Plant Disease
Epidemiology

Plant Disease Epidemiology

Population Dynamics and Management
Volume 1

Kurt J. Leonard
U.S. Department of Agriculture, Agricultural Research Service
North Carolina State University

William E. Fry
Cornell University

Macmillan Publishing Company
NEW YORK

Collier Macmillan Publishers
LONDON

Macmillan Publishing Company
866 Third Avenue, New York, NY 10022

Collier Macmillan Canada, Inc.

Printed in the United States of America

printing number   year
1 2 3 4 5 6 7 8 9 10   6 7 8 9 0 1 2 3 4 5

Library of Congress Cataloging-in-Publication Data
Main entry under title:

Plant disease epidemiology.

    Includes index.
    1. Plant-diseases—Epidemiology—Collected works.
2. Micro-organisms, Phytopathogenic—Host plants—
Collected works. 3. Micro-organisms, Phytopathogenic—
Control—Collected works. 4. Microbial populations—
Collected works. I. Leonard, Kurt J.   II. Fry,
William E.
SB731.P645   1986   632′.3   85-18906
ISBN 0-02-948820-6 (v. 1)

Contents

Preface

FROM ITS BEGINNINGS, the science of plant pathology has been based on a commitment to agriculture. It grew from the premise that the causes of destructive diseases of agricultural crops can be discovered, and that once they have been discovered and sufficiently understood, they can be controlled. With rare exceptions, plant pathologists are not concerned with providing therapy for individual diseased plants. Instead, they justify their research on the conviction that the knowledge gained will be useful in finding ways to reduce the size of pathogen populations, reduce their rates of population growth, or to otherwise inhibit the spread of pathogens through populations of plants.

Van der Plank's eloquent application of the simple exponential and logistic growth equations to the increase of disease in plant populations more than 20 years ago in his book *Plant Diseases: Epidemics and Control* set off a revolution in the ways plant pathologists regard epidemic increase of disease. What had previously been largely a descriptive science is rapidly developing mathematical precision in the expression of its concepts. The surge of interest and research in quantitative epidemiology and population dynamics of plant pathogens has brought a new orientation to the science of plant pathology.

Nevertheless, this new orientation has not come easily. Plant pathologists who have spent many years studying the biology of plant pathogens in relation to disease often find it difficult to think of disease in mathematical terms. They may even feel inadequate or intimidated by the modeling work of younger, more mathematically inclined colleagues. These experienced, descriptive biologists would agree with Goethe that "mathematicians are like Frenchmen; whatever you say to them they translate into their own language and, forthwith, it is something entirely different." The criticism is sometimes justified. Many of the younger modelers have achieved their high level of mathematical competence at the expense of thorough knowledge of biological constraints in the systems they attempt to model. We now need a major effort to ease the transition from descriptive to quantitative epidemiology.

This volume is intended as the first in a series of volumes designed to aid this transition. Our goal is to encourage and facilitate the integration of descriptive "field" epidemiology with the more abstract mathematical modeling of plant disease epidemics. Therefore, we have invited authors to cover a wide range of topics. Some chapters should enable classical plant pathologists to gain valuable insight into the methods and benefits of quantitative epidemiology. Other chapters will provide modelers with a better appreciation of the environmental and biological constraints and the importance of some of the questions facing plant pathologists involved in applied disease control work. In all cases we have encouraged the authors to describe their subjects in ways that will make them accessible to more than just the specialists in their own fields of research.

The chapters should be understandable to anyone, from graduate students to mature scientists, who has a serious interest in epidemiology and population dynamics of plant pathogens. Of course, we cannot absolve the readers from all responsibility to invest their own energies into the struggle to assimilate difficult material. As Albert Einstein put it, "Everything should be made as simple as possible but no simpler."

Any attempt to cover the entire range of important topics in plant disease epidemiology in a book this size would necessarily give only superficial coverage of many topics. Instead, we chose to emphasize more penetrating analyses of a restricted number of important topics. We asked the authors to present their topics in sufficient depth to prepare the readers to participate in research in that specific area of plant disease epidemiology.

The early development of quantitative epidemiology occurred more or less independently from the work on population biology in related disciplines. In recent years, however, plant pathologists have increasingly found new insights in gleaning techniques and concepts from related biological disciplines and adapting them to plant disease epidemiology. In turn, scientists from other disciplines have shown increased interest in plant pathology. Ecologists who study population dynamics of interacting species can find in plant disease epidemiology a fertile source of extensive data and observational information to apply to models of the population dynamics of host-parasite interactions. Soil microbiologists have drawn on the extensive data of population fluctuations of soilborne plant pathogens. Population geneticists have found the modeling of gene frequency changes and equilibria in plant host-parasite interactions to be a challenging area of research. The agricultural foundation of the science of plant pathology provides an important attraction to scientists from other fields. The importance of crop production gives a greater human relevance to their research as it can be applied to plant disease epidemics.

We have intended this volume and subsequent volumes in this series to be read and understood by plant pathologists who want to expand their own competence in research and teaching of quantitative epidemiology. This series is also intended for scientists in related fields such as agronomy, entomology, ecology, genetics, and meteorology who may discover previously unappreciated applications of their research to plant pathology. We have chosen population growth models as the foundation and starting point for this modern treatment of plant disease epidemiology. The chapters of Part 1 of this first volume deal with the use, interpretation, and statistical analysis of disease progress curves. Readers who have been discomforted by the variety of growth models that have been applied recently to plant disease progress curves and by the relative lack of biological rationale for their use will be reassured by Part 1. The unifying concepts in Part 1 demonstrate that quantitative plant disease epidemiology has moved beyond the phase of mere curve fitting as a descriptive exercise in categorizing epidemics.

Part 2 illustrates some specific ways in which physical and biological factors in the environment influence pathogen survival and reproduction. The section provides examples of the types of voluminous data that have been generated in attempts to understand the biology of important plant pathogens. These data represent both an invaluable resource and a challenge to modelers of population dynamics of pathogenic microorganisms. They illustrate why real epidemics deviate from theoretical population growth models, but they also provide the raw material for refining theory to more accurately represent real epidemics.

P. M. A. Bourke said that "taking to computers is generally regarded as a sign of maturity in a field of study in much the same way as experimentation with tobacco and alcohol is a sign that one's children are growing up." Eventually, of course, adolescent exuberance gives way to more sustained, serious contemplation of long-range objectives. As Part 3 shows, plant pathology is a maturing science in this sense; Part 3 includes two of the best recent examples of the development and use of simulation models for disease management as well as a general discussion of the application of systems analysis to the underlying goal of alleviating plant disease epidemics. In an era of economic hardship in agriculture, it is not enough to fight disease. The choice of disease control methods and the timing of their implementation need to be carefully refined. For a few important plant diseases, models have been developed to accurately assess the magnitude of the disease threat and to predict just how much effort and expense in disease control is justified for optimal economic returns.

Part 4 contains a single chapter that may be regarded as a link between the chapters in Part 3 and those in Part 5. Complex simulation models can be very useful in developing and testing practical disease management strategies, but as discussed in Part 4, they are not always the best tools for advancing the theory of quantitative epidemiology. A few years ago it would have seemed incongruous to speak of theoretical plant pathology as a field of study. Theories in plant pathology were distilled directly from experimental data. Little advance in the theoretical understanding of plant disease epidemiology was thought possible without the infusion of masses of new experimental data.

It became apparent that quantitative plant disease epidemiology was coming of age when theoretical papers based on the construction and analysis of mathematical models of disease increase and spread began to appear in our journals. The development of theoretical research in plant disease epidemiology owes much to earlier advances in the fields of quantitative ecology, population genetics, and meteorology. The influence of these fields can be seen in Parts 4 and 5. One can also see in those sections the great potential for a stimulating dialog between the experimental and theoretical approaches to the study of plant disease epidemiology. Theoreticians will increasingly share in advancing the frontiers of research in plant disease epidemiology as they become more adept at integrating experimental conclusions and predicting fruitful areas for experimental study.

The original idea for this book was conceived by Sarah Greene, who, like the two of us, was reared on the treeless but fertile plains of eastern Nebraska and western Iowa and was educated at Cornell University. Her early encouragement and continuing support and advice ensured the successful completion of the book. We are indebted to the members of the editorial board, who helped plan the contents and select the authors for the book. We are also indebted to them and to others for their helpful reviews of selected chapters. Those who reviewed chapters and provided valuable suggestions include Donald Aylor, Jerry M. Davis, John W. Duniway, Howard Ferris, Harvey J. Gold, James V. Groth, Kenneth P. Minogue, Erik V. Nordheim, and Gregory E. Shaner. We extend our sincere thanks to the authors whose timely and excellent manuscripts made this volume possible.

Kurt J. Leonard
William E. Fry

Contributors

Dr. C. Lee Campbell	Department of Plant Pathology, Box 7616, North Carolina State University, Raleigh, NC 27695–7616
Dr. James E. DeVay	Department of Plant Pathology, University of California, Davis, CA 95616
Dr. Howard Ferris	Division of Nematology, University of California, Davis, CA 95616
Dr. B. D. L. Fitt	Plant Pathology Department, Rothamsted Experimental Station, Harpenden, Hertfordshire AL5 2JQ, England
Dr. A. P. Gutierrez	Division of Biological Control, University of California, Berkeley, CA 94720
Dr. D. Haith	Department of Agricultural Engineering, 308 Riley-Robb Hall, Cornell University, Ithaca, NY 14853
Dr. M. J. Jeger	Department of Plant Pathology and Microbiology, Texas A & M University, College Station, TX 77843

Dr. A. L. Jones   Department of Botany and Plant Pathology,
                           Michigan State University, East Lansing, MI
                           48824

Dr. Laurence V. Madden   Department of Plant Pathology, Ohio Agricultural
                           Research and Development Center, Wooster, OH
                           44691

Dr. H. A. McCartney   Rothamsted Experimental Station, Harpenden,
                           Hertfordshire AL5 2JQ, England

Dr. Kenneth P. Minogue   Department of Botany and Plant Pathology,
                           Purdue University, West Lafayette, IN 47907

Dr. David E. Pedgley   Overseas Development Administration, Tropical
                           Development Research Institute, College House,
                           Wrights Lane, London W8 5SJ, England

Dr. Alan Roelfs   U.S. Department of Agriculture, Agricultural
                           Research Service, Cereal Rust Laboratory,
                           University of Minnesota, St. Paul, MN 55108

Dr. Robert C. Seem   Department of Plant Pathology, New York
                           Agricultural Experiment Station, Cornell
                           University, Geneva, NY 14456

Dr. James O. Strandberg   University of Florida, Institute of Food
                           and Agricultural Sciences, P.O. Box 909,
                           Sanford, FL 32771

Dr. Thierry C. Vrain   Agriculture Canada, Research Station,
                           6660 N.W. Marine Drive,
                           Vancouver, B.C. V6T 1X2, Canada

Dr. Paul E. Waggoner   Connecticut Agricultural Experiment Station,
                           P.O. Box 1106, New Haven, CT 06504

Part One

Disease Progress Curves

Chapter 1

Progress Curves of Foliar Diseases: Their Interpretation and Use

Paul E. Waggoner

Why and how does a plant pathologist plot the progress of an epidemic? A progress curve, which Kranz (1974) called "the graph of an epidemic," relates a change in the number of pathogens or symptoms to time. In this section I shall suggest some "why's" or reasons for plotting, analyzing, and even simulating the progress of a disease. Then I will examine the "how" or methods. Finally, I shall return to "why," examining whether the reasons that are proposed for curves of disease progress are, in fact, supported by their interpretation and usefulness.

WHY EXAMINE THE PROGRESS OF A FOLIAR DISEASE?

The reflex answer to "why?" is "to control disease" or "identify an effective fungicide or resistant variety." The reflective answer follows recollections that before we used graph paper, statistics, or computers, we controlled disease by using a Bordeaux mixture or a disease-resistant variety of wheat. An effective fungicide can be readily identified by a spotless yield and a resistant variety of a crop plant by a heavier yield. The reflective answer deals with more subtle subjects than the reflex answer to "Why?"

The ultimate reason for studying the progress of disease is to determine its cause. Since it is evident that cause precedes effect, we make a series of observations over a period of time and ask not only what level has disease attained but what level it came from. Only when we have "before" and "after" can we reason about causes.

At this point, asking why a *series* rather than a *pair* of observations is a reasonable question. If the environment were steady and the mechanism of increase were surely known to be, say, straightforward multiplication, a pair would suffice. The rate of multiplication would be precisely defined by the ratio of two observations, and interpolation and even extrapolation would be easy and accurate. However, a steady multiplication—generation after generation—is rare in the real world because the environment and the quality and amount of host available to the pathogen all change. Hence, the progress of the disease can only be revealed by a series, not a mere pair of observations.

"Progress" suggests graphs. A statistician, whom one might expect would begin with numbers rather than graphs, wrote instead,

Before starting to compute, we should examine the linearity . . . and the easiest test is a simple diagram. . . . Ideally, the plotted points will form a linear trend, with about the same vertical dispersion over the full range. . . . If the overall trend is non-linear, their pattern may suggest . . . the initial observations can be transformed. (Bliss 1970)

In other words, one figure is worth a thousand numbers. Nevertheless, we do use numbers.

I often say that when you can measure what you are speaking about, and express it in numbers, you know something about it; but when you cannot measure it, when you cannot express it in numbers, your knowledge is of a meagre and unsatisfactory kind. (William Thomson [Lord Kelvin] in a lecture to the Institution of Civil Engineers, 1883)

With numbers we can pass on to the mathematics of epidemics. According to Thompson (1942), Immanuel Kant (1782–1804) declared that the chemistry of his generation was a science, but not true science because the criterion of true science lay in its relation to mathematics. Earlier, Leonardo da Vinci (1452–1519) wrote in his notebooks, "No human investigation can be called real science if it cannot be demonstrated mathematically." Still earlier, Roger Bacon (1220–1292) wrote, "Mathematics is the door and the key to the sciences."

All this testimony could be put down to snobbery if we did not know that mathematics is the door to rigorous and formal reasoning, and its raw material is numbers. According to Koestler (1980), Galileo (1564–1642), delighting in unraveling the laws of order hidden in the puzzling diversity of phenomena, wrote: "The book of nature is written in the mathematical language. Without its help it is impossible to comprehend a single word of it." Koestler concluded that for Galileo quantitative measurements and formulations were simply the most effective tools for revealing the inherent rationality of nature.

To learn the causes underlying the inherent rationality of nature, a plant pathologist makes a series of observations of disease during its progress, plots the progress to reveal its gross features, and then analyzes the progress by the formal logic of mathematics. Symbols used in this chapter are as follows:

K	Maximum foliage as dimensionless number or sq cm area
Mu	Moment about the mean, time raised to a power
N	Diseased foliage as dimensionless number or sq cm area
R	Relative rate increase of host K, per day
RMS	Root mean square error of prediction of percent, dimensionless
S	Skewness, dimensionless
T	Temperature, °C
X	Distance, cm
a	Combined parameters
b	Combined parameters or change per day per day in r
c	Parameter for shape, Weibull equation, dimensionless
gompit x	$-\ln(\ln[1/x])$, dimensionless
k	Contagion, negative binomial equation, dimensionless
logit x	$\ln(x/[1-x])$, dimensionless
m	Parameter for shape, Richards equation, dimensionless
n	Number of lesions, dimensionless
r	Relative increase of disease per day
t	Time, days
t'	Days after disease appears or radians of annual period
x	N/K portion diseased, dimensionless

Subscripts:
a	Amplitude
d	Defoliation
g	Gompertz
h	Richards
m	Mean
n	Number of moment about mean
s	Simple interest
w	Weibull
x	Maximum
z	Zero
0	At time = 0
1	Time when disease first appears or disease at unit distance

HOW TO ANALYZE THE PROGRESS OF A FOLIAR DISEASE

The progress of foliar diseases is exemplified by the observations of epidemics by Fry (1975) of late blight in potatoes (Fig. 1), by Berger (1973) of early blight in two varieties of celery (Fig. 2), and by Imhoff et al. (1982) of rust in beans

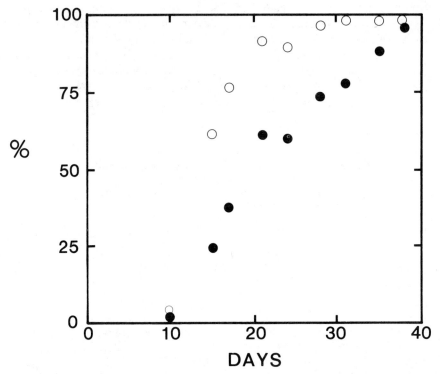

Figure 1 Progress of late blight of Russet Rural potatoes after 29 July 1974. Foliage was unsprayed (○) or sprayed (●) with 0.22 kg zinc ion and maneb per hectare. *(Data from Fry, 1975.)*

(Fig. 3). What inherent rationality can we find in these typical, sigmoid progressions of foliar diseases?

Logistic Model

In the mid-nineteenth century, Verhulst, considering the growth of human populations, proposed the simplest model of increase: its proceeds by multiplication and is limited by a maximum (Kingsland, 1982). Later and independently, M'Kendrick and Pai (1911) proposed the same mathematical rule for the population growth of bacteria confined in a test tube. Still later, Pearl and Reed (1920) rediscovered the logistic law that Verhulst had proposed decades earlier and wrote that a population would increase in proportion to the number of its members N at time t and to the difference between N and its maximum size K ($K - N$). The relationship can be expressed in a differential equation for the rate of increase dN/dt of the population N:

$$\frac{dN}{dt} = rN \frac{K - N}{K} \tag{1}$$

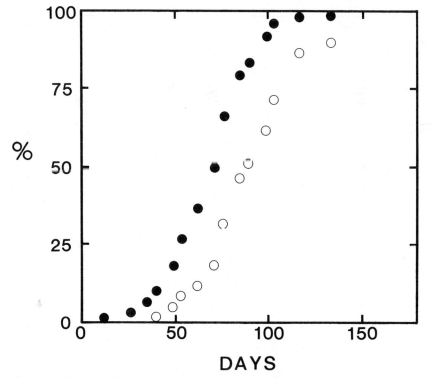

Figure 2   Progress of early blight of EES1624 (○) and Fla683 (●) celery after 2 February 1972. *(Data from Berger, 1973.)*

where r is the proportional rate. Plant pathologists customarily consider K the sum of healthy and diseased host and equate N/K with the fraction x of the host diseased. Since N and K must have the same dimensions, K is plants, leaves, or leaf area when N is diseased plants, leaves, or leaf area. The value x is dimensionless and Eq. 1 becomes

$$\frac{dx}{dt} = rx(1 - x) \tag{2}$$

If K and r are constant, Eq. 2 can be easily integrated to obtain the familiar logistic equation

$$x = \frac{1}{1 + (1/x_0 - 1) \cdot \exp(-rt)} \tag{3}$$

where x_0 is the portion x of the hosts diseased at time zero. Since the usual goal is estimating r (the proportional rate of increase per time), Eq. 3 is often converted into a linear form for the estimation of r by a graph or linear regression:

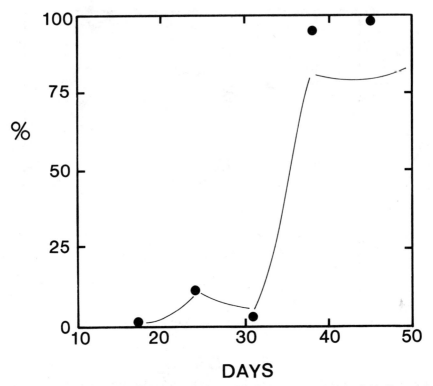

Figure 3 Progress of rust of beans in days after the beans emerged in the field. Observed disease (●) and the progress curve calculated from the observed growth of the host and from *r* varied weekly. *(Data from Imhoff et al., 1982.)*

$$\ln\frac{x}{1-x} - \ln\frac{x_0}{1-x_0} = rt \qquad (4)$$

where ln indicates natural logarithm. If $\ln[x/(1-x)]$ is the logit of x,

$$\text{logit}(x) - \text{logit}(x_0) = rt$$

Equation 3 traces a progress curve for x that is sigmoid. It is symmetrical around the inflection at $N = K/2$ or $x = 1/2$ and $t = -\text{logit}(x_0)/r$ where the absolute rate dN/dt or dx/dt stops increasing and starts decreasing. Since compound interest increases money as $(dN/dt)/N$, Van der Plank (1963) often used the term *compound interest disease* for infections that increased as Eq. 1, remembering that the compound interest was $r(1-x)$ rather than r. When disease is slight and $1-x$ is close to 1, Eq. 1 does cause compounding at a nearly constant rate r.

The question is whether the diseased hosts' N actually increases with a constant r, in proportion to N, in proportion to the opportunity $(K-N)$, and toward a constant K? The usual test is to transform the observed x into logits, plot them, and note whether the progress then delineates a straight line on a graph (Van der Plank, 1963). Although I shall later do this, I will first introduce a measure of the shape of the progress of the portions x themselves.

The logistic curve is defined by three parameters: (1) the initial number N_0 of infections, which determines where the curve rises, (2) the rate r, which determines how fast it rises, and (3) the maximum K, which determines how high it rises. The curve is always symmetrical around its inflection at $K/2$. Generally, however, a real progress curve is not precisely symmetrical. It may be negatively skewed with a long tail of small values, or positively skewed with a long tail of large values. Quantitatively, skewness S is the ratio of the third moment about the mean to the 1.5th power of the second moment. The second moment is, of course the variance, and its 1.5th power is the third power of the standard deviation SD. S is dimensionless and has a standard error SE of $(6/$ [number in the sample]$)^{1/2}$ (Geary and Pearson, 1938). I have taken the number of observations per time as the number in the sample. The skewness of the logistic curve is zero since it is symmetrical about the inflection.

Calculating S requires estimates of the variance and third moment. Because progress curves are repeated observations of a single population, the moments cannot be estimated from the usual averaging and summing of squares employed with a random sample from a population. Ferrandino (1985) has shown that the mean time t_m of a progress curve can be estimated as the area above the curve after the time t_1 when disease x first appears. The higher moments Mu_n about the mean are then estimated as

$$\text{Mu}_n = n\left[\int_{t_1}^{t_m} (t_m - t)^{n-1} (1 - x)\, dt + \int_{t_m}^{\infty} (t - t_m)^{n-1} x\, dt\right] \quad (5)$$

S can be calculated for disease progress curves by extrapolation. This was done for Figs. 1 through 3, and the results are given in Table 1. The S values were derived from the following extrapolations: The progression of late blight in the treated potatoes of Fig. 1 was extrapolated to 100% at 40 days, the progressions for early blight in Fig. 2 were extrapolated to 100% at 135 days in the susceptible celery and to 91% at 135 days and 100% at 170 days in the tolerant celery, and the progression for bean rust in Fig. 3 was extrapolated to 100% at 77 days.

In Table 1, the progress is summarized by skew S and its standard error SE. It is also summarized by regression of logits L or gompits G on time as shown by rates (per day) and coefficients of determination. The root mean square errors RMS are for the differences between observed percent and those expected from logits or gompits or from the Richards R equation with m shown. Number 1 is sample size for regression and number 2 is for RMS.

The skewness ranges from the positive value of the unsprayed potatoes with

Table 1   Characteristics of epidemics of *Phytophthora infestans* on potatoes, *Cercospora apii* on celery, and *Uromyces phaseoli* on beans

Parameters[a]	*P. infestans* on unsprayed potatoes	*P. infestans* on sprayed potatoes	*C. apii* on EES1624 celery	*C. apii* on FLA683 celery	*U. phaseoli* on bean
Skew, S	1.50	0.52	0.45	−0.22	−1.51
SE	0.20	0.20	0.37	0.37	0.07
Regressions:					
Rate L, % per day	37	25	7	9	32
Rate G, % per day	28	15	3	6	22
Coef Det % L	94	89	96	99	82
Coef Det % G	98	95	96	90	82
Number 1	9	10	18	17	5
Errors:					
RMS L	12.4	12.2	3.7	3.2	15.9
RMS G	3.6	5.8	5.4	11.7	24.4
m	0.75	0.99	1.50	2.25	3.00
RMS R	2.7	5.8	1.9	2.5	12.9
Number 2	11	11	19	19	6

[a] L = Logistic; G = Gompertz; RMS = root mean square; Number 1 = sample size for regression; Number 2 = sample size for RMS.

the lagging progress at a high percentage of blight, through the mixed positive and negative skewness of the celery, and on to the negative skewness for the beans with a "burst" of rust following a period of low percentage of disease. Remembering that the skewness of the logistic curve is zero, one concludes that, generally, the logistic law reproduces the sigmoid curves of epidemics and shows the inherent rationality of conceiving pathogens that simply increase in proportion to their population and the remaining opportunity—but not exactly. Is there a more rational and realistic law of growth?

We are not the first to forage for another equation. In the article, "The Refractory Model: The Logistic Curve and the History of Population Ecology," Kingsland (1982) tells how a barrage of criticism followed Pearl's claim that the logistic curve was a *law* of growth rather than a convenient model. As we search for a more useful equation, we must follow Pearl's advice. His advice was to examine qualitative resemblances between the natural mechanism and the concept underlying the model and not to be satisfied with mere quantitative resemblances between the outcomes of nature and the model. (Unfortunately, he disregarded his own advice and got in trouble.)

Gaussian Model

The equation of the normal curve of error of Gauss is too familiar to require restating here. Like the logistic equation it traces a symmetrical and sigmoid course and has three parameters which for our use are: (1) the maximum spread of the disease, (2) the mean or time of 50% progress of the disease, and (3) the standard deviation or rate of getting from 16% to 50% disease. Long ago it was used to plot the course of *Alternaria solani* blight on tomato (Barratt, 1943) and has often been used since then, notably by Large (1952).

Numerically, the logistic and normal curves are close when $r = 1.67/SD$. For example, if $r = 0.167$ and $SD = 10$, the two curves agree within about 0.5% from 0 to 10 days and 1% from 10 to 40 days. In logits, they agree within 0.2 from 0 to 10 days, 0.4 at 20 days, and 1.6 at 30 days.

Attempting to rationalize the gaussian as a growth curve, Thompson (1942) wrote, "It represents on either hand the natural passage from best to worst" conditions for growth. Why this rationale should make dN/dt precisely proportional to $\exp[-(t - t_m)^2/2]$, however, is far from clear. Since the rationale for the gaussian curve is less appealing and more complex than the rationale of the logistic Eq. 1, I shall abandon the gaussian curve.

Limitation without Multiplication

If the opportunity for infection decreases as infection increases while the source of infection is steady rather than increasing, Eq. 1 becomes, simply,

$$\frac{dN}{dt} = r_s(K - N) \tag{6}$$

and if r_s and K are constant,

$$x = 1 - (1 - x_0) \cdot \exp(-r_s t) \tag{7}$$

or in linear form,

$$-\ln(1 - x) + \ln(1 - x_0) = r_s t \tag{8}$$

Unlike the logistic curve where the inoculum multiplies with N and the absolute rate dN/dt increases fully to $K/2$, Eq. 6 incorporates the assumption that inoculum is constant, making the rate fastest when N is smallest and decreasing it steadily as N increases. The progress curve traced by Eq. 7, unlike the logistic, is neither sigmoid nor symmetrical.

The subscript s of r_s indicates "simple interest." Van der Plank (1963) called Eqs. 6 to 8 *simple interest* or *monomolecular* equations, and since then, pathologists have deduced that progress curves made linear by the transformation $\ln(1 - x)$ and Eq. 8 are caused by inoculum that does not increase with disease during the period of observation. Just as $r(1 - x)$, not r, is the compound rate of interest in Eq. 2, so $r_s(1 - x)$, not r_s, is the simple rate of interest in Eq. 6. Van der Plank illustrated simple interest with the increase of cotton wilt during a single season; the fungus sown on the field at the beginning of the season or already in the soil was likely the sole source of infection and was not likely augmented by spread from plant to plant. Since simple interest does not cause a sigmoid progress as many, perhaps most, pathogens do, I turn to a curve that is sigmoid like the logistic and gaussian but positively skewed like the curves implied in Fig. 1.

Gompertz Model

Although the Gompertz curve appeared in the phytopathological literature after the logistic, it actually antedates the logistic. Fully 20 years before Verhulst published his logistic equation, Benjamin Gompertz published in 1825 in the *Philosophical Transactions of the Royal Society*, "On the Nature of the Function Expressive of the Law of Human Mortality." He showed that if the average exhaustions of a man's power to avoid death were such that at the end of equal infinitely small intervals of time, he lost equal portions of his remaining power to oppose destruction then

$$\frac{dN}{dt} = r_g N \cdot \ln \frac{K}{N} \tag{9}$$

where r_g is the rate specific to the Gompertz model. Like the logistic, the Gompertz equation can be expressed in a linear form for the estimation of the rate r_g by linear regression:

$$-\ln \left(\ln \frac{1}{x} \right) + \ln \left(\ln \frac{1}{x_0} \right) = r_g t \tag{10}$$

or if $-\ln \left(\ln \frac{1}{x} \right)$ is the gompit of x,

$$\text{gompit}(x) - \text{gompit}(x_0) = r_g t$$

The Gompertz curve was resurrected by Wright (1926) during the controversy evoked by Pearl's apotheosis of the logistic law. Applying Eq. 9 to growth, one can use Gompertz's phrasing to say Eq. 9 states that in equal small intervals of time the organism loses equal proportions of its power to increase. Pearl's student, Winsor (1932) compared the logistic equations of his teacher to the resurrected Gompertz. He concluded

In each curve, the degree of skewness, as measured by the relation of the ordinate at the point of inflection to the distance between the asymptote(s), is fixed. It has been found in practice that the logistic gives good fits on material showing an inflection about midway between the asymptotes. No such extended experience with the Gompertz curve is as yet available, but it seems reasonable to expect that it will give good fits on material showing an inflection when about 37 per cent of the total growth has been completed. Generalizations of both curves are possible, but here again there appears to be no reason to expect any marked difference in the additional freedom provided.

The 37% of total growth at inflection, is, of course, K/e, where e is the base of natural logarithms. Windsor could have added that the logistic curve has a more transparent rationale in Eq. 1 than the Gompertz has in Eq. 9.

Gompertz vs. Logistic Model

Returning to the examples of Figs. 1 to 3, we can ask whether the logistic or Gompertz curve fits more closely. Theoretically, the skew of the logistic curve is zero, and I have found empirically that the skew of the Gompertz is about 1. Thus positively skewed progress of disease should be fit more closely by the Gompertz equation, whereas symmetrical or negatively skewed progress should be fit more closely by the logistic. In fact, Table 1 shows that the coefficients of determination for the regression of gompits on time are greater than for logits where the skew is positive, which is exactly as expected since the Gompertz curve is positively skewed and the logistic is not. Where the skew is negative, the coefficients of determination for the logistic are greater than for the Gompertz curve. Berger (1981) compared the coefficients of determination for the logistic and Gompertz curves and 113 epidemics, and the coefficients for the Gompertz curve were usually greater. Table 2 shows the calculations I have done using data from Berger as well as several other investigators.

The gompit and logit transformations magnify differences between observation and model at low and high percentages, and it is important to consider the differences between observation and model in the percentages themselves, which were used to examine skewness. This was done by first estimating parameters in Eqs. 4 and 10 by linear regression; then the percentages expected from these parameters were calculated at each observation time. Finally the differences or errors between the expected and observed percentages and the root mean squares RMS of these errors were calculated.

This procedure does not weight the low and high percentages heavily and shows the great differences between model and observation in the intermediate percentages where the logistic and Gompertz curves differ most. Since a logit or gompit cannot be calculated for an observation of 100%, whereas an expectation can be compared with an observation of 100%, the sample size for RMS, number 2, can be more than that for regression, number 1.

In Table 1 it is apparent that the RMS for the Gompertz equation is less when the skew is positive and, for the logistic, RMS is less when the skew is negative. The single exception in the five examples of Table 1 is for the tolerant celery EES1624 where the skew is only 0.45 with a standard error of 0.37 and the RMS for the Gompertz is slightly more than for the logistic. The RMS of the errors in percent differentiate the fit of each model more clearly than do the coefficients of determination. Kranz (1974) employed a similar criterion for differentiating the fit of models. In Table 1, sometimes one model fits better, sometimes the other, according to whether the epidemic lags and then explodes or explodes and then stagnates. Since we cannot accept as a natural law that all epidemics will progress exactly according to either Eq. 1 or Eq. 9, we would

Table 2 Relative rate *r* of increase of disease per day and coefficient of determination for fit of logistic and Gompertz models to disease progress curves

Pathogen	Host	N^a	$r \times 100$	Coefficient of determination, $\%^b$		Source
				Logistic model	Gompertz model	
Botrytis cinerea	Begonia	3	–0–18	3–98	3–98	Plaut and Berger (1981)
Cercospora apii	Celery	18	7–15	44–86	94–99	Berger (1981)
C. arachidicola	Peanut	3	5–12	85–91	89–96	Plaut and Berger (1981)
Cercosporidium personatum	Peanut	1	15	69		Plaut and Berger (1980)
Erysiphe graminis	Wheat	4	18–24	57–84	97–99	Berger (1981)
Helminthosporium maydis	Maize	3	10–11	91–98	96–98	Berger (1981)
H. turcicum	Maize	1	13	85	98	Berger (1981)
Hemileia vastatrix	Coffee	2	12	86–92	83–90	Kushalappa and Ludwig (1982)
Phytophthora infestans	Potato	2	25–37	89–94	95–98	Fry (1975)
Puccinia coronata	Oats	19	12–52			Luke and Berger (1982)
P. graminis	Wheat	1	41			Van der Plank (1963)
P. recondita	Wheat	80	7–12	71–95	94–98	Berger (1981)
P. striiformis	Wheat	1	11	94	97	Berger (1981)
Uromyces phaseoli	Bean	3	4–20	92–95	90–98	Plaut and Berger (1981), Kushalappa and Ludwig (1982), Imhoff et al. (1982)
Uromyces spp.	Yucca	3	5–6	71–92	94–96	Berger (1981)
Venturia inaequalis	Apple	1	19	34	74	Berger (1981)
Various	Various	59	1–30			Kranz (1968a)

Note: r is calculated using the logistic equation.
[a]*N* is the total number of epidemics studied.
[b]Percentage of variability of the dependent variable explained by the model.

like to have a general equation that encompasses logistic and Gompertz models, as well as shades in between them.

Richards Model: A Generalization

Beginning with an equation for the growth of animal populations proposed by L. von Bertalanffy, Richards (1959) wrote a general equation for growth:

$$\frac{dN}{dt} = r_h N \frac{(K/N)^{1-m} - 1}{1 - m} \tag{11}$$

where r_h is rate for the Richards model. Independently Turner et al. (1969) proposed the same equation and rationalized it nicely, writing that the number of empty spaces or opportunities for multiplication is conceived, not as the simple difference between the maximum and the present population, but rather as the difference between their $(m - 1)$th powers.

If $m = 0$, Eq. 11 becomes Eq. 6 for the monomolecular or simple interest case. If, on the other hand, $m = 2$, Eq. 11 becomes Eq. 1 for the logistic case. Although the proof is more difficult, making $m = 1$ converts Eq. 11 into Eq. 9 for the Gompertz equation. Thus we have in one general equation the range from the monomolecular with its steadily declining rate, through the Gompertz with its positive skew and inflection at K/e, to the symmetrical logistic with its inflection at $K/2$. In fact, when $m > 2$, Eq. 11 produces a curve with a negative skew, and when m is very large, it approaches unlimited exponential growth or compounding at a steady r_h/m per time.

Integration of Eq. 11 produces

$$x = [1 - b \cdot \exp(-r_h t)]^{1/(1-m)}$$

and

$$b = 1 - x_0^{1-m} \qquad \text{when } m < 1$$

or

$$x = [1 + b \cdot \exp(-r_h t)]^{1/(1-m)}$$

and

$$b = -1 + x_0^{1-m} \qquad \text{when } m > 1 \tag{12}$$

The inflection of Eq. 12 is at $x = m^{1/(1-m)}$, which is easily seen to occur at $x = 1/2$ for $m = 2$ using the logistic curve and, with more difficulty, is seen to occur at $1/e$ for $m = 1$ using the Gompertz curve. When m is small, the inflection is near m; as m increases, the inflection moves toward 1.

We come now to relating the generalization of the Richards curve to actual progress curves. The linear form of the Richards equation is illustrated for $m < 1$:

$$-\ln(1 - x^{1-m}) + \ln(1 - x_0^{1-m}) = r_h t \tag{13}$$

Because Eq. 13 has the parameter m as well as an intercept and slope to be estimated, linear regression alone will not fit the equation to a progress curve, and the required additional information is sought in the skewness S. Since S can be estimated from the moments of a progress curve, the S produced by different m values is a clue to the m value that might be selected to fit such progress curves as in Figs. 1 to 3. For $m = 0$ or the monomolecular case, S is approximately 2; for $m = 1$ or the Gompertz model, S is about 1; for $m = 2$ or the logistic model, S is 0; and for $m = 4$, S is about -0.5.

The positive S of the first three columns or epidemics of Table 1 suggest $m < 2$, and the negative of the last two suggest $m > 2$. The m values shown in the table, in fact, did produce Richards curves with RMS as small or smaller than either the logistic or the Gompertz models, and the sizes of m values relative to 2 and the symmetry are logical. The m values were selected by trial to minimize the RMS: an m was selected for trial, Eq. 13 was fit by linear regression, the expected x values were calculated from Eq. 12, the RMS values between the expected and observed percentages were calculated, and repeated trials produced the small RMS values of Table 1.

For 2 years, Analytis (1973) observed that apple scab increased quickly and then lagged, tracing a progress that was strongly and positively skewed with a maximum increase dx/dt when only about half the time had passed. The Bertalanffy equation equivalent to the Richards with $m = 0.5$ fitted these positively skewed progressions somewhat more closely than did the Gompertz model ($m = 1$) and much more closely than the logistic model ($m = 2$).

The progress of late blight in nine replications of three varieties of potato was measured on ten occasions in 1975 by Ion Cupsa (personal communication), who kindly allowed me to analyze these remarkably thorough observations. The x increased from $0 \leq x_0 \leq 8\%$ on 25 June to as little as 13% in the resistant variety and as much as 98% in the susceptible by 26 August. When Richards curves were fit, the correlation was greatest when m was 0 to 1.5 in the susceptible potato, 0.8 to 1.8 in the intermediate, and 1.0 to >5.0 in the resistant. Thus the same pathogen in the same environment and on the same host species produced both positively and negatively skewed progress curves.

A healthy attempt to interpret the constants of his general equation was made by Richards (1959). Obviously x_0 locates the curve at time zero, and K specifies how high it can rise. I have already mentioned that inflection from increasing to decreasing dx/dt and hence the fastest rate of infection or highest value for dx/dt will occur at $K \cdot m^{1/(1-m)}$. The meaning of the proportional rate r_h, which seems so transparent, is not. Although it is the rate of increase of the units or "richits" in the left-hand member of Eq. 13, their meaning is not easily grasped and depends on the shape of the curve. The relation between rate r_h and m is most easily grasped by considering the fastest rate, which is at the inflection and is simply r_h/m. The rate dx/dt and the ratio r_h/m is fixed by nature in a set of observations, whereas the estimate of r_h falls with our choice of a curve and accompanying m. Thus, to the extent that the dx/dt at the inflection represents the entire progress, fitting the same progress curve with the Gompertz equation ($m = 1$) produces a rate that is about half as large as fitting it with the logistic ($m = 2$). For example, Berger (1981) estimated rates that were 2 to 10 times larger for the logistic than the Gompertz in 113 epidemics. Similarly, Analytis (1973) estimated regularly smaller rates as m was decreased from the Gompertz curve to the even more positively skewed Richards curve. Our estimates of the proportional rate depend as much on how we look at the epidemic as on the epidemic itself!

Weibull Model: A Flexibility

In 1951 Weibull introduced a function for frequency distributions, which fits both negatively and positively skewed curves. Weibull dealt with probabilities of an entire system in which failure of one segment, like failure of a link in a chain, caused the entire system to fail. However, a notable use of his function is fitting distributions of wind speeds, which are limited at zero and include a few extremely fast gusts. In pathology, Pennypacker et al. (1980) introduced the Weibull function as a flexible curve to fit progress curves of all shapes, and wrote

$$\frac{dx}{dt'} = cr_w(r_wt')^{c-1} \cdot \exp[-(r_wt')^c] \tag{14}$$

and

$$x = 1 - \exp[-(r_wt')^c] \tag{15}$$

where time t' is measured from the moment t_1 when x is no longer zero. The t' is always greater than zero. Skewness is controlled by c.

Examination of Eq. 14 shows that the Weibull equation becomes the monomolecular Eq. 6 when $c = 1$. Pennypacker et al. showed empirically that when $c = 3.6$ the Weibull model behaves quantitatively much like the symmetrical logistic and Gaussian equations. I am unable to show that $c = 3.6$ makes Eq. 14 qualitatively similar to the differential Eq. 1, which embodies the rationale of the logistic model.

The rationale of the Weibull equation may, however, be irrelevant. For example, Campbell et al. (1980) fitted it to the progress of hypocotyl rot in 22 fields of snapbeans to summarize the plethora of observations of these 22 series in 2 years. They then used the manageable 22 c's to test whether the monomolecular or logistic equations (and presumably the accompanying rationales) fit the progress curves. None of the 22 could be attributed to the limitation without multiplication embodied in the monomolecular model, and half could not be attributed to the limitation and multiplication embodied in the logistic model.

Thus, in the end, we are left with the monomolecular and logistic models with clear rationales and the Gompertz model with somewhat turbid rationale. Sometimes all fail to fit real progress curves. Other times, first one fits and then the other, with little or no rhyme or reason. We can always retreat to the Richards model, which shades from one form and presumably one rationale to the other, or to the Weibull model, which has no evident rationale.

CHOOSING FROM THE CAFETERIA OF EQUATIONS

Although the statistical data of Table 1 could be expanded to include the other equations, we can use what is there for comparison. Imagining this comparison as a contest, the victorious equation would be the one that best fits an epidemic. In a wider field, the victor might be the equation that fits the most epidemics. In the long run, of course, victory will go to the equation with the most parameters. In fact, if the polynomial equation in powers of t is given as many fitted coefficients as there are observations, it will fit perfectly and always win. This reduction to the absurd raises the question, "Why use an equation?"

Our purpose, like that of Galileo, was to learn the inherent rationality of nature. We shall succeed only by examining qualitative as well as quantitative resemblances between nature and equation. This means using a differential equation for dN/dt in which we can, first, comprehend inherently rational biological and physical qualities and with which we can, second, calculate a progress curve that resembles the observations. It is likely that the comprehensible and rational equation will not fit all the wiggles of every progress curve as we go from crop to crop, disease to disease, and environment to environment. Rather we shall view this progress curve as the embodiment of the general, larger strategies of Nature. We shall consider the deviations from it as evidence of her variable, smaller tactics that are to be understood on a case-by-case basis.

The logistic model, Eq. 1, is a comprehensible and rational equation and it produces curves that resemble those of real epidemics. I shall discuss the interpretation of progress curves by seeing what can be done with the deviations from the logistic equation, beginning with deviations caused by changes in r with passing time.

HOW TO INTERPRET PROGRESS CURVES

Changing Host, Environment, and r

When the differential Eqs. 1 and 2 were introduced, integration was made easy by explicitly assuming a constant rate r and implicitly assuming that environment and susceptibility were also constant. In reality, a constant proportional increase of infection is rare during the advance of an epidemic in inconstant weather and on aging foliage. Reality, as Kranz (1968a) demonstrated, is plotted by progress curves that have retreats as well as advances of disease. In the long run, we must interpret declining as well as rising curves and cope with great fluctuations in r.

A general or "analytical" solution of Eq. 2 with variable r follows its definition as a function of time. If r changes because susceptibility increases or decreases linearly with time, it can be replaced in Eq. 2 by $r_0 + bt$ where r_0 is r at time t_0 and b is the increase or decrease in r per day. Then the logits of the left-hand member of Eq. 4 become equal to $r_0t + bt^2/2$ rather than simply rt. If $b > 0$ for increasing susceptibility, the progress in logits will curve up rather than be linear, and the progress of the percentages x will be negatively skewed. Alternatively, if $b < 0$ for decreasing susceptibility, disease will only increase to a maximum when time is $-(r_0/b)$ and then decrease; the progress of the percentages x will be positively skewed during the increase.

Environment, however, is the great cause of variation in r (Kranz, 1968a), and although weather might cause the linear increase or decrease in r that was just examined, a more likely course for an r affected by environment is controlled by a sinusoidal seasonal course. It is reasonable to express r first as a function of temperature T, namely,

$$r = r_x - \frac{(T - T_x)^2}{b} \tag{16}$$

where T_x is the temperature that makes r the maximum r_x and $\sqrt{br_x}$ is the degrees warmer or cooler than T_x that makes $r = 0$. Thus r declines parabolically as T becomes warmer or cooler than T_x. At $\sqrt{br_x}$ degrees warmer or cooler than T_x, $r = 0$. It becomes negative at more extreme temperatures and disease decreases. Climatologically in the northeastern United States, daily mean or extreme temperature is approximately $T_m + T_a \cdot \sin t'$, where t' is in radians or $2 \cdot \pi/365$ multiplied by the number of days after 20 April, T_m is the annual mean temperature, and T_a the annual amplitude of the daily value. If T in Eq. 16 is replaced by the sinusoidal function of time and r in Eq. 2 is replaced by Eq. 16, then

$$\text{logit}(x) - \text{logit}(x_0) = \left[r_x - \frac{(T_m - T_x)^2}{b} - \frac{T_a^2}{2b} \right] t'$$
$$+ \frac{2(T_m - T_x)T_a(\cos t' - 1)}{b} + \frac{T_a^2 \sin 2t'}{4b} \tag{17}$$

The sinusoidal variation of the temperature controlling r and the parabolic or quadratic decrease of r on each side of an optimum T_x can cause either "hesitations" or actual decreases in the standard logistic course.

Examples show progress curves produced by an r varying as shown in Eq.

17. The potatoes of northern Maine experience maximum temperatures during the day that can be represented by T_m and T_a of 10° and 16° C, whereas the same temperatures in, say, West Virginia can be represented by 20° and 12° C. The relation of the r of late blight of potatoes to this temperature is unknown, but Crosier (1934) observed the briefest incubation from inoculation to visible mycelium occurred near 20° C and became very long in the neighborhoods of 3° and 33° C.

The curve for Maine in Fig. 4 was produced by its climatic parameters above, given 5% infection on 20 April (day 0). The hypothetical fungus had a narrow tolerance with T_x at 20° C, and b adjusted to make $r = 0$ at 10° and 30° C. Since the maximum temperature specified for Maine is only 26° C, the progress looks much like a logistic curve with constant r, although it is negatively skewed by the slow progress of disease during the cool spring. In West Virginia, however, the temperature was above 30° C for a long time, and disease declined from a maximum at day 56 to a minimum on day 125. If the tolerance of the fungus is broadened by a T_x at 18° C, and $r = 0$ at 3° and 33° C, on the other hand,

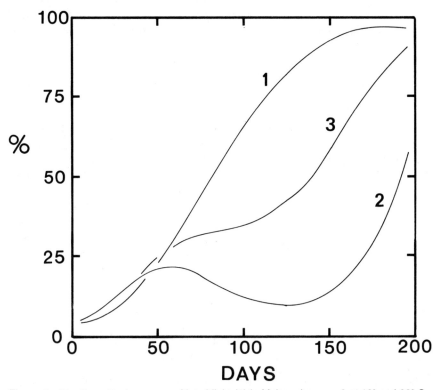

Figure 4 The theoretical progress of late blight (1) in Maine when $r = 0$ at 10° and 30° C and maximum at 20° C; (2) in West Virginia when r behaves as in Maine; and (3) in West Virginia when $r = 0$ at 3° and 33° C and maximum at 18° C.

disease does not decrease but only slows in its rise. Thus, we can interpret curves by analytical equations that incorporate variations in r that independent evidence suggests.

The use of independent evidence can also be exemplified by the observations of beans and their rust by Imhoff et al. (1982). They observed rust weekly for a month, and I assume the weekly increase in rust reflected the weather of the preceding week. The second week was the wettest of the four, and the long hours of wetness likely caused the great increase in rust in the third week. Taking the liberty of assuming the first week was unfavorable for the rust, as evidenced by its decrease in the second week, I calculated the progress depicted in Fig. 3 from r of 0.3, 0, 0.6, and 0.3 per day during the four weeks. The initial x was assumed to be 1%. The initial K and its growth rate were taken from observations. (Changing K will be discussed in the next section.) Arresting rust by an $r = 0$ while the host grew during the second week mimicked the decrease in x and gave the negatively skewed progress curve a logical explanation.

When the entire life cycle of a pathogen is computed by a simulator that employs actual weather observation, the changing rate and consequent starts and stops in progress can be simulated. Shaner et al. (1972), in fact, simulated the start, stop, and start again of southern corn leaf blight, using parameters for Indiana weather in a mathematical simulator, EPIMAY.

When pecans were inoculated at different times, several of their disease progress curves stopped rising, as if different inoculation times caused different K's (Gottwald and Bertrand, 1983). Early inoculation caused high maximums and later caused low maximums or apparent K's. Instead of assuming different K's, one can join the observers in interpreting the cause to be great susceptibility during midseason and the expansion of the nuts followed by their resistance. Changing r from moderate during the early season, to rapid during midseason, and to nil during the late season mimics the progress curves from a logical foundation.

Growing Host and K

Back at the transformation of Eqs. 1 and 2 to Eq. 3, I stated that integration was easy if K as well as r was constant. If, however, K is the total number of leaves in a crop, then during much of the season, K is increasing. In fact, a comparison of Tables 1 and 2 with Table 3 shows that hosts often grow with proportional rates R that are as large as the r of many diseases.

Sticking to the rational and reasonably realistic logistic equation, we can write a differential equation like Eq. 1 for the maximum opportunities K leaves or leaf area that grow with time, starting with K_0 at time zero and approaching a maximum K_x after a long time. Since this new seasonal K_x is a constant and r is still considered a constant, the new equation can be solved, producing an equation for K resembling Eq. 3. This equation for K can be substituted for the old K in Eq. 1:

Table 3 Growth rate R per day of hosts estimated for the logistic law

Host	Organ	R × 100/ day	Source
Bean	Leaf	23	Imhoff et al. (1982)[a]
Bean	Leaf	13–20	M. P. H. Gent (1981, personal communication)[a]
Cannabis gigantea	Weight	13	Blackman (1919)
Coffee	Leaf	8–24	Kushalappa and Ludwig (1982)
Helianthus spp.	Weight	11–17	Blackman (1919)
Maize	Weight	11	Thompson (1942)
Maize	Leaf	12–18	R. B. Curry (1974, personal communication)[a]
Potato	Leaf	6	Watson (1947)[a]
Tobacco	Leaf	13	R. B. Peterson (1984, personal communication)[a]

[a]R calculated from data presented in these studies.

$$\frac{dN}{dt} = rN\left\{1 - \frac{N}{K_x}\left[\left(\frac{K_x}{K_0} - 1\right) \cdot \exp(-Rt) + 1\right]\right\} \qquad (18)$$

The r is still the proportional rate of increase of N diseased leaves or leaf area, and the new rate R is the corresponding rate for the growth of opportunities K. In the appendix, F. J. Ferrandino integrates Eq. 18 to obtain Eq. 19 for the course of x that is now a ratio of N/K where both disease N and opportunities or host K change with time. When r does not equal R,

$$x = \frac{1 + b \cdot \exp(-Rt)}{1 + a \cdot \exp(-rt) + r/(r - R)\, b\, [\exp(-Rt - \exp(-rt)]} \qquad (19)$$

where $a = K_x/N_0 - 1$ and $b = K_x/K_0 - 1$.

Equation 19 is derived as Eq. A10 in the appendix, which includes a second equation for the case of $r = R$. Equation 19 is a generalization of the familiar logistic Eq. 3, which embodied Nature's usual, large strategy. It is the same as the equation of Turner et al. (1969) for a growing K and their exponent m equal to 1. With Eq. 19 we can analyze departures from the strategy of the logistic model caused by the tactic of a growing host.

A few examples remove the forbidding nature of Eq. 19. If the host initially has a size K_0 as large as its final size K_x and thus is not growing, Eq. 19 quickly reduces to the familiar logistic Eq. 3 with constant K, and it traces the familiar symmetrical and sigmoid curve 1 in Fig. 5. The larger that K_0 is relative to K_x, the slower the growth $RK(K_x - K)/K_x$ of the host K, and the smaller the impact of growth in general and a given proportional rate R in particular on the increase of disease.

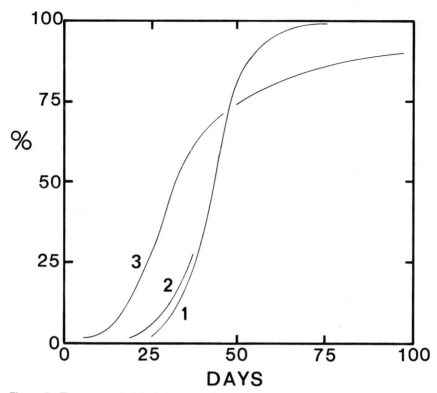

Figure 5   The symmetrical, logistic progress of disease when $r = 0.2$ and (1) $R = 0$, and the progress; (2) when $R = 0.20$, and (3) when $R = 0.05$.

If the host grows more rapidly than the disease N, x can decrease until the growth of the host slows near K_x, and the disease catches up. In fact, Kushalappa and Ludwig (1982) observed a decrease in coffee rust on a growing host. With K_0/K_x fully 0.4 and R only 2 to 3 times faster than r, however, the growth of the coffee accounted for a leveling but not the observed decrease in x. If growth of the host is logistic as conceived in Eq. 19 and R is much greater than r, Eq. 19 becomes approximately Kushalappa and Ludwig's Eq. 9 for the relation between x and r in a growing host.

Until now, I have been assuming a maximum K_x that is only about 50 times as large as the initial size K_0. If K_x is much greater than 50 times K_0, the host K will grow essentially exponentially for a while; and if the disease N increases with an r much faster than the R of the host, Eq. 19 becomes approximately Van der Plank's Eq. 8.2 for the relation between x and r in a growing host.

Exploring the effect of R, K_0, and K_x on the skewness of a progress of x is a logical alternative to empirically fitting, say, Eq. 12 or 15. Returning to a host with K_x, say, about 50 times as large as K_0 and both R and r equal to 0.20, one sees that the consequent curve 2 in Fig. 5 is negatively skewed. The rapid increase of the proportion x is slow until the host has nearly reached its maximum K_x and disease can catch up.

A positive skew is produced when the host grows with only $R = 0.05$, which is much slower than the $r = 0.20$ of the disease. The growth of the host is slow but persistent since K is less than K_x for a long time. The slow but prolonged growth of the host opens new opportunities for infection, x slowly increases and curve 3 in Fig. 5 has a positive skewness. Repeated calculation shows that the skewness, which is zero with no growth, increases abruptly to about 3 when R is only 0.02, falls to about 1 when $R = 0.05$, becomes zero near $R = 0.10$, reaches -0.7 near $R = 0.20$ and increases slowly toward zero as R becomes very fast.

Beginning with Van der Plank (1963), plant pathologists have been paying much attention to correcting the rates estimated from the logits of Eq. 4 for the the growth of the host. Although the correction is difficult using Eq. 19, it is straightforward by Van der Plank's Eq. 8.2 if K_x is much larger than K and r is faster than R. Alternatively, correction is straightforward by Kushalappa and Ludwig's Eq. 10 if R is much faster than r. In the end, however, we can ask why labor to estimate precisely a shifty r that is a characteristic of the changing environment and not an immutable characteristic of a pathogen-host pair (Kranz, 1968a)?

Interacting Host and Pathogen

Although a microbe that had no impact on its host would not be a pathogen, all the preceding equations have had no place for the impact called disease. As every pathologist knows by training and every farmer knows by experience, pathogens do have an impact. At the very least the host that would otherwise grow in proportion to its foliage K toward a limit K_x is slowed by defoliation removing the product of earlier growth and reducing photosynthesis for further growth. And defoliation decreases the number N of infected leaves outright as well as decreasing the K that limits N. These interactions are evident in the pair of equations:

$$\frac{dN}{dt} = rN\frac{K - N}{K} - \frac{N}{t_d}$$

$$\frac{dK}{dt} = RK\frac{K_x - K}{K_x} - \frac{N}{t_d}$$

where t_d is the average time from infection to defoliation. Since $dx/x = dN/N - dk/K$,

$$\frac{dx/dt}{x} = \left(r - \frac{1}{t_d}\right)(1 - x) - R\left(1 - \frac{K}{K_x}\right) \tag{20}$$

Equation 20 differs from the previous models we have been using for the percentage of the plant infected. This equation incorporates the new feature of a decrease by defoliation; if $1/t_d$ is faster than r, x actually decreases with time despite infection at r per day. This decrease in the percentage of diseased foliage is familiar to every pathologist who has crawled among crops, hoping to record a perfect logistic increase but being disappointed by the facts of defoliation outpacing infection in weather unfavorable to the disease. Further, the decrease in x caused by growth at R per day has already been introduced in Eq. 18 and must be included in Eq. 20 along with defoliation.

Up to now, N has been defined as the number of diseased units of the host and since the increase in N is proportional to N itself, there has been the implication that all N are producing propagules. If we changed the definition of N to sporulating units of the host, then t_d would become the average duration of sporulation. Since t_d removes units from K as well as N in Eq. 20, the equation implies that the lost sporulating units can be replaced by growth.

Equation 20 is a new model, calling attention to the impact of disease on the host. It can be interpreted as a statement that when a unit of foliage is infected it continues to function, contributing to growth for an average of t_d days. The leaf unit then falls and no longer contributes to growth. As the equation is written, the defoliation is not subtracted from K_x and does open an opportunity for future growth. Other equations could be written for an effect of disease on growth before the leaf falls off or for a K_x decreased by defoliation. Boote et al. (1980) observed that before defoliation, levels of 1% and 25% leafspot on peanut leaves decreased photosynthesis by 20% and 80%.

Accepting Eq. 20 for the present, however, we can surmise an effect of disease on the eventual size K. The results of several calculations with several rates of host growth and with disease beginning at $N = 0.02$ and increasing at a typical r of 0.20 per day for 100 days are given in Table 4. Integration was numerical. If R is zero and the host K neither grows nor shrinks by defoliation and is steadily 100, disease x increases by the familiar logistic path, reaching 100% by the end of the 100 days. If, on the other hand, infected foliage falls after an average t_d of 10 days, the final K is only 19, and x is only 81% of that 19.

If the host grows from $K = 2$ at R of 0.05 to 0.30 per day, K reaches 75 to 100 by day 100 if leaves are not lost. If, however, leaves fall after an average of $t_d = 10$, growth makes the host K only 23 to 85 and disease makes the portion x only about 50% by day 100.

Another calculation in Table 4 shows the effect of varying but reasonable t_d on the size K of the host and on disease when both K and N increase at 0.20/day, with initial values of $K = 2$ and $N = 0.02$. If t_d is only 5 days or $1/r$, infections are removed as fast as they occur, and K is 100 and x is less than 1%, even after 100 days. As t_d increases from 10 to 30 days, disease increases and, surprisingly, the final K is also greater because the infected leaves live longer.

Table 4 The effect of defoliation on leaf area K and percent disease x

R per day	No defoliation			Defoliation		
	x(%)	K, area	Skew	t_d	x(%)	K, area
0	100	100	0	10	81	19
0.05	92	75	1.0	10	52	23
0.20	100	100	−0.6	10	48	78
0.30	100	100	−0.3	10	47	85
0.20				5	<1	100
0.20				10	48	78
0.20				20	75	81
0.20				30	83	86
0.20				Infinite	100	100

Note: Defoliation occurs t_d days after leaves become infected. Percent disease, x (dimensionless), and leaf area, K (dimensionless), are given for day 100. K increases at the rate R (per day) and x increases at $r = 0.20$/day. On day 0, K is 2, N is 0.02, and x is 1%.

An examination of Eq. 20 for $dx/dt/x$ shows that it becomes zero when

$$\frac{1 - x}{1 - K/K_x} = \frac{R}{r - 1/t_d}$$

For example, when $R = 0.20$, $r = 0.30$ and $t_d = 10$, then K/K_x, and x do not change when they both reach 67%. Or, when $R = 0.10$, $r = 0.30$ and $t_d = 4$, equilibrium is reached when K/K_x is 58% and x is 17%. When there is no defoliation, K/K_x and x reach equilibrium when

$$\frac{1 - x}{1 - K/K_x}$$

equals R/r.

Defoliation has not been overlooked in pathology. Plaut and Berger (1980) found it proceeded faster than the visible spotting of peanut leaves by *Cercosporidium personata,* and Kushalappa and Ludwig (1982) carefully corrected their observations for fallen leaves before estimating the impact of growth on x.

Van der Plank (1963) corrected his basic infection rate for removals and called it R_c. If the host is not growing and the R in Eq. 20 is zero, then it becomes

$$\frac{dx/dt/x}{1 - x} = \frac{d(\text{logit } x)}{dt} = \frac{rt_d - 1}{t_d} \tag{21}$$

Equating t_d with Van der Plank's period during which tissue remains infectious quickly leads to his Threshold theorem: "No epidemic can start unless conditions of host, pathogen, environment and fungicide interact to make" $rt_d > 1$. Returning to Eq. 20 and a host growing R per day, one sees

$$\frac{d(\text{logit } x)}{dt} = \frac{rt_d - 1}{t_d} - R\frac{1 - K/K_x}{1 - x} \tag{22}$$

shows there is more to the Threshold theorem: no epidemic can start unless conditions of host, pathogen, environment, and fungicide interact to make r greater than $1/t_d + R(1 - K/K_x)/(1 - x)$. Thus, a host growing rapidly because of small size relative to K_x, a fast R, and (surprisingly) a large percentage of the host already diseased join a short t_d in setting a high threshold for an r that will increase disease. Of course when the host is large or mature and K is nearly K_x, r need only be greater than $1/t_d$ to start an epidemic.

Spreading Propagules

If the propagules of a pathogen are not nicely spread at random over a crop, the disease may not increase in proportion to the healthy fraction $1 - x$ of the crop, and the logistic Eq. 1 loses its foundation. Even casual observation of a field of diseased foliage often reveals foci of severe disease and regions of little disease, which is not surprising if the disease is contagious and spreading from early infections in the crop during several rapid generations of the pathogen within a season. On a smaller scale, using leaves rather than regions of a crop, one often finds more leaves with many infections and more with no infections than would be caused by a nicely random spread of propagules over the crop.

On the larger scale of foci in a field, Jeger (1983) derived equations for the decrease of disease x with distance X from the center of a focus. Where the increase of x at a spot is logistic and logit (x) decreases as $1/X^2$,

$$x = \frac{1}{1 + (1/x_{0,1} - 1) \cdot \exp(-rt) \cdot X^2} \tag{23}$$

The $x_{0,1}$ is the disease at zero time and unit distance. Integrating x over the circle with radius X produces an average x over space:

$$\frac{\ln(1 + bX^2)}{bX^2} = \frac{\ln x}{1 - 1/x} \tag{24}$$

where $b = (1/x_{0,1} - 1) \cdot \exp(-rt)$.

Thus if logit(x) varies as the reciprocal of the square of distance, the average x inside a circle around a focus can be determined from the x on the circle alone.

If the mean $x/(1 - x)$, not x or logit(x), is calculated over an annulus from, say, unit distance to X around the focus, the log of the mean increases linearly with time. Thus the average $x/(1 - x)$ over foci can produce progress curves that are logistic, and average x over foci can produce approximately the same result.

On the small scale of lesions on leaves rather than foci in fields, the outcome could also be simple. Commonly x is the fraction of the leaves that are diseased because they have one or more lesions. If n lesions are randomly distributed over K leaves, the fraction $(1 - x)$ of the leaves that are healthy will be the probability $\exp(-n/K)$ of no lesions on a leaf according to the Poisson function with mean n/K lesions per leaf, and the relation between n lesions on K leaves and the fraction x with any lesions will be the familiar multiple infection transformation (Gregory, 1948). The frequency of 0, 1, 2, 3, . . . lesions on leaves, however, often differs from the Poisson function that fits random distributions as Gregory anticipated, and instead is fit by the "contagious" negative binomial function, making the relation between the number n of lesions and fraction x of leaves diseased

$$n = Kk\,[(1 - x)^{-1/k} - 1] \tag{25}$$

where k is positive. The value of k is infinite when the distribution is random or Poisson, decreases as contagion or overdispersion increases and is near 1 in many foliar diseases (Waggoner and Rich, 1981).

To incorporate this evidence into a growth law, dN/dt is broken down to $dn/dt \cdot dN/dn$. The rate of production of propagules is dn/dt if effective propagules and subsequent lesions are equal, and dN/dn is the number of leaves infected per effective propagule.

In the logistic Eq. 1, dn/dt is rN, and dN/dt is $(K - N)/K$, as it must be for random distribution. If, however, the relation between n and N is contagious, as in Eq. 25, dN/dn is $(1 - x)^{1 + 1/k}$, which can be substituted for $(K - N)/K$ in Eq. 1. When $k = 1$, the new equation produces a positively skewed curve, which is convex upward when plotted as logits vs. time. Although this contagion and equation could be advanced as a cause of the positive skew often encountered and fit by the Gompertz and the Richards ($m < 1$) curves, its skewness is about 3, exceeding the skewness usually encountered.

The new equation for the increase of contagious disease can be modified further, making the rate dn/dt of propagule production proportional to the number of lesions rather than the number of diseased leaves. Since a leaf with two lesions would logically produce twice as many propagules as a leaf with one lesion, the rN for dn/dt is logically replaced in Eq. 1 by rn. With $dn/dt = r$ multiplied by the right-hand side of Eq. 25, and $dN/dn = (1 - x)^{1 + 1/k}$,

$$\frac{dN}{dt} = rKk[(1 - x)^{-1/k} - 1](1 - x)^{1 + 1/k} \tag{26}$$

Thus a new equation for growth has been derived that still follows the logistic concept that the organism multiplies in proportion to itself and to the opportunities. The new features are (1) multiplication in proportion to the number n of lesions rather than to the number N of leaves with one or many lesions and (2) the contagious distribution of propagules and consequent lesions to a degree indicated by the k of the negative binomial function.

Among 112 frequency distributions of lesions on plants, including many with foliar infections, fully 98 were not random, whereas only 5 were not fitted well by the negative binomial function. The k of the function ranged from about 4 for ash rust to 0.01 for two bacterial diseases of leaves. Since the *Puccinia peridermiospora* rusting ash leaves comes from another host and does not multiply on ash, its moderately large k is a logical indication of only moderate contagion. The smaller k is a logical indication of great contagion in bacterial pathogens spread by splashing (Waggoner and Rich, 1981).

The skewness of the progress curve traveled by the new function is zero when $k = 1$ because then the new function reduces to the classic logistic model proposed by Verhulst. When k is larger, say 4, however, skewness is -0.9; and when $k = 0.2$, skewness is $+1.7$. That is, multiplication in proportion to lesions and contagion to degree $k = 1$ produce the same curve as multiplication in proportion to infected leaves—regardless of lesions per leaf—when there is no contagion. When, however, contagion is slight, making $k = 4$ (as in the example of ash rust) and multiplication is proportional to lesions, the progress of infected leaves is negatively skewed. Finally, when contagion is great as with the $k = 0.2$ of the bacterial disease, the progress of infected plants is positively skewed.

Because much was written earlier about the inflection point or level of x where the progress curve passed from increasing to decreasing dN/dt, I determined that the inflection of Eq. 26 is at $1 - x = [k/(1 - k)]^k$. Thus Eq. 26 has its inflection at the same $1/e$ as the Gompertz equation when $k = 1/3$, and one might be tempted to conclude that infections are aggregated in a pattern of considerable contagion whenever a progress curve is better fitted by the Gompertz than the logistic model. Since other factors can cause positive skewness, however, we must have independent evidence of contagion and multiplication in proportion to lesions before we can interpret the skewness of a progress curve as a consequence of contagion.

HOW TO USE PROGRESS CURVES

Although distinguishing the uses of progress curves from their interpretation, which was discussed at length in the preceding section, is arbitrary, the question of "Why use curves?" is not fully answered until use has been explicitly examined. Sigmoid progress has a rate, a location plus skewness, and an area beneath that can be used.

Rate

Van der Plank's great theme is "In the long run, a high rate of interest is more important than a large balance in the bank today." In other words, a modest decrease in r may prevent an epidemic. Importantly, the characteristic time $1/r$

of multiplication must be shorter than the average duration t_d of infectivity for disease to increase and for an epidemic to ensue, even on a host that is not growing. In this context, r is the rate inherent in the production and spread of propagules and their incubation, whereas $(r - 1/t_d)$ is the rise of the progress curve if the host is not growing. If the host is growing, then Eq. 22 must be consulted. Thus a knowledge of progress curves is used in weighing whether estimates of r, t_d, and R imply the rise or fall of the curve and the explosion or control of an epidemic.

Since so many factors affect progress, it is foolhardy to use the progress curve alone to infer whether r, t_d, or R caused an observed slowing of progress. It is, then, doubly foolhardy to infer from the curve alone that a single factor of genes of the host, dryness of the weather, or action of a fungicide decreased r and slowed the progress of an epidemic. And to infer from a curve alone that, say, a gene initially decreased sporulation and thereby decreased r and thereby slowed progress is triple foolhardiness. But all is not lost.

Although we cannot infer causes from curve, we can deduce the curve from causes. This is possible because the parameters in Eqs. 20 and 26, for example, are logical. Because the rate r reflects the daily arrival at new hosts of effective propagules per source, and we can deduce the changed progress following halved production, viability, or flight of propagules. Because r is commonly expressed in units of days^{-1} and incubation is often a long stretch of days, the effect of doubling incubation time can also be deduced, as can the effects of shorter infectivity t_d and faster growth R of the host. Van der Plank's (1963) book is filled with examples showing the power of these deductions.

Finally, a combination of a progress curve with other observations of an epidemic give insight into causes that the curve alone cannot. For example, if the weather turns dry, no propagules are produced on expanding foliage, and the percentage x of leaves diseased declines, then there seems little doubt about either the cause of the altered progress or the effect of the weather and growth. Similarly, if propagules appear on leaves in proportion to lesions, the frequency distribution of lesions per leaf follows the negative binomial function with $k = 1$, and the logits of disease increase linearly, there seems little doubt that multiplication and contagion conspired to fool us with a logistic curve, as Eq. 26 predicted. A simple progress curve reveals little, but the theory that it represents allows us to deduce the consequences of changed rates, and with other evidence, we can both infer causes and confirm their impact.

Location and Skewness: Extrapolation

Although the ends of curves are less certain than their middles as the variances of estimates from regression lines demonstrate, the ends are important. The front end of a progress curve is the inoculum, sometimes overwintering, and its reduction is the goal of sanitation. The front end is the first date t_1 of appearance or x_1, which Kranz (1968b) found was the best predictor of ensuing disease. The tail end of a curve is the disease that is harvested or carried through the winter to inoculate the crop next spring.

Scarce initial infection may be overlooked until an epidemic reminds us that we should have looked, and it is always difficult to find. Extrapolation of the progress curve to an earlier time estimates the missing or inconspicuous x_1. Imprecision of our knowledge of r affects x_1 greatly, of course, because it is multiplied by the long time between x_1 and the mean to obtain the imprecise x_1. The progress curve can also be extrapolated forward to predict when the epidemic will reach a given level (McCoy, 1976).

Skewness also affects extrapolation. Berger (1981) cautioned that choosing, say, the symmetrical logistic rather than the skewed Gompertz equation changes the extrapolation profoundly. Consider the outcome of sanitation that decreases x_1 from 10 to 0.1% before multiplication ensues at $r = 0.2$ for 30 days. According to the logistic model, this theoretical sanitation decreases disease from 98 to 29%. According to the Gompertz equation, when r is half as fast, as Richards (1959) showed it must be for comparing logistic and Gompertz models, the same theoretical sanitation only decreases disease after 30 days from 89 to 71%. Berger expressed the difference between the two equations for the increase of disease as a far briefer postponement of 50% disease when the Gompertz rather than the logistic model ruled. The curve that fits reality exactly must be used for extrapolation because the trouble accumulates in the ends of curves.

Area under the Progress Curve

Although we exert ourselves to grow crops only because they yield, evaluations of the relation of disease progress curves to yield are surprisingly infrequent. In Table 4, I examined the effect of disease on K after 100 days. This might pass for the relation of disease progress to yield if K were the foliage and the foliage were the yield, as in tobacco or hay. In many crops, such as wheat and potatoes, however, yield is not related to the leaf area at a single time but rather to the integral of leaf area over time, which is called the *leaf area duration* (Watson, 1947). Since yield is largely photosynthate (since photosynthesis occurs in leaves and since products accumulate over time), it is not surprising that leaf area duration is the relevant parameter.

A parameter reminiscent of leaf area duration is used in plant pathology: ADPC or area under the disease progress curve is the integral over time of the percentages x of foliage diseased. It is considered a better index of effectiveness of fungicides and resistance than r because:

ADPC uses all data available and does not obscure the variation in rate of disease development because of transformations. Moreover, minor differences in disease severity early in the season have little effect on ADPC. (Shaner and Finney, 1977)

The apparent infection rates for the two treatments were not significantly different, although the ADPC [were]. The ADPC reflected both the onset as well as the rates of epidemic development. (Fry, 1978)

These are, of course, testimonials to the ability of ADPC for distinguishing between the disease following treatments and not claims that it is related to yield.

And the integral ADPC of a ratio of N to K is neither an integral of leaf area that is healthy and producing food nor of the leaf area that is diseased.

Despite the paucity of studies relating disease progress to yield, we can reason what the relationships might be, using the foundation that leaf area duration determines yield. At this primitive stage, I have continued to neglect both (1) the impact of infection on photosynthesis and growth until a leaf falls and (2) the effect of defoliation on K_x. That is, I have used Eq. 20 to calculate the growth of the host K and increase of disease N. Assuming K and N are areas, I calculated ADPC as the integral of x for 100 days. Seeking an analog of Watson's leaf area duration that would be proportional to yield, I also calculated the healthy area duration HAD by integrating the difference between K and N over a 100-day period, using Eq. 20.

When the host K and the rate r of disease are constant and no leaves fall, x is simply related to time by the familiar Eq. 3, and this x is easily integrated over time, showing that ADPC is $t + \ln(x_0/x)/r$, which approaches $t + \ln(x_0)/r$ after a long time. When disease is progressing on a host that is growing or losing leaves, however, the relation of x to time is complex. To calculate HAD, I used numerical integration (Table 5). Because Fry, in the previous quotation, found ADPC advantageous in reflecting the time of onset of disease, the effect of initial infection N_0 was examined. The ADPC for a crop that is not infected (i.e., $N_0 = 0$) is zero, and the integral HAD of healthy leaf area is 80. The

Table 5   Relationship to ADPC and HAD of initially diseased area N_0, rate of increase of disease r, growth of host R, and mean time for defoliation t_d

Parameter and value		ADPC (portion-days)	HAD (area-days)
N_0	= 0	0	80
	0.02	57	24
	0.10	65	17
	0.20	69	14
	2.00	80	9
r (per day) =	0.02	<1	80
	0.10	17	64
	0.20	57	24
R (per day) =	0.02	68	1
	0.10	57	11
	0.20	57	24
t_d (days) =	5.00	<1	80
	10.00	12	65
	20.00	35	43
	30.00	42	37
	Infinite	57	24

Note: Unless otherwise specified, $N_0 = 0.02$; $r = 0.2$; $R = 0.2$; and there is no defoliation. Host K begins at 2 and reaches a maximum K_x of 100.

increase of N_0 increases ADPC from 57 to 80 as N_0 increases 100-fold from 0.2 to 2 while $r = 0.2$. The decrease in yield indicated by HAD is, however, relatively greater, falling from 24 to 9. Interestingly, in the mathematical crop, some HAD is accumulated even when $N_0 = 2$ and 100% of the initial K of 2 because dK/dt, which is limited by the relatively large K_x, outdistances dN/dt, which is limited by the relatively smaller K.

If disease increases from $N_0 = 0.02$ at a rate $r = 0.02$, which is much less than the $R = 0.2$ of the host, x is very small for a long time and ADPC is less than 1, while HAD is fully the 80 of the healthy host. Increasing r to 0.2 increases ADPC to many times its value at $r = 0.02$, but healthy leaves still exist during the progress of the disease, and HAD is decreased by only two-thirds.

Whether the host grows at only $R = 0.02$ or fully 0.20/day, disease increasing at 0.20/day produces a relatively invariant ADPC of 57 to 68. The faster-growing crop, nevertheless, produces a substantially larger HAD and presumably larger yield than the slower-growing crop.

Finally, the impact of defoliation is examined in Table 5. If infected leaves have an average life t_d of only 5 days while infection increases at 0.20, ADPC is less than 1, and ADPC increases to many times this value if the leaves hang on long after infection. A longer life (>5 days) t_d of the infected leaves, on the other hand, only decreases HAD by two-thirds.

All in all, ADPC does summarize the seasonal course of disease better than r, but being the integral of a ratio of diseased to healthy plant material, it does not vary exactly as HAD. Rather, HAD, being close to the integral of net photosynthesis, seems the logical parameter to relate to yield. Testing whether HAD or another parameter better shows the impact of disease seems a worthy assignment if epidemiology is to go beyond watching pathogens to become an effective tool for laying bare the inherent rationality of how crops grow.

APPENDIX: Modification of the Logistic Spread of Disease for a Growing Crop

If both the number of infected leaves N and the total number of leaves K within a canopy obey a logistic law of growth, then the course of the ensuing epidemic is described by a pair of coupled, first-order, nonlinear differential equations

$$\frac{dN}{dt} = r N \frac{K - N}{K} \tag{A1}$$

and

Appendix by Francis J. Ferrandino of the Connecticut Agricultural Experiment Station.

$$\frac{dK}{dt} = R K \frac{K_x - K}{K_x}$$ (A2)

where r is the time rate of the spread of the disease, R is the time rate of leaf production, and K_x is the upper asymptotic limit to the value of K. Both Eqs. A1 and A2 are of the same form

$$\frac{dX}{dt} = A X \frac{Y - X}{Y}$$ (A3)

and have the same general solution.

To solve Eq. A3 for general $Y(t)$, make the substitution $Z = 1/X$, which yields

$$-\frac{1}{Z^2} \frac{dZ}{dt} = \frac{A}{Z} \left(1 - \frac{1}{YZ}\right)$$

or (A4)

$$\frac{dZ}{dt} + AZ = \frac{1}{Y}$$

Multiplication of both sides of Eq. A4 by the integrating factor e^{At} renders it exact. The result is

$$e^{At} \frac{dZ}{dt} + A e^{At} Z = \frac{d}{dt} (e^{At}Z) = \frac{e^{At}}{Y}$$ (A5)

which can be directly integrated to yield the general solution to Eq. A3:

$$Z(t) = \frac{1}{X(t)} = \frac{e^{-At}}{X(0)} + e^{-At} \int_0^t \frac{e^{At'}}{Y(t')} dt'$$ (A6)

Comparing Eqs. A2 and A3, identify $X = K$, $A = R$, and $Y(t) = K_x$ (a constant). Insertion of these values into Eq. A6 gives the solution for $K(t)$ once the integration is performed. The result is

$$K(t) = \left[\frac{e^{-Rt}}{K(0)} + \frac{1}{K_x} \left(1 - e^{-Rt}\right)\right]^{-1}$$ (A7)

which is the logistic equation. Similarly in comparing Eqs. A1 and A2, identify $X = N$, $A = r$, and $Y(t) = K(t)$. Insertion of these values into Eq. A6, using the solution for $K(t)$ from Eq. A7 yields

$$N(t) = \left\{\frac{e^{-rt}}{N(0)} + e^{-rt} \int_0^t e^{rt'} \left[\frac{e^{-Rt'}}{K(0)} + \frac{1}{K_x} \left(1 - e^{-Rt'}\right)\right] dt'\right\}^{-1}$$ (A8)

Factoring out K_x and performing the integration, we see there are two solutions depending on whether or not $r = R$:

$$N(t) = K_x \left\{ 1 + \left(\frac{K_x}{N(0)} - 1 \right) e^{-rt} \right.$$
$$\left. + \frac{r}{r - R} \left(\frac{K_x}{K(0)} - 1 \right) \left(e^{-Rt} - e^{-rt} \right) \right\}^{-1} \quad \text{for } r \neq R$$

$$N(t) = K_x \left\{ 1 + \left(\frac{K_x}{N(0)} - 1 \right) e^{-rt} + r \left(\frac{K_x}{K(0)} - 1 \right) t \, e^{-rt} \right\}^{-1} \quad \text{for } r = R$$

\hfill (A9)

Defining $x = N(t)/K(t)$, $a = K_x/N(0) - 1$ and $b = K_x/K(0) - 1$ and using Eqs. A7 and A9 we obtain

$$x(t) = (1 + be^{-Rt}) \left[1 + ae^{-rt} + \frac{r}{r - R} b \, (e^{-Rt} - e^{-rt}) \right]^{-1} ; \quad r \neq R$$

$$x(t) = (1 + be^{-rt}) \left[1 + ae^{-rt} + r \, bt \, e^{-rt} \right]^{-1} ; \quad r = R$$

\hfill (A10)

as given in the text.

REFERENCES

Analytis, S. 1973. Methodik der Analyse von Epidemien dargestellt am Apfelschorf (*Venturia inaequalis* (Cooke) Aderh.) *Acta Phytomedica* 1:7–75.

Barratt, R. W. 1943. The relationship between *Alternaria* blight and "physiological" maturity in the tomato plant. M. S. Thesis, University of New Hampshire, Division of Biological Sciences, June, 1943.

Berger, R. D. 1973. Infection rates of *Cercospora apii* in mixed populations of susceptible and tolerant celery. *Phytopathology* 63:535–537.

———. 1981. Comparison of the Gompertz and logistic equations to describe plant disease progress. *Phytopathology* 71:716–719.

Blackman, J. H. 1919. The compound interest law and plant growth. *Ann. Botany* 33:353–360.

Bliss, C. I. 1970. *Statistics in Biology*, vol. 2. McGraw-Hill Book Co., New York, 639 pp.

Boote, K. J., Jones, I. W., Smerage, G. H., Barfield, C. S., and Berger, R. D. 1980. Photosynthesis of peanut canopies as affected by leafspot and artificial defoliation. *Agron. J.* 72:247–252.

Campbell, C. L., Pennypacker, S. P., and Madden, L. V. 1980. Progression dynamics of hypocotyl rot of snapbean. *Phytopathology* 70:487–494.

Crosier, W. 1934. Studies in the biology of *Phytophthora infestans*. New York (*Cornell*) *Agr. Exp. Sta. Mem.* 155. 40 pp.

Ferrandino, F. J. 1985. A mathematical model of biological development. *J. Theor. Biol.* 112:609–626.

Fry, W. E. 1975. Integrated effects of polygenic resistance and a protective fungicide on development of potato late blight. *Phytopathology* 65:908–911.

————. 1978. Quantification of general resistance of potato cultivars and fungicide effects for integrated control of potato late blight. *Phytopathology* 68:1650–1655.

Geary, R. C., and Pearson, E. S. 1938. Tests of normality. Biometrika Office, Univ. College, London.

Gottwald, T. R., and Bertrand, P. F. 1983. Effect of time of inoculation with *Cladosporium caryigenum* on pecan scab development and nut quality. *Phytopathology* 73:714–718.

Gregory, P. H. 1948. The multiple infection transformation. *Ann. Appl. Biol.* 35:412–417.

Imhoff, M. W., Leonard, K. J., and Main, C. E. 1982. Analysis of disease progress curves, gradients, and incidence-severity relationships for field and phytotron bean rust epidemics. *Phytopathology* 72:72–80.

Jeger, M. J. 1983. Analyzing epidemics in time and space. *Plant Pathol.* 32:5–11.

Koestler, A. 1980. *Bricks to Babel*. Random House, New York. 697 pp.

Kingsland, S. 1982. The refractory model: The logistic curve and the history of population ecology. *Quart. Rev. Biol.* 57:29–52.

Kranz, J. 1968a. Eine Analyse von annuellen pilzlicher Parasiten. II. Qualitative und quantitative Merkmale von Befallskurven. *Phytopathol. Z.* 61:171–190.

————. 1968b. Eine Analyse von annuellen Epidemien pilzlicher Parasiten. III. Uber Korrelationen zwischen quantitativen Merkmalen von Befallskurven und Ahnlichkeiten von Epidemien. *Phytopathol. Z.* 61:205–217.

————. 1974. Comparison of epidemics. *Annu. Rev. Phytopathol.* 12:355–374.

Kushalappa, A. C., and Ludwig, A. 1982. Calculation of apparent infection rate in plant diseases: Development of a method to correct for host growth. *Phytopathology* 72:1373–1377.

Large, E. C. 1952. The interpretation of progress curves for potato blight and other plant diseases. *Plant Pathol.* 1:109–117.

Luke, H. H., and Berger, R. D. 1982. Slow rusting in oats compared with the logistic and Gompertz models. *Phytopathology* 72:400–402.

McCoy, R. E. 1976. Comparative epidemiology of the lethal yellowing, Kaincope, and Cadang-Cadang diseases of coconut palm. *Plant Disease Reporter* 60:498–502.

M'Kendrick, A. C., and Pai, M. K. 1911. The rate of multiplication of micro-organisms: A mathematical study. *Proc. R. Soc. Edin.* 31:649–655.

Pearl, R., and Reed, L. J. 1920. Rate of growth of the population of the United States since 1790 and its mathematical presentation. *Proc. National Acad. Sci.* 6:275–288.

Pennypacker, S. P., Knoble, H. D., Antle, C. E., and Madden, L. V. 1980. A flexible model for studying plant disease progression. *Phytopathology* 70:232–235.

Plaut, J. L., and Berger, R. D. 1980. Development of *Cercosporidium personatum* in three peanut canopy layers. *Peanut Sci.* 7:46–49.

————. 1981. Infection rates in three pathosystem epidemics initiated with reduced disease severities. *Phytopathology* 71:917–921.

Richards, F. J. 1959. A flexible growth function for empirical use. *J. Exp. Bot.* 10:290–300.

Shaner, G., and Finney R. E. 1977. The effect of nitrogen fertilization on the expression of slow-mildewing resistance in Knox wheat. *Phytopathology* 67:1051–1056.

Shaner, G. E., Peart, R. M., Newman, J. E., Stirm, W. L., and Loewer, O. L. 1972. EPIMAY: An evaluation of a plant disease display model. Purdue Univer. Agr. Exp. Sta. RB-890. 15 pp.

Thompson, D. W. 1942. *Growth and Form*. MacMillan Co., New York, 1116 pp.

Turner, M. E., Blumenstein, B. A., and Sebaugh, J. L. 1969. A generalization of the logistic law of growth. *Biometrics* 25:577–580.

Van der Plank, J. E. 1963. *Plant Diseases: Epidemics and Control*. Academic Press, New York, 349 pp.

Waggoner, P. E., and Rich, S. 1981. Lesion distribution, multiple infection, and the logistic increase of plant disease. *Proc. Natl. Acad. Sci. USA* 78:3292–3295.

Watson, D. J. 1947. Comparative physiological studies on the growth of field crops. I. Variation in net assimilation rate and leaf area between species and varieties, and within and between years. *Ann. Bot. NS* 11:41–76.

Weibull, W. 1951. A statistical distribution function of wide applicability. *J. Appl. Mechanics* 18:293–297.

Winsor, C. P. 1932. Gompertz curve as a growth curve. *Proc. Nat. Acad. Sci.* 18:1–8.

Wright, S. 1926. Review of the biology of population growth. *J. Am. Stat. Soc.* 21:493–497.

Chapter 2

Interpretation and Uses of Disease Progress Curves for Root Diseases

C. Lee Campbell

Root diseases are economically important in most parts of the world. Losses due to root diseases often go unnoticed, however, because aboveground symptoms may be indistinct or attributed to other stresses on the plants. The study of root disease epidemiology is still in an early stage of development compared with the epidemiology of foliar diseases. Not only has the need been less obvious, but also the quantitative characterization of root disease epidemics has been more difficult because of the relative inaccessibility of the roots. Pathogen populations and numbers of infections are less easily monitored in the soil than in leaves or other aerial plant parts. Furthermore, conceptual interpretations developed for epidemics of foliar diseases have not proved satisfactory for direct application to root disease epidemics. New conceptual approaches are required.

MONITORING ROOT DISEASE EPIDEMICS

Quantifying Populations of Pathogen Propagules

Models of root disease epidemics all depend on careful assessment of initial inoculum. Numerous techniques have been developed for this purpose. Some involve direct counting of large propagules extracted from the soil by wet sieving, elutriation, or differential centrifugation. Others depend on counts of colonies that develop on selective media. For any of these methods, careful attention must be given to sampling design, because soilborne inoculum typically occurs in clustered or aggregated spatial patterns (Campbell and Noe, 1985).

Quantifying Disease in Host Populations

Symptom Evaluation   The progress of root disease epidemics may be monitored as the increase in incidence or severity of either shoot or root symptoms. For verticillium or fusarium wilts the shoot symptoms are often distinct enough to be easily recognized. In those cases shoot symptoms may be used as a measure of disease incidence. It is risky, however, to try to infer root disease severity from the severity of shoot symptoms of individual plants. For diseases such as Rhizoctonia hypocotyl and root rots, shoot symptoms may be poorly defined, but root and hypocotyl lesions may be characteristic enough to allow accurate estimates of disease severity as well as incidence.

The lag time between infection and the appearance of symptoms on roots and shoots further complicates the monitoring of root disease progress. In particular, the time required for development of shoot symptoms may be extended, depending on variable environmental conditions and changes in host resistance. The delay in the appearance of shoot symptoms, which may be variable, must be taken into account in interpreting disease progress curves based on shoot symptoms.

The advantage of relying on shoot symptoms is that they can be nondestructively evaluated, which allows repeated assessments on the same plants. Root symptoms usually provide a more accurate assessment of disease severity, but in most cases their assessment involves destructive sampling. In addition, removal of roots from the soil and washing them before evaluation requires much more time and effort than visual evaluation of shoot symptoms. Destructive sampling by removing entire plants from the experimental area may also alter the course of the epidemic. Assessment of root disease severity should be done in comparison with roots of healthy plants in order to reveal the effects of root pruning by the pathogen.

Sampling Considerations   The absolute minimum number of sampling dates necessary for adequate statistical analysis of a disease progress curve is five. Greater numbers of sampling dates increase the resolution of the true course of the epidemic over time and enhance the accuracy of the estimated rate of disease progress.

Formulas are available to calculate the optimum sample size (Karandinos, 1980). If the propagules or diseased plants occur randomly throughout the sample area, systematic sampling is as suitable as other procedures and is easy to perform. For root diseases, however, inoculum and disease usually occur in clustered patterns. If the area of individual clusters is relatively large, it may be possible to map the locations of clusters prior to extensive sampling so that a stratified sampling scheme can be devised. For small clusters, a simple random sample or a two-stage or multistage sample may be appropriate. In a two-stage sample, quadrats are selected randomly and samples are taken within each quarter. See Gilligan (1982) for considerations involved in the choice of size and shape of sampling units.

ANALYSIS OF DISEASE PROGRESS CURVES

The various models described by Waggoner (Chap. 1) for analyzing foliar disease epidemics have also been applied to root diseases. Studies of root disease progress curves and the models employed are summarized in Table 1. The list is small, reflecting the limited development of the study of epidemics of root diseases.

Initial Conceptual Basis for Analysis

The analysis of disease progress curves in phytopathology has its origins in the work of J. E. Van der Plank (1963). Chester (1946) wrote of the "tempo" of cotton wilt development and noted differences in the shape of progress curves for this disease. It was Van der Plank's synthesis of previous knowledge, however, that provided the techniques and conceptual framework necessary for the meaningful analysis of disease progression.

Van der Plank (1963) presented two biological models—monomolecular and logistic—for describing disease progression. These models are based on the assumption that disease occurrence and pathogen spread are random. This is not

Table 1   Root pathosystems in which disease progression has been described using biological or statistical models

Host	Pathogen	Model utilized[a]	Reference
Celery	Fusarium	L	Welch (1981)
Cotton	*Verticillium dahliae*	Linear	Pullman and DeVay (1982)
Lettuce	*Sclerotinia minor*	M, L	Jarvis and Hawthorn (1972)
Pea	*Aphanomyces euteiches* f. sp. *pisi*	L	Pfender and Hagedorn (1983)
Snapbean	*Rhizoctonia solani*	Polynomial	Campbell et al. (1980a;1980b)
Tobacco	*Phytophthora parasitica* var. *nicotianae*	M, L	Kannwischer and Mitchell (1978)
Tobacco	*P. parasitica* var. *nicotianae*	M, L, W	Campbell and Powell (1980)
Tobacco	*P. parasitica* var. *nicotianae*	M, BR, G, L, W	Campbell et al. (1984)
Tobacco	*P. parasitica* var. *nicotianae*	L	Ferrin and Mitchell (1984)
Wheat	*Cochliobolus sativus*	M	Verma et al. (1974)
Wheat/barley	*Cochliobolus sativus*	L, W	Stack (1980)

[a]BR = Bertalanffy-Richards; G = Gompertz; L = logistic; M = monomolecular; W = Weibull.

always the case for foliar diseases (Waggoner and Rich, 1981) and is probably even less likely for root diseases. The monomolecular model can be written as:

$$\frac{dy}{dt} = k(1 - y) \tag{1}$$

where dy/dt is the absolute rate of disease increase, k is the rate parameter, and y is the proportion of disease at time t. The maximum value that y can reach is assumed to be 1. The model assumes that the absolute rate of disease increase is proportional to the amount of healthy tissue remaining at any given time. The logistic model can be written as

$$\frac{dy}{dt} = ky(1 - y) \tag{2}$$

where the terms are as previously defined. In this case the absolute rate of disease increase is assumed to be proportional to the amount of disease and the amount of healthy tissue present.

Van der Plank (1963) provided a biological basis for each model in describing progress of appropriate types of disease. The monomolecular model describes "simple interest" diseases and the logistic model, "compound interest" diseases. In simple interest diseases, there is only one cycle of infection and inoculum production per growing season (i.e., they are monocyclic). In compound interest diseases, cycles of infection and inoculum production occur repeatedly and thus many disease cycles occur during a growing season (i.e., they are polycyclic).

The monomolecular and logistic models are biological models based on prior assumptions about the mechanisms of disease increase (Madden, 1980, and Chap. 3). These models differ from statistical models which are mathematical formulas with values of parameters chosen to adequately describe disease progress for specific data sets, but which lack a precise biological interpretation. Attempts to use and interpret statistical models as "biological" models have caused confusion in the recent literature on the analysis of root disease progress curves.

The concept that diseases induced by soilborne pathogens are monocyclic was, for a time, firmly established in the literature (Campbell and Powell, 1980). As Pfender (1982) noted, this is probably the result of Van der Plank's (1963) use of fusarium wilt as a typical example of a simple interest disease caused by a soilborne pathogen. The concept was promoted (Bald, 1969) and used to interpret experimental results (Verma et al., 1974).

In analyzing disease progression of common root rot of wheat (*Cochliobolus sativus*), Verma et al. (1974) assumed the biological system to be monocyclic. Consequently, they used the monomolecular model to describe disease progression. Subsequently, Stack (1980) found that the logistic model was more statistically suited for describing epidemics of common root rot of wheat than was the monomolecular model. He concluded that common root rot was a compound interest disease. Similar types of conclusions were drawn from work by Campbell et al. (1980b), and Campbell and Powell (1980) for several root pathosystems.

The error in such interpretations (Huisman, 1982; Pfender, 1982) is in inferring the nature of the disease cycle from the description of the disease progress curves. In the epidemics of common root rot of wheat (Morrall and Verma, 1981; Stack, 1980) and bean hypocotyl rot (Campbell et al., 1980a, 1980b), the data constitute valid disease progress curves; however, these curves represent a composite of infection rates and pathogen growth rates that vary over time and do not necessarily indicate whether the disease is of the simple or compound interest type.

The models utilized in the above cited examples and those listed in Table 1 (e.g., Gompertz, logistic, monomolecular) were originally developed as biological models, but are applied in the sense of a statistical or empirical model. The valid purposes in using the models are to characterize the epidemic and provide a basis for comparisons among epidemics. Biological determinations of infection cycles, however, must be based on other more direct experimental evidence.

Model Selection

Many of the biological and statistical models used to describe disease progress curves for foliar pathosystems (Madden, 1980, Waggoner, Chap. 1) may also be useful in analysis of root disease progress curves. The choice of which model to use may be based on an examination of the plot of dy/dt or y vs. t (or other independent variable). If the principal task is to simply describe disease progression, the model that best fits the observed data is sufficient. Statistical criteria for evaluation of model fit are discussed by Madden (Chap. 3).

A flexible model such as the Weibull or Richards model is likely to provide the best description of a range of disease progress curves. These models gain flexibility with an increase in the number of parameters. The greater number of parameters requires use of relatively new techniques for model fitting (Madden, Chap. 3, and Thal et al., 1984) and extreme care in interpretation of the parameters.

One problem is the difficulty of comparing rate parameter k values from the Richards model for curves with different estimates of m, the shape parameter. Comparisons among curves can be made by calculating ρ, the weighted mean rate parameter (Richards, 1959),

$$\rho = \frac{Ak}{2m + 2} \tag{3}$$

where k and m are as previously defined and A is the asymptotic value y_{max} for the system. The use of ρ combines the power of a flexible model for describing disease progression and the mechanism for comparing rates of progression of disease among epidemics.

Assumptions Concerning y_{max}

In analyzing progression of root disease, it is commonly assumed that the final or maximum amount of disease is equal to 100% (i.e., y is assumed to approach a value of 1 asymptotically). Root disease progress curves have been published, however, for which y_{max} is considerably less than 1. The models used to describe disease progress include provisions for varying the asymptote value. Inserting an unrealistic value of $y_{max} = 1$ in a model for a disease in which a much lower asymptote is appropriate may cause the rate parameter to be underestimated (Hau and Kranz, 1977).

Selection of an Independent Variable

As stated previously, time in days or weeks is usually used as the independent variable for disease progress curves. This allows comparisons among epidemics over years with regard to the timing of an event or sequence of events, but it does not take into account the developmental or physiological stage of the host. It also fails to account for variation in environmental factors, such as temperature and moisture, that may significantly influence the rate of disease progress.

The use of environmental independent variables such as degree days (temperature) or centibar days (soil moisture) may provide a better biological basis for comparing epidemics. Pullman and DeVay (1982) used degree days (>53.5°F) as a physiological time scale for modeling progression of incidence of verticillium wilt of cotton. Ferrin and Mitchell (1984) determined that increase of tobacco black shank was positively correlated with centibar days. Pfender and Hagedorn (1983) regressed incidence of Aphanomyces root rot against soil temperature degree days (>4°C).

INTERPRETATION OF ROOT DISEASE PROGRESS CURVES

There is an extensive body of knowledge concerning the theoretical interpretation of progress curves of foliar diseases based on the accumulation of a large base of empirical data (Van der Plank, 1963; Waggoner, Chap. 1; Zadoks and Schein, 1979). As a result, theoretical interpretations have been derived for the influence of many biological and environmental variables on disease progression. This has not been true for root disease progress curves for reasons that have already been discussed. A rational approach to the development of such theoretical interpretations is to examine the three basic components of the pathosystem—pathogen, host, and environment—individually with the goal of eventually combining the interpretations of component effects into a general interpretation.

What follows is a presentation of ideas and hypotheses concerning the components of root pathosystems. The ideas and interpretations may not be equally applicable to all systems, but they will provide a basis for further experimentation and discussion.

Pathogen Component

Inoculum Density (ID) The ID for fungi is usually given as propagules per gram (ppg) dry soil, while for nematodes, density may be expressed as eggs or larvae of a certain stage per 500 cm^3 soil. Whatever the units for expression of ID, some description of the accuracy of the method of ID determination is essential for reliable interpretation of the data.

Variation in initial pathogen ID can alter the time of epidemic onset, the rate of disease progression, the shape of the disease progress curve, and the observed final disease (y_{max}), depending on the disease and the interaction of the pathogen with host and environment. Specific examples will illustrate these effects.

Onset of Phymatotrichum root rot of cotton was apparently delayed 4 weeks in deep chiseled plots compared to plots with conventional tillage (Rush and Lyda, 1984). Deep chisel plowing reduced the effective ID of *Phymatotrichum omnivorum*. It is possible that disease onset was not actually delayed by the reduction in ID, but rather that the appearance of foliar symptoms was delayed by the reduced number or severity of root infections. Griffin and Tominatsu (1983) found that only 0.02 to 1.03% of observed root infections by *Cylindrocladium crotalariae* on peanut resulted in visually detectable shoot symptoms. Numerous independent infections of *Fusarium oxysporum* f. sp. *pisi* were necessary to produce severe wilt symptoms in peas (Nyvall and Haglund, 1972).

Van der Plank (1963) provided a theoretical basis for an increase in the measured rate of disease progression with increased ID for monocyclic diseases. He presented the monomolecular model for simple interest diseases as

$$\frac{dy}{dt} = QR(1 - y) \tag{4}$$

where Q is a measure of the initial inoculum and R is the rate parameter. QR is equivalent to k in Eq. 1. It follows, therefore, that the regression parameter k of Eq. 1 will be increased by an increase in Q and that this will result in an increase in dy/dt.

Empirical evidence for a relationship between rate of progression and ID has been provided for root diseases caused by *Verticillium dahliae* (Ashworth et al, 1979a, 1979b; Evans and McKeen, 1975; Pullman and DeVay, 1982). *Verticillium dahliae* survives in soil as microsclerotia (ms) and verticillium wilt is probably a monocyclic disease. For wilt of eggplant, the rate increased as ID varied from 1 ms/g soil to 10 to 13 ms/g (Evans and McKeen, 1975). In the cotton wilt system, slope values (rate) of linear disease progress curves increased as ID increased to 4 ppg soil, but decreased from the maximum rate when ID was greater than 40 ppg soil (Pullman and DeVay, 1982). Rate of disease progression for verticillium wilt of tomato also increased with increased ID (Ashworth et al., 1979a). Similar differences in time of onset and rate of disease increase at several inoculum population densities were found for three kinds of fusarium wilt (Komada, 1975).

Production, Spread, and Effectiveness of Inoculum   There has not been sufficient research to either confirm or deny Bald's (1969) view that "during one season, or the growth of an annual crop, most soil-borne pathogens will cause 'simple interest' diseases." It is clear, however, that some soilborne pathogens can cause polycyclic disease. The monocyclic/polycyclic character of a disease will, of course, influence the interpretation of progress curves for epidemics of that disease.

The importance of latent period and infectious period, as these concepts are generally applied to foliar diseases (Van der Plank, 1963), has not been documented in the epidemiology of root diseases (Gilligan, 1983). It is known, however, that for several diseases the production of inoculum by root pathogens may lead to repeated disease cycles during a growing season. Examples of secondary inoculum production for fungi include: *Aphanomyces euteiches* in soils where peas were grown (Pfender and Hagedorn, 1983); *Phytophthora cinnamomi* following soil flooding in the presence of *Abies fraseri* (Kenerley et al., 1984); *P. parasitica* var. *nicotiana* in the rhizosphere of susceptible tobacco plants (*Nicotiana tabacum*) increasing from 0.75 to 250 ppg soil within 13 to 46 days after planting (Kannwischer and Mitchell, 1978); and in soils of various texture for *P. cryptogea* (Duniway, 1976). Other phycomycetous fungi such as *Pythium* spp., which cause seedling damping-off, may also have secondary cycles of inoculum production and infection (Burdon and Chilvers, 1975a).

Certain plant parasitic nematodes complete several life cycles, and thus disease cycles, during a growing season. *Meloidogyne incognita*, which induces root knot on a number of hosts, completes several life cycles in a season and may have high population levels by harvest (Barker and Campbell, 1981). *Pratylenchus penetrans* can complete its life cycle within 30 days at 30°C and thus could induce a polycyclic disease within a growing season (Mai et al., 1977).

For many root pathogens, inoculum production probably occurs over a growing season. Inoculum may, however, be released from host material into soil only once annually, e.g., at harvest or when soil is tilled. Examples of predominantly monocyclic root pathogens are *Verticillium dahliae* (Ashworth et al., 1979a, 1979b), *Macrophomina phaseolina* (C. L. Campbell, unpublished), and certain *Heterodera* spp. (Norton, 1978).

The factor that determines the importance of a polycyclic nature for a pathogen is spread of that pathogen within a season. A pathogen can induce a disease that is truly polycyclic (i.e., new reproduction structures and subsequent infections are produced several times during a season) or effectively polycyclic (i.e., propagules may or may not be produced repeatedly but mycelium spreads through soil or via root contact and induces disease.

The distance and time required for the movement varies for different diseases. Zoospores of *P. cryptogea* swam 25 to 35 mm in surface water or through a coarse-textured soil mix at a matric potential of ≥ -1 millibar (mb), but no detectable zoospore movement occurred at a matric potential of ≤ -100 mb (Duniway, 1976). In the *A. euteiches*–pea system (Pfender and Hagedorn, 1983), the pathogen spread from an initially infected plant to that plant's second to fifth nearest neighbors in each direction within rows and to plants in the next adjoining rows. Maximum distance of spread from the initial source was 18 cm. *Sclerotinia sclerotiorum* infected underground tissue of sunflower and spread from a primary infection focus by root contact from plant to adjacent plant when plants were separated by 30 cm or less (Huang and Hoes, 1980). Spread was less likely when plant distances ranged from 31 to 40 cm. Mean time for plant-to-plant spread of *S. sclerotiorum* ranged from 1.5 to 4.3 weeks with plant spacing of 10 to 30 cm, respectively.

The demonstration of production and spread of secondary inoculum with subsequent disease production may indicate that root pathogens induce "compound interest" diseases sensu Van der Plant (1963). The disease progress curves (percent disease incidence vs. time) for the *A. euteiches*–pea system (Pfender and Hagedorn, 1983) are nearly S-shaped and the logistic model adequately described most of the curves. Although it is tempting to conclude from this that Aphanomyces root rot of pea is a typical compound interest disease, the conclusion would be unjustified. Such a conclusion fails to take into account other important factors, especially root growth, that influence the course of the epidemics (Huisman, 1982).

Spatial Pattern of Inoculum   Clustering or spatial aggregation of inoculum, which is common for soilborne pathogens (Barker and Campbell, 1981; Campbell and Noe, 1985; Hau et al., 1982; Noe and Campbell, 1985; Shew et al., 1984; Taylor et al., 1981), can have at least two significant effects on an epidemic. If there are no secondary cycles of infection, the final level of disease may be lower if the inoculum is highly clustered than if it is less aggregated or randomly dispersed. Also, aggregation of inoculum may increase the variance for disease severity levels (Elliott, 1977). Similarly, the maximum amount of disease that could occur, y_{max}, may be reduced by aggregation of inoculum.

Some plants will be exposed to large amounts of inoculum within clusters while others may not contact inoculum at all. Ferrin and Mitchell (1984) found a highly significant negative correlation between final incidence of tobacco black shank and the value of Lloyd's index of patchiness for inoculum at time of planting (greater values of this index indicate increased aggregation of inoculum).

For some diseases, y_{max} may be a function of the pathogen's ability to spread among adjacent plants as well as initial spatial pattern of inoculum. As noted by Pfender and Hagedorn (1983), if inoculum density of *A. euteiches* is fairly uniform and high, many pea plants will contact inoculum early in the growing season and y_{max} would be expected to be 100%. If inoculum density were uniform but lower, the number of primary infections could be increased by pathogen spread and, again, y_{max} could equal 100%. If, however, very few disease foci were initiated as a result of extremely low ID or highly aggregated inoculum, even with secondary spread of the pathogen, a low y_{max} could result at season's end.

Disease Complexes   The soil environment is biologically diverse. It is not surprising, then, that plant roots may be affected by more than one pathogen concurrently. This fact complicates the interpretation of root disease epidemics, but it should be acknowledged. Most investigations into the epidemiology of root diseases have been confined to diseases with a single, recognized causal agent. Interactions among pathogens have been studied (Powell, 1971), but not generally in the framework of quantitative epidemiology.

Kirkland and Powelson (1981) presented empirical evidence that the composition of mixed pathogen populations in a complex can influence the rate of disease progression in the potato "early dying" disease. This disease complex involves *Verticillium dahliae* and a complex of other pathogens which include fungi, a bacterium, nematodes, and a virus (Kotcon et al., 1984). Rate of disease progress was greater for potato plants infected with *V. dahliae* and *Erwinia caratovora* var. *atroseptica* than for plants infected with either pathogen alone. In plants infected with *V. dahliae* and *E. caratovora* var. *caratovora* and/or *Colletotrichum atromentarium*, rate of disease progression was slower than for plants infected only with *V. dahliae* (Kirkland and Powelson, 1981). These results indicate the potential impact that disease complexes may have on the progression of root diseases and on the subsequent interpretation of progression curves.

Host Component

Root Growth   Root growth is a primary determining factor in the progression of root disease epidemics. The importance of this concept has been recognized (Last, 1971); however, as Huisman (1982) correctly states, "little work has been done to quantitate and refine this concept beyond its recognition level." There is, then, a significant need for plant pathologists to utilize the existing studies on root growth (Bohm, 1979; Carson, 1974; Kotcon et al., 1984; Lungley, 1973; Page and Gerwitz, 1974; Russell, 1977; Torrey and Clarkson, 1975) and to undertake new studies of the effects of root growth on progression of root diseases. Initial advances in incorporating data on root growth into calculation of rates of root infections have been made by Tominatsu and Griffin (1982) and Smith and Walker (1981).

Root growth may affect epidemics in several ways. Where pathogens have limited mobility, root growth provides increased opportunity for root-pathogen contacts. As roots grow through soil the probability of infection increases. The effective pathogen ID on a per plant basis increases with root growth. Even for mobile pathogens, root growth may increase disease incidence by providing increased numbers of sites available for infection.

In interpreting progress curves for root disease incidence, it is significant that root growth, like growth of most plant organs, exhibits a typical sigmoidal or S-shaped curve (Huisman, 1982). This sigmoidal shape of root growth curves provides one possible explanation for observations of a sigmoidal shape for some root disease progress curves (Campbell et al., 1984; Chester, 1946; Rush and Lyda, 1984). Both root length of Douglas fir seedlings and seedling mortality due to *Fusarium oxysporum* had sigmoidal curves when plotted against time (Bloomberg, 1979a). Bloomberg's model to describe root rot of Douglas fir seedlings assumes that the dynamics of root growth is the main factor determining host-pathogen contact. If the probability of root contact with the pathogen is determined principally by root growth, it is reasonable that progress curves for disease incidence should parallel curves for root growth (Huisman, 1982). This can occur if the pathogen-root interaction distance and the probability of infection over time remain constant. Empirical evidence to support this hypothesis is needed.

Data on root growth dynamics such as those presented by Kotcon et al. (1984) are essential to our understanding and interpretations of root disease systems. With the recent development of reliable methods for quantifying root length (Bohm, 1979; Tennant, 1975; Voorhees et al., 1980), studies on dynamics of root growth should be facilitated.

Plant Spacing and Density   The effect of plant density on disease progression varies with the pathosystem. Incidence of verticillium wilt of cotton was reduced when plant density was increased (Longenecker et al., 1970). Minton et al. (1972) determined that the disease incidence was related to area occupied by individual host plants, i.e., more disease resulted when plants occupied a greater area. The reduction in disease incidence with reduced plant spacing indicates that plant-to-plant spread is unimportant and that there is sufficient inoculum in the soil to infect only a limited number of plants.

When the pathogen can grow and spread from plant to plant, rate of disease increase will probably vary inversely with plant spacing. For Sclerotinia wilt of sunflower, increased plant spacing increased the time needed for plant-to-plant spread and decreased the number of new infections developing from each infection focus (Huang and Hoes, 1980). For damping-off of garden cress seedlings induced by *Pythium irregulare*, varying the density of the host population or inoculum resulted in changes in number of primary infection foci (Burdon and Chilvers, 1975b). At low population or inoculum densities, the number of primary infection foci was proportional to density. Rate of advance of the disease front and apparent infection rate of damping-off showed a negative linear relationship with the mean distance between adjacent host plants (Burdon and Chilvers, 1975a).

Host stand also affected the appearance of the disease progress curve (Burdon and Chilvers, 1976). Disease progress occurred in steps of rapid increase of disease when pathogen spread was predominantly within plant clumps followed by slower increase when spread was mainly between clumps. Rate of spread of *Sclerotium cepivorum* between onion plants was slower if plants were 8 or 13 cm apart than when plants were only 5 cm apart (Scott, 1956); spread was only possible when plant roots were in contact with one another.

Symptomatology   The incubation period between infection and expression of root symptoms is usually relatively short, but the incubation period between root infection and expression of shoot symptoms (e.g., wilt) may be variable. This variability may confound interpretation of rates of disease progression.

The site at which infection occurs within a root system can influence rate of development of shoot symptoms. Infection of small-diameter, distal roots may have little or no effect on shoot symptomatology. Conversely, if a major root or tap root is infected, the effect may be significant. This is a topic that should be investigated further to refine our understanding of the progression of disease caused by soilborne pathogens.

The age of plants at the time of infection should also be taken into account. Seedlings or juvenile tissue may readily exhibit symptoms of some root diseases, whereas symptoms of others may develop more rapidly in senescent host tissue in another system.

Environmental Component

The examples of the effects of environment on disease development are too numerous to review here. Leach's (1947) studies relating the coefficient of velocity of seedling emergence to damping-off are a good illustration of the importance of environment on the development of disease induced by a soilborne pathogen. The effects of various environmental factors on root growth and their subsequent support on the epidemiology of root diseases have been reviewed by Huisman (1982). The importance of environment is also exemplified by the development of simulators for root diseases (Bloomberg, 1979b; Gutierrez et al., 1983; Reynolds, 1984).

The way in which environmental variation may influence the interpretation of root disease progression can be illustrated for black shank of tobacco. Jacobi et al. (1983) proposed that the disease progress curve may be sigmoidal because plants are initially infected by *P. parasitica* var. *nicotianae* during relatively cool, moist conditions, but do not express shoot symptoms until 6 to 8 weeks after transplanting when moisture and temperature stress are important. Thus, the onset of hot, dry weather causes a rapid increase in the proportion of infected plants that show shoot symptoms. Other factors such as host age and pathogen population dynamics may also contribute to the rapid increase of visibly infected plants at this time.

Of the three pathosystem components—pathogen, host, and environment— the effects of environment on root disease progress curves are the least understood. One of the greatest challenges to root disease epidemiologists will be to increase our knowledge and interpretive power in this area.

USES OF ROOT DISEASE PROGRESS CURVES

Disease progress curves serve to integrate a large number of complex, interacting effects into a singular interpretable expression. The number of uses for such curves is limited only by the ingenuity and needs of researchers. If we can escape what Last (1971) referred to as the "root pathologists predilection for a single end-of-season observation," we can use our disease progress curves in a multitude of theoretical and practical applications.

Understanding Root Pathosystems

Our understanding of interactions among root pathosystem components, particularly at the population level, is minimal. Through the analysis and comparison of root disease epidemics, we may hope to derive principles concerning the operation of these pathosystems. Such analyses will provide a basis for developing plausible biological explanations for the interactions that make up root disease epidemics.

Disease progress curves for root diseases cannot be used alone to imply the operation of specific biological processes. As discussed earlier, the appropriateness of the logistic model for describing root disease progression does not imply that the pathogen is multiplying, spreading, and infecting additional plants within the growing season. It is appropriate, however, to provide explanations for the sigmoidal shape of a root disease progress curve on the basis of sound, empirical biological data. A knowledge that root growth is usually described by a sigmoidal curve over time (Huisman, 1982) is an example of such an explanation for a sigmoidal disease progress curve.

The analysis and comparison of progress curves within or among pathosystems when certain factors are held constant can provide intriguing insights into the reasons for differences among disease progress curves. For instance, if inoculum density and environment are held constant and host cultivars are compared, testable, biological hypotheses may be developed concerning the ways in which differences in host phenology and resistance affect disease progression. Similar comparisons can be made for many other components of root pathosystems. Such comparisons will help fill in gaps in our knowledge and determine directions for further research.

Root Disease Management

Root disease management has not progressed as rapidly as the management of foliar diseases. The reasons for this include: a lack of perceived need for the short-term or within-season management of root diseases; a lack of available management strategies and delivery systems; and a paucity of knowledge concerning the course or progression of root diseases and the factors that affect this progression.

As our knowledge of the losses caused by root diseases increases, the need for management strategies within a growing season of an annual crop will become more apparent. We have not done a good job of determining actual losses to root diseases within fields, counties, states, or nations. Increased emphasis on measuring yield losses resulting from root disease will make the need for root disease management much clearer.

Perhaps the greatest benefit that can be gained from the development and interpretation of root disease progress curves is the identification of factors within root pathosystems that are amenable to management. The effects of inoculum density on disease progression are fairly well documented (Ashworth et al., 1979a, 1979b; Evans and McKeen, 1975; Pullman and Devay, 1982). This provides a basis for determining thresholds of inoculum density below which unacceptable disease losses do not occur.

The further study of the relationships between root growth and development of root disease epidemics should provide information that will be useful in root disease management. Resistance to many root diseases may lie in the ability of plants to regenerate roots during the course of the epidemic. Perhaps cultivars can be developed with root patterns and growth rates that will minimize the impact of root disease.

Although nematicides have been used extensively, the use of pesticides to manage root diseases caused by other pathogens has not been as extensive as for the management of foliar diseases. Analysis of root disease epidemics should provide a clearer view of the economics of the use of pesticides to reduce pathogen populations or the rate of epidemic development.

REFERENCES

Ashworth, L. J., Jr., Huisman, O. C., Harper, D. M., and Stromberg, L. K. 1979a. Verticillium wilt disease of tomato: Influence of inoculum density and root extension upon disease severity. *Phytopathology* 69:490–492.

———. 1979b. Verticillium wilt of cotton: Influence of inoculum density in the field. *Phytopathology* 69:483–489.

Bald, J. G. 1969. Estimation of leaf area and lesion sizes for studies on soil-borne pathogens. *Phytopathology* 59:1606–1612.

Barker, K. R., and Campbell, C. L. 1981. Sampling nematode populations, in: *Plant Parasitic Nematodes*, vol. 3. (B. M. Zuckerman and R. A. Rohde, eds.), pp. 451–474. Academic Press, New York.

Bloomberg, W. J. 1979a. A model of damping-off and root rot of Douglas fir seedlings caused by *Fusarium oxysporum*. *Phytopathology* 69:74–81.

———. 1979b. Model simulation of infection of Douglas-fir seedlings by *Fusarium oxysporum*. *Phytopathology* 69:1072–1077.

Bohm, W. 1979. *Methods for Studying Root Systems*. Springer-Verlag, Berlin.

Burdon, J. J., and Chilvers, G. A. 1975a. Epidemiology of damping-off disease (*Pythium irregulare*) in relation to density of *Lepidium sativum* seedlings. *Ann. Appl. Biol.* 81:135–143.

———. 1975b. A comparison between host density and inoculum density effects on the frequency of primary infection foci in *Pythium*-induced damping-off disease. *Aust. J. Bot.* 23:899–904.

————. 1976. The effect of clumped planting patterns on epidemics of damping-off disease in cress seedlings. *Oecologia Berlin* 23:17–29.

Campbell, C. L., Jacobi, W. R., Powell, N. T., and Main, C. E. 1984. Analysis of disease progression and the randomness of occurrence of infected plants during tobacco black shank epidemics. *Phytopathology* 74:230–235.

Campbell, C. L., Madden, L. V., and Pennypacker, S. P. 1980a. Structural characterization of bean root rot epidemics. *Phytopathology* 70:152–155.

————. 1980b. Progression dynamics of hypocotyl rot of snapbean. *Phytopathology* 70:487–494.

Campbell, C. L., and Noe, J. P. 1985. The spatial pattern analysis of soilborne pathogens and root diseases, *Annu. Rev. Phytopathol.* 23:129–148.

Campbell, C. L., and Powell, N. T. 1980. Progression of diseases induced by soilborne pathogens: Tobacco black shank. *Prot. Ecol.* 2:177–182.

Carson, E. W. 1974. *The Plant Root and Its Environment.* Univ. Press of Virginia, Charlottesville.

Chester, K. S. 1946. The loss from cotton wilt and the tempo of wilt development: A study of new uses for old data. *Plant Dis. Rep.* 30:253–260.

Duniway, J. M. 1976. Movement of zoospores of *Phytophthora cryptogea* in soils of various textures and matric potentials. *Phytopathology* 66:877–882.

Elliott, J. M. 1977. *Some Methods for the Statistical Analysis of Samples of Benthic Invertebrates,* 2d ed. Freshwater Biological Assoc., (Sci. Publ. No. 25.). Titus Wilson and Son, Ltd., Kendal, U.K.

Evans, G., and McKeen, C. D. 1975. Influence of crops on number of microsclerotia of *Verticillium dahliae* in soils and the development of wilt in southwestern Ontario. *Can. J. Plant Sci.* 55:827–834.

Ferrin, D. M., and Mitchell, D. J. 1984. The influence of density and patchiness of inoculum on the epidemiology of tobacco black shank. (Abstr.) *Phytopathology* 74:839.

Gilligan, C. A. 1982. Size and shape of sampling units for estimating incidence of sharp eyespot, *Rhizoctonia cerealis,* in plots of wheat. *J. Agr. Sci. Cambridge* 99:461–464.

————. 1983. Modeling of soilborne pathogens. *Annu. Rev. Phytopathol.* 21:45–64.

Griffin, G. J., and Tominatsu, G. S. 1983. Root infection pattern, infection efficiency, and infection density-disease incidence relationships of *Cylindrocladium crotalariae* on peanut in field soil. *Can. J. Plant Pathol.* 5:81–88.

Gutierrez, A. P., DeVay, J. E., Pullman, G. S., and Friebertshauser, G. E. 1983. A model of Verticillium wilt in relation to cotton growth and development. *Phytopathology* 73:89–95.

Hau, B., and Kranz, J. 1977. Ein Vergleich verschiedener Transformationen von Befallskurven, *Phytopathol. Z.* 88:53–68.

Hau, F. C., Beute, M. K., and Campbell, C. L. 1982. Inoculum distribution and sampling methods for *Cylindrocladium crotalariae* in a peanut field. *Plant Dis.* 66:568–571.

Huang, H. C., and Hoes, J. A. 1980. Importance of plant spacing and sclerotial development of Sclerotinia wilt of sunflower. *Plant Dis.* 64:81–84.

Huisman, O. C. 1982. Interrelations of root growth dynamics to epidemiology of root-invading fungi. *Annu. Rev. Phytopathol.* 20:303–327.

Jacobi, W. R., Main, C. E., and Powell, N. T. 1983. Influence of temperature and rainfall on the epidemiology of black shank of tobacco. *Phytopathology* 73:139–143.

Jarvis, W. R., and Hawthorne, B. T. 1972. *Sclerotinia minor* on lettuce: Progress of an epidemic, *Ann. Appl. Biol.* 70:207–214.

Kannwischer, M. E., and Mitchell, D. J. 1978. The influence of a fungicide on the epidemiology of tobacco black shank. *Phytopathology* 68:1760–1765.

Karandinos, M. G. 1980. Quantification of disease progression. *Prot. Ecol.* 2:159–176.

Kenerley, C. M., Papke, K., and Bruck, R. I. 1984. Effect of flooding on development of Phytophthora root rot in Fraser fir seedlings. *Phytopathology* 74:401–404.

Kirkland, M. L., and Powelson, M. L. 1981. The influence of pathogen interactions on apparent infection rates in potatoes with "early dying" disease. (Abstr.) *Phytopathology* 71:886.

Komada, H. 1975. Development of a selective medium for quantitative isolation of *Fusarium oxysporum* from natural soil. *Rev. Plant Protec. Res.* 8:114–125.

Kotcon, J. B., Rouse, D. I., and Mitchell, J. E. 1984. Dynamics of root growth in potato fields affected by the early dying syndrome. *Phytopathology* 74:462–467.

Last, F. T. 1971. The role of the host in the epidemiology of some nonfoliar pathogens. *Annu. Rev. Phytopathol.* 9:341–362.

Leach, L. D. 1947. Growth rates of host and pathogen as factors determining the severity of preemergence damping-off. *J. Agr. Res.* 75:161–179.

Longenecker, D. E., Thaxton, E. L., Jr., Hefner, J. J., and Lyerly, P. J. 1970. Variable row spacing of irrigated cotton. *Texas Agr. Exp. Sta. Bull.* 1102, College Station.

Lungley, D. R. 1973. The growth of root systems in a numerical computer simulation model. *Plant Soil* 38:145–149.

Madden, L. V. 1980. Quantification of disease progression. *Prot. Ecol.* 2:159–176.

Mai, W. F., Bloom, J. R., and Chen, T. A., eds. 1977. Biology and ecology of the plant-parasitic nematode *Pratylenchus penetrans, Pennsylvania State Univ. Agr. Exp. Sta. Bull.* 815, University Park.

Minton, E. B., Brashears, A. D., Kirk, I. D., and Hudspeth, E. B., Jr. 1972. Effect of row and plant spacings on Verticillium wilt of cotton. *Crop Sci.* 12:764–767.

Morrall, R. A. A., and Verma, P. R. 1981. Disease progress curves, linear transformations, and common root rot of cereals. (Letter to the Editor.) *Can. J. Plant Pathol.* 3:182–183.

Noe, J. P., and Campbell, C. L. 1985. Spatial pattern analysis of plant parasitic nematodes. *J. Nematol.* 17:86–93.

Norton, D. C. 1978. *Ecology of Plant-parasitic Nematodes.* Wiley-Interscience, New York.

Nyvall, R. F., and Haglund, W. A. 1972. Sites of infection of *Fusarium oxysporum* f. *pisi* Race 5 on peas. *Phytopathology* 62:1419–1424.

Page, E. R., and Gerwitz, A. 1974. Mathematical models, based on diffusion to describe root systems of isolated plants, row crops and swards. *Plant Soil* 41:243–254.

Pfender, W. F. 1982. Monocyclic and polycyclic root diseases: Distinguishing between the nature of the disease cycle and the shape of the disease progress curve. (Letter to the Editor.) *Phytopathology* 72:31–32.

Pfender, W. F., and Hagedorn, D. J. 1983. Disease progress and yield loss in Aphanomyces root rot of peas. *Phytopathology* 73:1109–1113.

Powell, N. T. 1971. Interactions between nematodes and fungi in disease complexes. *Annu. Rev. Phytopathol.* 9:253–274.

Pullman, G. S., and DeVay, J. E. 1982. Epidemiology of Verticillium wilt of cotton: A relationship between inoculum density and disease progression. *Phytopathology* 72:549–554.

Reynolds, K. M. 1984. The epidemiology of Phytophthora root rot of Fraser fir. Ph.D. thesis, North Carolina State Univ., Raleigh, 215 pp.

Richards, F. J. 1959. A flexible growth function for empirical use. *J. Exp. Bot.* 29:290–300.

Rush, C. M., and Lyda, S. D. 1984. Evaluation of deep-chiseled anhydrous ammonia as a control for Phymatotrichum root rot of cotton. *Plant Dis.* 68:291–293.

Russell, R. S. 1977. *Plant Root Systems: Their Function and Interactions with the Soil.* McGraw-Hill, London.

Scott, M. R. 1956. Studies on the biology of *Sclerotium cepivorum* Berk. II. The spread of white rot from plant to plant. *Ann. Appl Biol.* 44:584–589.

Shew, B. B., Beute, M. K., and Campbell, C. L. 1984. Spatial pattern of southern root rot caused by *Sclerotium rolfsii* in six North Carolina peanut fields. *Phytopathology* 74:730–735.

Smith, S. E., and Walker, N. A. 1981. A quantitative study of mycorrhizal infection in *Trifolium:* Separate determination of rates of infection and of mycelial growth. *New Phytol.* 89:225–240.

Stack, R. W. 1980. Disease progression in common root rot of spring wheat and barley. *Can. J. Plant Pathol.* 2:187–193.

Taylor, J. D., Griffin, G. J., and Garren, K. H. 1981. Inoculum pattern, inoculum density-disease incidence relationships, and population fluctuations of *Cylindrocladium crotalariae* microsclerotia in peanut field soil. *Phytopathology* 71:1297–1302.

Tennant, D. 1975. A test of a modified line intersect method of estimating root length. *J. Ecol.* 63:995–1001.

Thal, W. M., Campbell, C. L., and Madden, L. V. 1984. Sensitivity of Weibull model parameter estimates to variation in simulated disease progression data. *Phytopathology* 74:1425–1430.

Tominatsu, G. S., and Griffin, G. J. 1982. Inoculum potential of *Cylindrocladium crotalariae:* Infection rates and microsclerotial density-root infection relationships on peanut. *Phytopathology* 72:511–517.

Torrey, J. G., and Clarkson, D. T., eds. 1975. *The Development and Function of Roots.* Academic Press, London.

Van der Plank, J. E. 1963. *Plant Disease: Epidemics and Control.* Academic Press, New York.

Verma, P. R., Morrall, R. A. A., and Tinline, R. D. 1974. The epidemiology of common root rot in Manitou wheat: Disease progression during the growing season. *Can. J. Bot.* 52:1757–1764.

Voorhees, W. B., Carlson, V. A., and Hallaver, E. A. 1980. Root length measurement with a computer-controlled digital scanning microdensitometer. *Agron. J.* 72:847–851.

Waggoner, P. E., and Rich, S. 1981. Lesion distribution, multiple infection, and the logistic increase of plant disease. *Proc. Natl. Acad. Sci.* 78:3292–3295.

Welch, K. E. 1981. The effects of inoculum density and low oxygen tensions on Fusarium yellows on celery. Ph.D. thesis, Univ. California at Berkeley.

Zadoks, J. C., and Schein, R. D. 1979. *Epidemiology and Plant Disease Management.* Oxford University Press, Oxford.

Chapter 3

Statistical Analysis and Comparison of Disease Progress Curves

L. V. Madden

An epidemic can be defined as an increase or change in disease intensity with time. Epidemics are described in various ways, but most commonly with a series of disease intensity estimates collected over time. A plot of these disease values vs. time—a disease progress curve—summarizes the effects of host, pathogen, and environment on epidemic development. The key to understanding an epidemic is to quantify disease progression over time. Quantification is conducted through the use of models.

There are two basic types of models for disease progression. The first is based on prior assumptions about the mechanisms of disease increase. These models generally start with a differential equation describing the absolute rate of disease increase which may then be represented in an integrated form (Madden, 1980). For example, Van der Plank (1963) proposed the logistic model based on the assumption that the absolute rate of disease increase is jointly proportional to the level of disease intensity and the level of healthy tissue. Many other models of this type have been proposed for growth of organisms or populations, where growth is defined as the change in magnitude of any measurable characteristic such as weight, and disease intensity (Madden, 1980; Richards, 1969). Growth models based on prior biological assumptions have been commonly termed "theoretical" and "biological."

The second type of model is based on the observed set of data with few assumptions about the mechanisms of disease increase. Simple and polynomial regression models are common forms of this type of model. Growth models based on the observed data are termed "empirical."

Empirical and theoretical models represent the extremes of a continuum of model types. There are many situations in which some prior information is used within the constraints of the observed data to develop a disease progression model. Often, the logistic and other theoretically derived models do not accurately describe observed disease progression data; yet the investigator does not wish to use a purely empirical model with no biological interpretation. In these cases, a model such as the Richards or Weibull could prove to be effective. Both of these models are flexible enough to fit a large number of disease progression patterns yet they can be developed, in certain circumstances, by using prior biological assumptions (Madden, 1980; Richards, 1969). The goal of model development should be a balance between prior assumptions and empirical descriptions of the data (Mandel, 1969).

All models are simplifications of reality (Edminster, 1978). In this context, no model will completely describe disease progress data. Statistical versions of theoretical and empirical models are used to handle the differences between predictions and observations. The degree of these differences, i.e., unexplained variability, is used to determine the accuracy of a particular model and also to compare different epidemics.

Often an investigator wishes to compare epidemics without specifying a model for disease progression. If disease intensity is estimated at different times, analysis of variance (ANOVA) can be used for such comparisons. There are special statistical problems with ANOVA when disease is measured over time and these are discussed in this chapter.

Data collected over time in scientific investigations are called sequential, longitudinal, vertical, or time series, of which disease progress data are a special case. Zadoks (1978) stated that "epidemiologists can do a much better job in sequential analysis; they have plenty of good data, but have not put these to their best use." The goal of this chapter is to show how to put epidemic data to good use. First, I discuss how to fit statistical versions of empirical and theoretical models to disease progress data. Means of evaluating the appropriateness of a particular model are presented followed by techniques for statistically comparing epidemic data. Limited multivariate statistical techniques are presented for analyzing and comparing epidemics.

I assume that the reader has a basic understanding of regression analysis and ANOVA. No attempt can be made here to teach these statistical subjects. Readers can gain most of the background necessary for this chapter by reading either Chap. 1 in Draper and Smith (1981) or Chap. 2 in Neter et al. (1983). This chapter stresses the special considerations or modifications that must be made to analyze properly disease progress data.

STATISTICAL MODELING OF DISEASE PROGRESS DATA

Empirical and Theoretical Models

An epidemic can be represented as

$$y = f(t) \tag{1}$$

in which y is a random variable representing disease intensity, t is time, and $f(t)$ is some function of time. Random variables are measured or observed quantities that cannot be known with absolute certainty; random variables are often described with means and variances. In general, y is expressed as a proportion and t is in convenient units. Typical forms of y include the proportion of plants in a population that are infected or the proportion of total plant tissue that is diseased. The function of time is of two general forms: linear and nonlinear. A linear model version can be written as

$$y = b_0 + b_1 t + b_2 t^2 + \cdots + b_q t^q \tag{2}$$

in which the b's are unknown parameters estimated from the data. Models of this type are called polynomials. Although parameters are constants, their estimates are random variables; the estimated parameters characterize epidemics. The highest power of t seldom exceeds 3 or 4. When the highest power of t is 1, Eq. 2 simplifies to a model for a straight line with slope b_1 and y intercept (y when $t = 0$) of b_0. Equation 2 is a linear model, in part, because the parameters are not multiplied or divided by other parameters.

Wishart (1938) introduced this type of empirical model for describing growth and it has been used extensively in many disciplines. It is a convenient approach, because growth curves can be reduced to a discrete number of parameters, and treatments can be compared through an analysis of the estimated parameters. Epidemiologists have not used polynomial models to a great extent because the parameters often do not have easily understood biological interpretation and also because there has been an effort to describe epidemics with theoretical models whenever possible. A further difficulty with polynomials is that extrapolating beyond the last value of t can produce nonsensical predictions. For example, predictions can reach a maximum and then decline even if this is impossible biologically. When predictions past the last recorded time are not needed and no theoretical model seems to fit the data, polynomial models become good alternatives. Griggs et al. (1978) analyzed and compared fusiform rust epidemics of slash pine with this approach.

A nonlinear version of Eq. 1 is the logistic model that can be written as

$$y = \frac{1}{1 + \exp(-\{\ln[y_0/(1-y_0)] + rt\})} \tag{3}$$

in which $\exp(\cdot)$ is e (2.718. . .) raised to a specific power, and the parameter y_0 represents the level of disease when $t = 0$, and r reflects the rate of disease increase; r has been termed the apparent infection rate (Van der Plank, 1963). The model is nonlinear because the parameters appear as exponents. Equation 3 is the integrated form of a differential equation describing the absolute rate of disease increase in meaningful biological terms. This model therefore can be considered theoretical. This and related models have been discussed numerous times (Jowett et al., 1974; Madden, 1980; Madden and Campbell, 1985; Van der Plank, 1963). Waggoner (Chap. 1) discusses many of the relevant models for disease progression. Not all nonlinear models are theoretical or biological; indeed, an infinite number of nonlinear models could be written to describe disease progression without a theoretical basis.

The logistic and many other nonlinear models can be transformed into a linear form. In the general case, Eq. 1 can be represented as

$$g(y) = f(t) \tag{4}$$

in which $g(y)$ is a function of y that enables $f(t)$ to be written in linear form; $f(t)$ is linear but not necessarily of the same form as in Eq. 2. For the logistic model, $g(y) = \ln[y/(1 - y)]$, and the transformed model is written as

$$\ln \frac{y}{1-y} = \ln\left(\frac{y_0}{1-y_0}\right) + rt \tag{5}$$

which is an equation for a straight line with two parameters, intercept $\ln[y_0/(1 - y_0)]$ and slope r. In this chapter, $g(y)$ will be written as y^*. With the logistic model, y^* is known as the logit of y. Note that not all nonlinear models can be transformed into linear models.

Table 1 contains disease progress data for a generated logistic epidemic with statistical variability. In addition to t and y, y^* is calculated assuming that disease progression is logistic.

Statistical Model Versions

Unexplained Variability   Disease progress models have been introduced under the assumption that a model completely describes the data. In reality, there will be differences between what the model predicts and the observations. To handle these differences, both empirical and theoretical models must be expanded into statistical forms. A statistical version of Eq. 1 can be written as

$$y_i = f(t_i) + u_i \tag{6}$$

Table 1 Example time t (in days), disease intensity y, and transformed diseases ($y^* = \ln[y/(1 - y)]$) data for a logistic disease progress curve[a]

	Ordinary least squares regression				Autocorrelation regression			First-difference regression		
t_i	y_i	y_i^*	$\hat{y}_i^*$	a_i^*	$t_i' =$ $t_i - 0.792t_{i-1}$	$y_i^{*'} =$ $y_i^* - 0.792y_{i-1}^*$	$\hat{y}_i^{*'}$	$t_i' =$ $t_i - t_{i-1}$	$y_i^{*'} =$ $y_i^* - y_{i-1}^*$	$\hat{y}_i^{*'}$
0	0.004	−5.517	−6.020	0.502	—	—	—	—	—	—
5	0.012	−4.410	−5.293	0.883	5.00	−0.041	−0.745	5	1.107	0.738
10	0.022	−3.794	−4.567	0.772	6.04	−0.301	−0.560	5	0.616	0.738
15	0.026	−3.623	−3.840	0.217	7.08	−0.618	−0.375	5	1.171	0.738
20	0.026	−3.623	−3.114	−0.509	8.12	−0.754	−0.189	5	0.000	0.738
25	0.025	−3.663	−2.387	−1.276	9.16	−0.794	−0.004	5	−0.040	0.738
30	0.064	−2.682	−1.661	−1.022	10.20	0.219	0.181	5	0.981	0.738
35	0.189	−1.456	−0.934	−0.522	11.24	0.668	0.366	5	1.226	0.738
40	0.322	−0.744	−0.208	−0.537	12.28	0.409	0.551	5	0.712	0.738
45	0.635	0.553	−0.519	0.035	13.32	1.143	0.736	5	1.298	0.738
50	0.761	1.158	1.245	−0.087	14.36	0.720	0.921	5	0.604	0.738
55	0.903	2.231	1.972	0.259	15.40	1.314	1.107	5	1.073	0.738
60	0.950	2.944	2.698	0.246	16.44	1.177	1.292	5	0.713	0.738
65	0.978	3.794	3.425	0.370	17.48	1.462	1.477	5	0.850	0.738
70	0.992	4.820	4.151	0.669	18.52	1.815	1.662	5	1.026	0.738

[a]Data were analyzed with ordinary least squares (Eq. 9), autocorrelation (Eq. 28), and first-difference (Eq. 30) regression. See text for description for all symbols.

59

in which the subscript i refers to the ith time and u_i is the unexplained variability or error term at the ith time. The subscript also indicates that there can be different disease values if there are multiple observations of y at a single t. The error term is the difference between the observed y_i and the expected or theoretical value $f(t_i)$, assuming that $f(t_i)$ represents disease progress.

Regression analysis is the most common method used to estimate parameters given a statistical model such as Eq. 6. To distinguish between regression models in the general case and epidemic models, Eq. 1 is written with uppercase Y to represent any "dependent" random variable and X to represent any predictor variable:

$$Y_i = f(X_i) + u_i \qquad (7)$$

The error term or unexplained variability of statistical models can be considered from many viewpoints. Since every model is a simplification of reality, the error can be considered to represent the discrepancy between reality (observed values) and the model (predicted values). Since many model assumptions do not hold (e.g., constant environment for the logistic model) (see Murray, 1979), the error can represent the discrepancy between the assumptions of the model and reality. Disease progression also can be considered a stochastic or random process which means that *average* disease progress can be described by a model such as Eq. 3, but the actual y values for any given epidemic will differ from the average. Finally, there are errors in measuring y. Measurement error causes special problems unless the error is independent of t, and therefore y; measurement error is not always independent of y, but this problem is not discussed further.

The manner in which the error term is incorporated into the disease progress model is open to debate. For empirical, linear models, incorporation is straightforward; the error term is simply added to $f(t)$ (see Eq. 6). For nonlinear models, there are different ways in which the error term can be incorporated. As with linear models, the u's can be added to $f(t)$. Using the logistic model as an example, the statistical model can be written as

$$y_i = \frac{1}{1 + \exp(-\{\ln[y_0/(1-y_0)] + rt_i\})} + u_i \qquad (8)$$

where u_i represents the difference between observed and predicted y's. The error term can also be added to the transformed model, which with the logistic can be written as

$$\ln \frac{y_i}{1-y_i} = \ln \left(\frac{y_0}{1-y_0} \right) + rt_i + u_i^* \qquad (9)$$

Here, u_i^* represents the difference between observed and predicted y_i^*'s. Unlike the case when no error is considered (Eqs. 3 and 5), Eq. 9 cannot be obtained from Eq. 8 using a transformation. Equation 9 can only be obtained from

$$y_i = \frac{1}{1 + \exp(-\{\ln[y_0/(1-y_0)] + rt_i + u_i\})} \qquad (10)$$

The error term is incorporated into the exponent. Most epidemiologists have not appreciated the differences among Eqs. 8 through 10. Transforming the y's to obtain a linear model implies that the error is not additive but logistic. Transforming y of Eq. 10 not only results in a linear equation, but also produces an additive error term.

Statistical Assumptions of Regression Analysis

Assuming the correct model is chosen, the usual assumptions of ordinary least squares regression are that the u_i's are normally and independently distributed (NID) random variables with mean 0 and constant variance σ^2 (Neter et al., 1983; Draper and Smith, 1981). As a consequence of this, the Y's are normally and independently distributed with mean $f(X_i)$ (or μ_i) and constant variance σ^2. There is a whole population of Y's at each level of X (Fig. 1). Furthermore, the Y's at each level of X are *independent* (i.e., Y at X_i is independent of Y at $X_{i'}$).

None of these regression assumptions is fully satisfied with disease progression. If y is a proportion, one would expect it to have a nonnormal distribution and have a variance that is directly related to the level of y. For instance, when

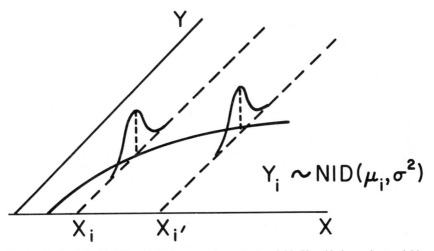

Figure 1   Statistical relationship between a dependent variable Y and independent variable X. (*Adapted from Neter et al., 1983.* Applied Linear Regression Models, *Richard D. Irwin, Inc., Homewood, IL.*)

y represents disease incidence (proportion of plants infected), an argument could be made that y has a binomial distribution with $\sigma^2 = y(1 - y)/n$, in which n is the number of data points used in calculating y (Fulton, 1979; Neter et al., 1983). If a transformation of y (y^*) is used to obtain a linear model, then y^* is the variable of interest. Even if y had a normal distribution with constant variance, y^* would not. Fulton (1979) gave the appropriate variance of y^*, if y was normally or binomially distributed and a logistic model described disease increase. Fortunately, regression results are not very sensitive to lack of normality or constant variance. Only when the variance is highly dependent on the level of y will the results be in serious error.

Independence of the u's is the most important regression assumption. Unfortunately, it is difficult to justify this based on biological reality. The level of y at any arbitrary time (i) can be represented by

$$y_i = y_{i-1} + (y_i - y_{i-1}) \tag{11a}$$

$$y_i = y_{i-1} + \Delta y_i \tag{11b}$$

The level of y at the ith time trivially is equal to the level of y at time $i - 1$ plus the difference. The error term associated with y at $i - 1$ (u_{i-1}) is incorporated into the change in y_i between the two times. The y's are not independent; indeed, they are highly dependent. This dependency of the y's can lead to high correlation of the u's. Although most investigators understand this obvious fact, they incorrectly use ordinary least squares regression to estimate model parameters. Estimating parameters with ordinary least squares regression is incorrect, especially when the correlation between u_i and u_{i-1} is high. At least, the variances and covariances of the estimated parameters will be wrong; and for small data sets, the estimates themselves might be incorrect (Thiel and Nagar, 1961).

Least Squares Regression Analysis

Regression analysis is a very powerful tool for estimating model parameters, describing a "dependent variable" (e.g., disease intensity) as a function of one or more "independent variables" (e.g., time), and for predicting future levels of the dependent variable. Various forms of regression are useful for working with disease progression models. Ordinary least squares regression estimates parameters by minimizing the sum of the squared differences between the observed and predicted values. This can be represented by the following function corresponding to Eq. 7:

$$Q = \sum_{i=1}^{n} (Y_i - f(X_i))^2 \tag{12}$$

in which Q is the sum of squares for any set of values for the parameters of $f(X)$. One could calculate Q with any arbitrary values for $f(X)$. If $f(X)$ is linear (e.g., Eq. 2), a unique minimum value for Q can be determined with calculus. The value of Q at the least-squares solution is termed the sum of squares for error (SSE) or the residual sum of squares and is written as

$$SSE = \sum_{i=1}^{n} (Y_i - \hat{Y}_i)^2 \tag{13}$$

in which $\hat{Y}$ is the predicted Y when $f(X)$ has the least-squares estimates of the unknown parameters. The difference between an observed and predicted Y value is called the residual and is represented by $\hat{u}$; $\hat{u}$ is the estimate of the unknown error term. Equations exist for defining the parameters of $f(X)$ at SSE. Empirical polynomial models for disease progression have the advantage of having statistically well-understood and unique parameter estimates. To use Eq. 12 with disease progression, substitute y for Y, and $f(t)$ for $f(X)$.

Parameters of linearized, nonlinear models can also be estimated with ordinary least squares regression. Equations 12 and 13 can be written as

$$Q = \sum_{i=1}^{n} (y_i^* - f(t_i))^2 \tag{14}$$

$$SSE = \sum_{i=1}^{n} (y_i^* - \hat{y}_i)^2 \tag{15}$$

In this case, the squared differences between the observed and predicted transformed y's (i.e., y^*'s) are minimized. Note that y^* is an artificial representation of disease intensity that has different distributional properties from y; y is the level of disease in a field; and y^* simply represents a method to obtain a linear model.

Ordinary least squares regression was used to fit the logistic model (Eq. 9) to the example disease progress data (Table 1). SSE equaled 5.89. Estimates of the intercept $\ln[y_0/(1 - y_0)]$ and slope r parameters were -6.02 and $0.145/\text{day}$, respectively (Table 2). The residuals also are listed in Table 1.

When $f(X)$, or in particular $f(t)$, is nonlinear, no unique minimum solution to Eq. 12 can be found with calculus. Instead, an iterative approach is necessary that starts with initial estimates (guesses) of the parameters supplied by the investigator, and then proceeds through an iterative scheme to find the "best" estimates. Best estimates are those that result in what appears to be a minimum Q. Obtaining initial estimates can be difficult. One approach is to transform the nonlinear model into a linear form, if possible, and then use ordinary least squares regression to derive parameter estimates; these estimates could then be used in the iterative nonlinear procedure. Another choice is simply to use reasonable values based on prior experience. For example, r and y_0 of the logistic model have typical values for many diseases (Kranz, 1978) that could be used.

Table 2 Estimated intercept and slope parameters, and their standard deviations when Eqs. 9, 28, and 30 are fitted to the data in Table 1

Also presented is the covariance of the estimated parameters, coefficient of determination R^2, standard deviation about the regression line s, degrees of freedom (df = number of observations minus number of parameters), autocorrelation parameter ρ, and the Durbin-Watson statistic D.

Regression	Intercept parameter[a]		Slope parameter (r)		Covariance	R^2	s	df	ρ	D
	Estimate	Standard deviation	Estimate	Standard deviation						
Ordinary least squares (Eq. 9)	−6.020	0.331	0.145	0.008	−0.0023	0.962	0.673	13	0.792	0.42
Autocorrelation (Eq. 28)	−1.635 (−7.861)	0.322 (1.548)	0.178	0.026	−0.078	0.799[b]	0.404[b]	12	0.224	1.29
First-difference (Eq. 30)	—	—	0.148	0.023	—	0.756[b]	0.435[b]	13	0.333	1.24

[a]Intercept equals $\ln[y_0/(1 - y_0)]$, except for the autocorrelation regression in which it equals $\ln[y_0/(1 - y_0)](1 - \rho)$. The estimate of $\ln[y_0/(1 - y_0)]$ and its standard deviation for the autocorrelation regression are presented in parentheses.
[b]One cannot directly compare R^2 or s from the different models.

There are four common iterative nonlinear procedures: (1) Taylor series expansion or Gauss-Newton iteration, (2) steepest descent, (3) Marquardt compromise of (1) and (2), and (4) the derivative-free method (Draper and Smith, 1981; Statistical Analysis System, 1982). These procedures often do not result in the same parameter estimates. Indeed, some or all of them may not work for certain data sets. Success depends on a large sample size (many time periods of disease assessment), low variability (as measured by SSE), and good initial parameter estimates. Estimation difficulty also increases with an increase in number of parameters in the model. With two- or three-parameter disease progress models, five time periods are the absolute minimum for attempting these procedures. Ideally, 10 or more times are needed. With smaller data sets, it is often not possible to successfully estimate parameters with nonlinear regression. The Taylor series expansion and Marquardt compromise appear to be the most reliable procedures for estimating parameters of nonlinear disease progress models (Thal et al., 1984).

With the example data set, all methods except the steepest descent method gave estimates of $y_0 = 0.0003$ and $r = 0.189$/day. The estimate of y_0 corresponds to $\ln[y_0/1 - y_0)] = -8.11$. The steepest descent method gave estimates of 0.0025 and 0.139/day for y_0 and r, respectively.

Confidence Intervals Although parameters are constants, their estimates are random variables with associated variances and covariances. With linear models, parameter estimates are said to be unbiased, i.e., an estimate such that the mean of a population of parameter estimates is equal to the unknown parameter. The variance of a parameter estimate is a function of both SSE and the spacing of the X's or t's. A simple confidence interval can be calculated for any parameter estimate using

$$\hat{\theta} \pm t[P;n - p]s[\hat{\theta}] \tag{16}$$

in which $\hat{\theta}$ is the estimate of an arbitrary parameter θ, $s[\hat{\theta}]$ is the estimated standard deviation of $\hat{\theta}$ (square root or the variance of $\hat{\theta}$; also called the standard error of $\hat{\theta}$), p is the number of estimated parameters in the model, and $t[P; n - p]$ is a critical value from a t distribution with $n - p$ degrees of freedom (df) and significance level of P. This significance level corresponds to a $100(1 - P)\%$ confidence interval for θ.

Estimated parameters of a single model are, in general, not independent of each other. The degree of association between two estimated parameters is their covariance $\sigma[\hat{\theta}_1, \hat{\theta}_2]$. The correlation of two parameters is given by

$$\psi = \frac{\sigma[\hat{\theta}_1, \hat{\theta}_2]}{\sigma[\hat{\theta}_1] \cdot \sigma[\hat{\theta}_2]} \tag{17a}$$

which is estimated by:

$$\hat{\psi} = \frac{s[\hat{\theta}_1, \hat{\theta}_2]}{s[\hat{\theta}_1] \cdot s[\hat{\theta}_2]} \qquad (17b)$$

When estimated parameters are highly correlated, simple confidence intervals (Eq. 16) will be misleading; the size of the error will depend on the degree of correlation of $\hat{\theta}$ with the other parameter estimates. A positive covariance (or correlation) indicates a tendency for the values of $\hat{\theta}_1$ and $\hat{\theta}_2$ to be jointly too high or too low; a negative covariance indicates that joint errors tend to be in opposite directions (Neter et al., 1983). Confidence *regions* are necessary for the set of parameter estimates in these circumstances, especially when correlations are greater than 0.90.

Investigators should be cautious in giving exact epidemiological meaning to any estimated parameter or its confidence interval when the estimate is highly correlated with other parameters. For instance, the covariance of the estimated intercept $\ln[y_0/(1 - y_0)]$ and slope r from Eq. 9 equals $-\bar{t}\sigma^2[\hat{r}]$, in which $\bar{t}$ is the average time; this implies that these parameter estimates will err in opposite directions. This is intuitive considering that an overestimate of r (i.e., a slope that is too steep) will result in an intercept that is too low. Researchers have shown the high negative correlation between the slope and intercept with experimental data (Rouse et al., 1981; Gregory et al., 1981). These results are not surprising given the statistical properties of their estimates.

Returning to the example, estimated standard deviations of the estimated parameters, as well as their covariance, are given in Table 2. The estimated correlation between the intercept and slope equaled -0.87. A 95% confidence interval for r equaled 0.145 ± 0.017.

Variances and covariances of estimated parameters from nonlinear models (e.g., Eq. 8) are only approximate for typical sample sizes. The exact significance levels for all tests also are not known. The estimated parameters may also be biased with small samples, and correlations between estimated parameters may be very high (Draper and Smith, 1981; Thal et al., 1984; L. V. Madden, unpublished). Indeed, correlations exceeding 0.95 can occur with even a two-parameter model. When the nonlinear Eq. 8 was fit to the data in Table 1, the correlation of the two estimated parameters equaled 0.99.

Model Evaluation and Corrections

Procedures for selecting a model and evaluating its performance have been reviewed numerous times in the statistical and phytopathological literature (e.g., Butt and Royle, 1974; Draper and Smith, 1981; Hau and Kranz, 1977; Madden, 1980). Formal tests and graphic procedures can be used to evaluate statistical assumptions. Of course, with nonlinear models the significance levels of the tests are only approximate. Most methods center on the residuals to determine normality, constant variance, independence, and whether an appropriate model was chosen. Important evaluation criteria include randomness of the residuals, realistic and significant estimated parameters, the coefficient of determination R^2, and the standard deviation about the regression line ($s = \sqrt{SSE/df}$).

Nonnormality and nonconstant variance are often associated with each other and correction for one often corrects the other. With variances proportional to y, a transformation of the y values would work in many circumstances but would also obscure the theoretical relationship between y (or y^*) and t. Suppose, for example, that Eq. 9 was used and an appraisal of the residuals revealed that the variance was increasing with y^*. Transforming y^* (equivalently, double transforming y) might stabilize the variance but the new model would no longer describe logistic increase in disease intensity.

Weighted ordinary least squares regression is the alternative to transformations when the variances are not constant. The least squares solution corresponds to minimizing Eq. 12 altered to

$$Q = \sum_{i=1}^{n} W_i (Y_i - f(X_i))^2 \tag{18}$$

in which W_i is the weight given to Y_i. The weight should be the inverse of the variance of u_i (or Y_i). Greater weight is given to observations with lower variance than to observation with higher variance. Weights can be determined empirically and theoretically (Neter et al., 1983).

Independence of the y_i's (or u_i's) should always be evaluated with disease progression models since there is good biological reason why the level of disease at successive times will be correlated. Correlation between successive times is called autocorrelation or serial correlation. Failure to adjust for this correlation results in erroneous estimates of the variances and covariances of the estimated parameters, and possibly of the parameters themselves. There are several methods to correct for serial correlation.

Autocorrelation   Instead of assuming that u_i's are independent, often it is reasonable to assume that u_i can be expressed as

$$u_i = \rho u_{i-1} + v_i \tag{19}$$

in which ρ is the *autocorrelation parameter* ($-1 < \rho < 1$) and v_i is a "new" error term that is NID with mean 0 and constant variance σ^2. When Eq. 19 is valid, the error is said to be first-order autoregressive. The error at each time is equal to a fraction of the previous error term (ρu_{i-1}) plus a new error. Of course when $\rho = 0$, Eq. 19 reduces to the traditional error term. Equation 19 has been successively used to represent serial correlation in many fields. With disease progression, one expects ρ to be positive.

Equation 19 can be expanded to show the relationship between u_i and u_{i-2}:

$$u_i = \rho(\rho u_{i-2} + v_{i-1}) + v_i$$
$$= \rho^2 u_{i-2} + \rho v_{i-1} + v_i \tag{20}$$

This equation could also be extended back s time periods to find the relation between u_i and u_{i-s}. When $0 < \rho < 1$, which is the typical case, the correlation between successive u's (i.e., u_i and u_{i-1}) is *greater* than more separated u's (u_i and u_{i-2}).

When the errors are serially correlated as represented in Eq. 19, the variance of u_i (or y_i) is given by

$$\sigma^2[u_i] = \frac{\sigma^2}{1 - \rho^2} \tag{21}$$

Note that the denominator is less than 1. Although the variance is a constant, it is greater than the variance of the ordinary least squares regression model. The covariance between any two successive u's is given by

$$\sigma[u_i, u_{i-s}] = \rho^s \frac{\sigma^2}{1 - \rho^2} \tag{22}$$

Because ρ^s gets smaller as s gets larger, Eq. 22 further supports the assertion in Eq. 20 that close u's have a higher correlation than well-separated u's.

A test can be constructed for serial correlation based on the following null H_0 and alternative H_a hypotheses:

$$H_0: \rho = 0$$
$$H_a: \rho > 0 \tag{23}$$

One first estimates model parameters with ordinary least squares regression or nonlinear regression assuming that $\rho = 0$. The autocorrelation parameter can be calculated as

$$\hat{\rho} = \frac{\sum_{i=2}^{n} (\hat{u}_{i-1} \hat{u}_i)}{\sum_{i=2}^{n} (\hat{u}_{i-1}^2)} \tag{24}$$

One could use a standard t-test to determine if ρ was significantly greater than 0, but Durbin and Watson (1951) have developed a special test for autocorrelation:

$$D = \frac{\sum\limits_{i=2}^{n} (\hat{u}_i - \hat{u}_{i-1})^2}{\sum\limits_{i=1}^{n} \hat{u}_i^2} \tag{25}$$

Small values of D lead to rejecting H_0 in favor of H_a because adjacent u's will have similar values when ρ is positive. Tables of the critical D values are available in most regression texts (e.g., Neter et al., 1983). The test requires a minimum of 15 datum points and there are certain intermediate values of D where the test is inconclusive, i.e., one cannot accept or reject H_0. I find that the Durbin-Watson test serves as a guide but that the decision on a positive correlation has to be made on additional information. In my experience, values of $\rho < 0.40$ or <0.50 will have little impact on the regression results, whereas $\rho > 0.70$ can have a substantial impact. Whenever ρ is much greater than 0.50, I feel that this autocorrelation should be accounted for in the analysis.

After fitting Eq. 9 to the data in Table 1, ρ was estimated to be 0.792. D equaled 0.42 which indicated significant positive autocorrelation ($P < 0.01$) of the residuals.

With linear models there are two basic ways of utilizing autocorrelation in the analysis. First, a form of weighted regression is used to estimate model parameters (Draper and Smith, 1981). Here the weight W_i of Eq. 18 is generalized to the inverse of the covariance matrix of the u_i's (or y_i's), $\mathbf{S}$. Matrix methods can then be used to reestimate model parameters. The approach of using a covariance matrix is called generalized least squares regression. Unfortunately, common computer regression programs do not have an option to perform this type of analysis.

An easier method for linear models is to transform the dependent variable y_i to

$$y_i' = y_i - \rho y_{i-1} \tag{26}$$

Substituting Eq. 26 into the statistical version of the linear model of Eq. 2, one obtains

$$y_i' = b_0 (1-\rho) + b_1 (t_i - \rho t_{i-1}) + \\ b_2 (t_i^2 - \rho t_{i-1}^2) + \cdots + b_q (t_i^q - \rho t_{i-1}^q) + v_i \tag{27}$$

which is a linear model that satisfies the statistical assumption of independence over time. Transformations of y to obtain linear models (y^*) can be transformed a second time in the form of $y^*{}_i' = y_i^* - \rho y_{i-1}^*$ to correct for autocorrelation. The linear logistic model (Eq. 9) can be written in the autocorrelation form as

$$y_i^{*\prime} = \left(\ln \frac{y_0}{1 - y_0} \right) (1 - \rho) + r \, (t_i - \rho t_{i-1}) + v_i \qquad (28)$$

In both cases, the estimate of ρ is substituted into the appropriate equation. With autocorrelation, the dependent and independent variables as well as the intercept parameter have different meaning from the original model. The original intercept can be obtained by $\hat{b}_0 = \hat{b}_0'/(1 - \hat{\rho})$; its standard deviation is calculated as $s[\hat{b}_0] = s[\hat{b}_0']/(1 - \hat{\rho})$.

Fitting Eq. 28 to the example data results in estimated parameters that are different from the ordinary least squares regression case. The estimate of ln $[y_0/(1 - y_0)]$ was reduced by 1.84 to -7.86; the estimate of r was increased by 0.033 to 0.178/day. Even greater relative changes in the standard deviations of the estimated parameters occurred when autocorrelation was corrected for. For instance, there was a greater than 3-fold increase in $s[\hat{r}]$ compared to that estimated with uncorrected regression. A 95% confidence interval for r equaled 0.178 ± 0.056. It is important to note that one cannot directly compare R^2 or s of the autocorrelation regression with R^2 and s from ordinary least squares regression. Both D and ρ were recalculated to verify that $y^{*\prime}$ corrected for autocorrelation (Table 2).

Incorporation of autocorrelation into nonlinear regression analysis is exceedingly more difficult than with the linear case. There are no simple transformations like Eq. 26 to eliminate the correlation. A form of generalized least squares regression for nonlinear models must be employed. The problems with the usual nonlinear least squares is compounded by incorporating a correlation structure to the errors. Gallant and Goebel (1976) and Glasbey (1979) independently developed techniques for using autocorrelation with nonlinear regression analysis. These techniques are not available with common computer regression packages, although Glasbey (1979) has a computer program for estimating parameters. Both procedures require at least 50 time periods to have a reasonable chance of deriving meaningful results. I doubt that these procedures as they now stand will have much use in most botanical epidemiological studies.

First Difference   It often seems reasonable to assume that ρ is close to 1 with data collected over time. Empirically, this is reasonable if a confidence interval for the estimate of ρ includes 1. Theoretically, this is reasonable if one can assume that *differences* of the u_i's (or y_i's) are independent of each other. When $\rho = 1$, Eq. 19 can be written as $u_i = u_{i-1} + v_i$.

The variances and covariances are of a different form when $\rho = 1$ (Starkey, 1977) than when $|\rho| < 1$. Equations 21 and 22 do not hold because the denominators equal 0. It can be shown that the variance of the u's and hence the y's increases with time (Starkey, 1977). The covariance between times decreases as the intervals increase.

With linear models, both generalized least squares regression and transformation approaches can be used for parameter estimation. For generalized least squares regression, the first y value (y_1) is subtracted from all other y_i's prior to analysis. A covariance matrix for these differences is determined, and a matrix version of Eq. 18 is used. A much easier approach is to calculate $y'_i = y_i - y_{i-1}$ as a special case of Eq. 26 (the so-called first differences) and substitute y'_i into the appropriate linear model. The first-difference versions of Eqs. 27 and 28 can be written as

$$y'_i = b_1(t_i - t_{i-1}) + b_2(t_i^2 - t_{i-1}^2)$$
$$+ \cdots + b_q (t_i^q - t_{i-1}^q) + v_i \tag{29}$$

$$y_i^{*'} = r(t_i - t_{i-1}) + v_i \tag{30}$$

In all cases, there is no intercept term in the model, which makes sense since the expected change in y or y^* (i.e., y' or $y^{*'}$) is zero when there is no change in t. The average y at the first assessment time can be used as the estimate of the intercept when it is necessary to have such a value.

The first-difference estimate of r with the example data equaled 0.148, slightly larger than the estimate from uncorrected regression. As with the autocorrelation regression approach, $s[\hat{r}]$ was considerably larger than with ordinary least squares regression.

The nonlinear methods of Gallant and Goebel (1976) and Glasbey (1979) likely could be extended to the first-difference situation. These researchers, however, did not consider this special case of autocorrelation.

Multivariate Growth Curves When replications of y at each time period exist, multivariate linear growth-curve models can be used. This is basically a form of generalized least squares regression in which a covariance matrix is calculated directly from the replicated data in contrast to the previous models where ρ was calculated and then used to generate the covariance matrix. The technique is very general because no prior assumptions about the variances and covariances, other than normality, are required. At least 6 replications per time, and preferably 12 or more, are needed to obtain a precise estimate of the covariance matrix. Khatri (1966), Rao (1965), Potthoff and Roy (1964), and Grizzle and Allen (1969) have all been instrumental in developing this regression procedure. Morrison (1976) provides the most readily available textbook description of the technique. Epidemiologists will find three problems with growth curve analysis: matrix methods are absolutely necessary, many replications are needed, and common statistical programs do not perform this relatively specialized analysis. Programs with matrix manipulation procedures (e.g., Minitab, SAS) can be used by the diligent epidemiologist.

Comparison of the Statistical Techniques There has only been one in-depth comparison of autocorrelation, first-difference, and multivariate growth curve models for epidemic data (Starkey, 1977). First-difference was the preferred method due to its simplicity, ease of computation, and high accuracy in predicting the final level of disease. For the prediction evaluation, the final level of y was not used to estimate parameters.

There has been no evaluation of the nonlinear forms of these models for disease progression data. Numerous growth studies (see Berkey, 1982) demonstrate that the autocorrelation declines markedly when the spacing between times increases. Weekly disease assessment values would have a lower ρ than daily values. Investigators who wish to use nonlinear regression should therefore use well-separated times to avoid or reduce the influence of serial correlation on the results. Increasing the spacing between times could substantially reduce the total number of observations, thus making nonlinear regression very difficult. If after conducting a nonlinear regression, the estimated autocorrelation is high ($\hat{\rho} < 0.70$) and the standard errors of the estimated parameters are to be used for comparisons, then the model should be transformed, if possible, to obtain a linear form. First-difference or autocorrelation models can then be used. Investigators who do not feel prepared to handle nonlinear regression should only use linear models with corrections for autocorrelation.

STATISTICAL COMPARISON OF DISEASE PROGRESS DATA

Although a description of a single disease progress curve can be very informative, investigators generally wish to compare several curves. Disease progression data can describe experimental treatments or naturally occurring epidemics. In either case, the goal is to determine if two or more disease progress curves are significantly different from each other, based on predetermined characteristics. Parameters of disease progress models comprise useful characteristics for comparisons as do the levels of disease at various times. There are numerous approaches for the comparison of longitudinal data, but I will limit this discussion to the most common.

Comparison of Model Parameters

Standard Error of the Parameter Estimates   Each estimated parameter of a disease progress model has an associated standard error and these can be used for statistical testing. Let $\hat{\theta}_1$ and $\hat{\theta}_2$ represent the estimated parameters for two different epidemics, e.g., the estimated apparent infection rates $\hat{r}_1$ and $\hat{r}_2$. A confidence interval for the difference between the two estimates is given by

$$(\hat{\theta}_1 - \hat{\theta}_2) \pm t[P; n_1 + n_2 - (2p)] \cdot s[d] \tag{31}$$

in which p is the number of parameters in each model; n_1 and n_2 are the number of observations for the two different disease progress curves; $t[\cdot]$ is the critical value from a t table with significance level P and $n_1 + n_2 - (2p)$ degrees of freedom; and $s[d]$ is the standard error of the difference which is calculated as the square root of the sum of the variances of the two estimated parameters:

$$s[d] = (s^2[\hat{\theta}_1] + s^2[\hat{\theta}_2])^{1/2} \tag{32}$$

If the confidence interval does not include zero, then one can reject the null hypothesis of parameter equality in favor of the alternative hypothesis of inequality. Essentially, Eq. 31 is a partial t test from regression analysis (Neter et al., 1983).

To use this confidence interval approach to compare two epidemics, the same model must be used for each epidemic. For example, the rate parameter of the logistic model cannot be directly compared to the rate parameter of the Gompertz. When the same model is used, each parameter estimate can be compared using the confidence interval of Eq. 31. This approach could still be misleading if the estimated parameters within each model are highly correlated and results are borderline.

The accuracy of confidence intervals of Eq. 31 depends on accurate estimates of the standard errors. When the u_i's or y_i's are highly correlated over time, calculated standard errors of the parameters will be underestimated, perhaps by 50% or more, if the autocorrelation is not utilized in the regression analysis. For example, if ordinary least squares regression is used and $\rho > 0$, then the standard error of the difference in Eq. 31 will be too low and the interval might not include zero when it should. In this situation significant differences could be found when, in reality, the differences are not significant. Using one of the previously described modifications of ordinary least squares regression will help correct this problem.

Assume that a second epidemic with 15 time periods was analyzed with ordinary least squares regression and that the estimate of r was 0.182/day with $s[\hat{r}] = 0.009$. If the data in Table 1 were also analyzed with ordinary least squares regression, $s[d] = s[\hat{r}_1 - \hat{r}_2] = 0.012$, and Eq. 32 would equal $(0.145 - 0.182) \pm 2.056(0.012)$ or -0.037 ± 0.025 because $t[0.05; 15 + 15 - 2(2)] = 2.056$. The interval does not include zero; thus, one would conclude that the values of r are significantly different at $P = 0.05$. Now assume that correction for autocorrelation was made with the second epidemic using first-difference regression. The estimate of r and $s[\hat{r}]$ equaled 0.19 and 0.027, respectively. Using the first-difference results for the first epidemic, $s[\hat{r}_1 - \hat{r}_2) = 0.036$. The confidence interval for $r_1 - r_2$ equals $(0.148 - 0.190) \pm 2.056(0.036) = -0.042 \pm 0.073$, which includes zero. One would thus conclude that the values of r for the two epidemics were not different at $P = 0.05$.

Standard errors of estimated parameters from nonlinear regression are only approximate; in general, the standard errors are too large. Significant differences might not be detected when in reality the differences are significant. Unfortunately, there is no simple way of correcting these values from nonlinear regression.

The confidence interval approach can be expanded to three or more disease progress curves. Confidence intervals for all pairwise differences of the appropriate parameters can then be calculated. For example, if there are four epidemics, all described by the logistic model, one could compare all four estimates of r in a pairwise fashion. With four epidemics there are $4(4 - 1)/2 = 6$ pairwise comparisons for each parameter. If each interval is performed at $P = 0.05$, then the overall significance level for the six tests (intervals) could be as high as $1 - (1 - P)^6 = 0.26$. Some statisticians recommend reducing the individual P to P divided by the number of comparisons (e.g., $P/6$) to achieve an overall (family) significance level equal or less than P. This approach is very cumbersome when there are many comparisons and can result in few significant differences. Currently, many statisticians argue that the individual comparisons are the test of interest and that controlling an overall significance level serves no purpose (Carmer and Walker, 1982). Epidemiologists should be aware that this area is highly controversial.

Analysis of Variance (ANOVA)   When designed experiments are divided into blocks or replications, there are several approaches for comparing disease progress curves. Instead of using a single value of y at each time, one could use the levels of y for all blocks at each time in the regression analysis. Alternatively, one could determine the mean y across blocks at each time, and then proceed with regression as performed above. With these two approaches, estimated parameters are compared with confidence intervals.

An alternative approach is to perform a regression analysis of disease progress on each block of each treatment separately, and then conduct an ANOVA on the estimated parameters. Estimated parameters, which are random variables, are handled in the same way as variables measured directly in the field. A separate ANOVA is performed for each parameter of the epidemic model. Some errors may result due to correlation of the estimated parameters. Multivariate ANOVA (MANOVA) can be used to analyze the set (vector) of estimated parameters when they are highly correlated. ANOVA works particularly well with complicated experimental designs consisting of several factors. For example, ANOVA permits direct testing of the effect of fungicide type, host cultivar, as well as their interaction on apparent infection rates. ANOVA has the disadvantage of not calculating the inherent variability of a single estimated parameter (e.g., $s[\hat{\theta}]$); ANOVA does estimate the variability among and within factors (e.g., treatments). Confidence intervals and ANOVA may not give the same results, especially if variability is large.

Comparison of Disease Levels

Often it is not possible to describe disease progression data with any regression model. Plots of y vs. t may show cyclical trends, discontinuities, or other unusual features that do not lend themselves to being described by the typical epidemic models. At other times, a single model may not describe the disease progress curves of every treatment. One epidemic might be described by the logistic model while another by the monomolecular. Sometimes, a flexible model such as the Weibull or Richards can be used in these situations. However, the investigator may not have access to a nonlinear regression program or there may not be enough data points to successively estimate model parameters. Other approaches are clearly needed, especially when the dynamics of disease increase are less important than merely evaluating experimental treatments.

Repeated-Measures ANOVA   Repeated-measures ANOVA is useful for analyzing disease progress data in designed experiments. The experimental design can consist of any number of factors. Experimental units (e.g., plants, plots) may or may not be divided into blocks. For the remaining discussion, I assume that the experiment does have b blocks and a treatments. Disease intensity is measured at t times on each experimental unit. It is tempting to consider this a randomized complete block design (RCBD), with treatment and time as factors of interest. Unfortunately, analyzing the experiment as an RCBD would be incorrect. There is a restriction on randomization when the same experimental unit (e.g., plot) is measured at more than one time. The randomized situation occurs only when there are t different experimental units of each treatment in each block to be measured only at one of the times. For example, if one were to measure disease at five times, one would need five plots for each treatment and block. The disease progress curve would be produced by assessing disease in only one of the plots at each time. This latter design would almost always be unwieldy and expensive, if not impossible to use.

Measurements of y at t times on all experimental units should be analyzed with repeated-measures ANOVA. Repeated-measures designs are similar to split-plot, split-unit, and split-block designs (Gill, 1978). Restriction on randomization over time is analogous to restriction on randomization in space of the "subplot" in agronomic studies. Many authors use these terms interchangeably, whereas others assign precise meaning to each of them (Monlezun et al., 1984). All involve restrictions on randomization. Some statisticians prefer to use the term *split-block* when blocks are used, *split-plot* when the experimental design is completely randomized without blocks (Gill, 1978), and *repeated-measures* only when measurements over time are taken. Repeated-measures differ from the other designs by having no randomization of the time factor; split-plot and split-block designs permit randomization of the subject *within* the whole plot factor. Use of the design terms interchangeably, nevertheless, is quite common and should cause no problems as long as the correct analysis is performed.

Gill (1978) provides a thorough description of the many types of repeated measures experimental designs. With all such designs, there are at least two different error variances (mean square errors) for the F test of significance. For our example, there is one error variance for the treatments (whole-plot) and

Table 3 Sources of variation, degrees of freedom *df*, and critical *F* values for a repeated-measures analysis of variance (ANOVA) when the experiment consists of *b* = 4 blocks, *a* = 5 treatments, and *t* = 5 times

		Example		
Source of variation	*df*	*df*	*F(P* = 0.05)	Conservative *F(P* = 0.05)
Block	$b - 1$	3		
Treatment	$a - 1$	4	3.26	3.26
Block × treatment (error for treatment)	$(a - 1)(b - 1)$	12		
Time	$t - 1$	4	2.52	4.54[a]
Time × treatment	$(t - 1)(a - 1)$	16	1.82	3.06[a]
Error for time, and treatment × time	$a(b - 1)(t - 1)$	60		

[a]The $t - 1$ term is replaced by 1.

another error variance for time and the interaction of time and treatment. Table 3 contains the sources of variation and degrees of freedom for the repeated measures example. The sum of squares and expected mean squares for each effect (e.g., treatment) can be found in Gill (1978), Monlezun et al. (1984), and elsewhere. Significance is determined with a *F* test, which entails dividing the mean square for each effect by the corresponding error mean square (variance) and comparing this value with a critical value from an *F* table, i.e., $F[P; df_1, df_2]$. Degrees of freedom from the numerator (df_1) and denominator (df_2) of the *F* tests correspond to the experimental effect and error variance, respectively, and are used in determining the critical *F* value.

When an investigator incorrectly analyzes a repeated-measures experiment as a RCBD with treatment and time as experimental factors, errors in opposite directions are likely to occur. The significance level of the treatment *F* test will be too low, or equivalently, treatment might be found significant when in actuality it is not. The significance level of the time and the treatment-time interaction *F* tests may be too high, or equivalently, time and the interaction might be found nonsignificant when in actuality they are significant. A significant treatment effect indicates that there are some significant differences among treatment level means; a significant time effect indicates that there is a significant change in *y* over time. A significant interaction indicates that the change over time is not the same for each treatment (i.e., the disease progress curves are not parallel), or equivalently, the differences among the treatment means are not the same at every time.

When the interaction is significant, one can compare the treatment means at each time separately. For example, one would compare the mean *y* for treatments 1 and 2 at time 1, then at time 2, and so on. Multiple comparisons or linear contrasts could be used to determine differences among the means. The error means square from the interaction F test is used. A two-way ANOVA of the blocks and treatments could also be performed at each time separately.

With a significant interaction, one could also evaluate the change in disease over time at each treatment level (e.g., fungicide or cultivar). Orthogonal polynomials (Hicks, 1973) could be used to determine if there are linear, quadratic, or higher-order changes in y for each treatment. This approach is analogous to modeling disease progress with Eq. 2. Recall that a reason for comparing actual disease levels instead of the estimated parameters from the various models is that the common models might not describe disease progress data. Therefore, use of orthogonal polynomials might be nonproductive.

Repeated-measures contrasts are useful alternatives to orthogonal polynomials. These contrasts consist of comparing differences of the means at *successive* times. If $\bar{y}_{ij}$ is the mean disease level for the jth treatment at the ith time, then a repeated measures contrast is based on the difference in disease between i and $i + 1$ (i.e., $\bar{y}_{ij} - \bar{y}_{(i+1)j}$). All successive pairs of means can be compared in this manner. It is important to state that many statisticians do not approve of pairwise testing of means when the experimental factor is quantitative and continuous, such as time (Chew, 1976). However, due to nonconstant environment or other factors, y may not change continuously over time and, therefore, will not be adequately described by one of the empirical or theoretical models discussed in a previous section. Under these conditions, there appears to be no reason why successive pairs of means cannot be compared.

When the interaction is not significant, one can compare the treatment means averaged across all times ($\bar{y}_{.j}$) with multiple comparisons or linear contrasts. The error variance for testing treatment (whole plot) effects should then be used (and not the error variances) for testing interaction. Likewise, with no interaction, disease averaged across all treatments ($\bar{y}_{i.}$) can be analyzed with orthogonal polynomials or repeated-measures contrasts. The error variance is the same as for the significant interaction case.

Serial Correlation Structure   There are several statistical assumptions that are necessary for valid F tests with ANOVA. Considerations of these assumptions is outside the realm of this chapter except for the serial correlation of the y's. I assume that the y's have constant variance and are normally distributed. As mentioned in a previous section, y will not have a constant variance if it measures the proportion of plants infected. Transforming y to $y^* = $ arcsin $\sqrt{y}$ will stabilize variances; the logit transformation, however, will not produce a constant variance. Weighted ANOVA can be used if variance inequality is a serious problem, but under most situations, unequal variances do not present major difficulties. A valid analysis of data from repeated-measures experiments requires that the covariance (or correlation) of the y's at *any* two times be homogeneous (Gill, 1978). This means that, regardless of the treatment, the correlation between y_{ij} and $y_{i'j}$ equals a constant ρ, whether i and i' are successive times or well-separated times. As discussed at length in previous sections, growth measurements at adjacent periods generally have a higher correlation than those more distantly separated in time.

Simple modifications of the F tests can be made when there is reason to believe that the correlations are described by Eq. 19. The critical F values for testing time and treatment-time interactions are replaced by the Geisser-Greenhouse critical F values (Gill, 1978). With our example, the critical $F[P; (t - 1), a(b - 1)(t - 1)]$ for testing time effects is replaced with $F[P; 1, a(b - 1)]$; $F[P; (t - 1)(a - 1), a(b - 1)(t - 1)]$ for interaction testing is replaced by $F[P; (a - 1), a(b - 1)]$. The actual F statistic is still calculated in the usual way without changing the degrees of freedom; only the critical value from the F table is altered. In general, the $t - 1$ component in the degrees of freedom is replaced by 1. The reduced degrees of freedom are used in all multiple comparisons and other contrasts of the means. The modified critical F value is considered conservative because, due to the reduced number of degrees of freedom, a larger calculated F statistic is needed to get a significant result (see example in Table 3). This approach is simple, can easily be handled by most researchers, and will prevent most serious errors in data analysis.

Conservative critical F values reduce the probability of determining significant differences when no differences exist. However, conservative values may also reduce the probability of determining significant differences when, in fact, they do exist. When the correlation structure of the y's is precisely known, other less conservative testing approaches are known. For correlations that can be described by Eq. 19, for instance, Box (1954) and Geisser and Greenhouse (1958) presented precise modifications of the critical F values to test for time and interaction effects. Recently, a technique was proposed to estimate ρ without using regression, and then to use $\hat{\rho}$ to correct for autocorrelation with ANOVA (Milliken and Johnson, 1984). For correlations that are represented by a more complicated second-order model, Bjornsson (1978) proposed a generalized ANOVA for significance testing. Determining the configuration of the correlations requires multivariate techniques. Likelihood ratio tests can be used to test whether the covariances are homogeneous or described by Eq. 22 (Hearne et al., 1983; Morrison, 1976). Dealing with complicated correlation structure is tedious and requires many observations. Most epidemiologists will be satisfied in using the described conservative critical F values with repeated measures experiments.

Multivariate Analysis of Variance (MANOVA)

ANOVA deals essentially with the analysis of a single random variable (e.g., disease intensity) for various combinations of factors (e.g., treatment, time, and block). MANOVA, on the other hand, deals with the analysis of a set or vector of random variables (e.g., disease intensity at each time) for various combinations of factors (e.g., treatment and block). Typically, the variables are highly correlated.

Consider an experiment in which the epidemic in each treatment and block is described with the two parameters of the logistic model, i.e., $\ln[y_0/(1 - y_0)]$

and r. Final disease intensity (y_f) and area under the disease progress curve (ADPC) can also be calculated for statistical comparisons. Epidemiologists typically analyze each variable separately with ANOVA. Because these variables are often highly correlated, misleading results can be obtained from this collection of ANOVAs (i.e., some variables might "compensate" for one another and obscure true treatment differences). Principal component and factor analysis are two multivariate techniques for formally assessing the relationships among epidemic variables. More information on these procedures can be found in Campbell et al. (1980), Madden and Pennypacker (1979), and Morrison (1976). MANOVA can be used to analyze the vectors of variables to determine if treatment differences exist. Contrasts of the mean vectors can then be utilized to separate treatments. Unfortunately, interpreting these multivariate contrasts is very difficult. I find that a satisfactory procedure is to first perform MANOVA, where appropriate, and if certain factors or interactions are significant, then to use ANOVA and contrasts of the means for each variable separately. MANOVA calculations were once very tedious, but now they can be readily performed by the major statistical computer packages.

If all statistical assumptions for both ANOVA and MANOVA are met, then ANOVA is the superior procedure. This is because ANOVA is more powerful than MANOVA in detecting differences. When certain repeated-measures ANOVA assumptions are violated, e.g., independent y's over time, conservative critical F values can be used with ANOVA. These new critical values might make ANOVA less powerful than MANOVA. The computational and interpretation difficulties of MANOVA will probably discourage most epidemiologists. Nevertheless, MANOVA is a useful and general technique for analyzing disease progress.

FUTURE PROSPECTS

Few, if any, of the currently used statistical techniques were developed specifically for epidemic data. Epidemiologists owe much to the research conducted in the fields of statistics, biometrics, and growth curve analysis. Some of the other procedures that have potential for analyzing disease progress data are briefly discussed.

Unexplained Variability

The unexplained variability term u can be incorporated into disease progress models in three ways. Adding u to either the untransformed (Eq. 8) or the transformed (Eq. 9) models have already been discussed. Because epidemics are dynamic processes, which are ultimately described by rates, it makes sense to incorporate the unexplained variability directly into the rate equation. The absolute rate of disease increase dy/dt can be represented by the derivative of the function of time in Eq. 1, $d(f(t))/dt$. With biological models, this derivative can be written in terms of y and the parameters, without t being in the expression

(Madden, 1980). Models for the rate of growth that incorporate unexplained variability are called stochastic differential equations and are of two forms:

$$\frac{dy_i}{dt_i} = \frac{d(f(t_i))}{dt_i} + u_i \tag{33a}$$

$$\frac{dy_i}{dt_i} = \frac{d(f(t_i))}{dt_i} + u_iy_i \tag{33b}$$

In Eq. 33a, errors are additive to the rate, whereas in Eq. 33b the errors are additive but proportional to y.

Although this approach is theoretically superior to adding the error term to either y or y^*, there are some difficulties. The mathematics for handling stochastic differential equations are exceedingly more difficult than for the other cases; often there is no unique analytical solution when integrating models of the form of Eq. 33 (Sandland and McGilchrist, 1979). White and Brisbin (1980) presented solutions to certain growth models (e.g., logistic) under strict conditions. The solutions are very complicated and parameters are difficult to estimate. Under most situations, rates of growth between successive times have to be estimated and then the stochastic differential equations fit directly to the estimated rates of increase. This requires a large number of assessment times. Jeger (1982) used a form of stochastic differential equations for studying an apple powdery mildew epidemic. Perhaps in a few years when statisticians develop improved methods for handling these types of models, epidemiologists will make use of stochastic differential equations.

Alternatives

When some of the assumptions for repeated-measures ANOVA are not met, a nonparametric statistical analysis can be performed (Koch et al., 1980). Nonparametric procedures are not based on a particular underlying statistical distribution and are therefore generally applicable. These procedures unfortunately are limited to relatively simple experimental designs and are not very useful with many experimental factors. Additionally, these procedures are sensitive to nonconstant error variance. Using revised and conservative critical F values is one way of handling certain violations of repeated-measures ANOVA assumptions. With simple experimental designs, however, there are no reasons why an investigator should not use nonparametric analysis when the assumptions for other techniques are violated.

Other approaches are possible for analyzing disease progress curves. The u's (or y's) might be described by models that are more complicated than Eqs. 19 or 20 (Gallant and Goebel, 1976; Bjornsson, 1978). One can also use time series analysis sensu Box and Jenkins (1970) to describe disease progress curves. Unfortunately, observations generally need to be taken at regular intervals, and at least 50 time periods are required.

CONCLUSIONS

It has been over 20 years since Van der Plank (1963) showed the value of monitoring epidemics over time and calculating rates of disease increase. A great deal of progress has been made in quantifying and modeling epidemics since 1963. One area that has not received the attention it deserves is the statistical analysis and comparison of disease progress curves. Van der Plank did not propose any special statistical techniques and most epidemiologists since then have used standard regression analysis and ANOVA. Unfortunately, there are special statistical problems when data are collected over time; failure to account for these problems can cause serious errors with statistical testing. Utilization of statistical procedures developed specially for sequential data can alleviate many likely errors. Research reported in the statistics, biometrics, and growth curve analysis literature needs to be used by plant pathologists monitoring disease progression. Several conclusions should be considered by epidemiologists wishing to monitor and compare disease progress curves.

   1.   Many empirical and theoretical models are available for fitting to disease progress data (Hau and Kranz, 1977; Jowett et al., 1974; Madden, 1980; Waggoner, Chap. 1). At least five observation times are necessary for accurately fitting these models to data. Because there will not be perfect agreement between observed values and those values predicted by any model, statistical versions of the appropriate model must be used. It is unlikely that all of the statistical assumptions will be met by a model for disease progression. The most important assumption of independence of the error terms or unexplained variability over time is least likely to be met; the serial correlation can, however, be accounted for in several ways. With nonlinear models, a difficult, generalized least squares can be used with large data sets. With linear models, or linearized nonlinear models, versions of generalized least squares regression as well as autocorrelation and first-difference transformations can be made. The latter transformations comprise the easiest corrections for serial correlation and can be performed most readily with common statistical computer programs or even with calculators.

   2.   Once parameters are estimated, epidemics can be compared using either the standard deviations of the estimated parameters or analysis of variance of the parameters separately estimated for each replication of each treatment. The latter approach is advantageous to use when dealing with many treatments or with complicated experimental designs.

   3.   An investigator wishing to compare epidemics without fitting a specific model to the progression data should use repeated-measures ANOVA or MANOVA. Inspection or testing of the covariance of y values will suggest whether or not modified critical F values should be used in determining significance levels of the various tests. Without testing the covariances, a safe procedure is to use conservative critical F values in testing for main effects and their interactions.

   4.   There are several techniques in addition to regression and ANOVA, such as nonparametric statistics, that have potential for analyzing and comparing disease progress curves. Use of these techniques may require changes in the way data are collected.

5. Investigators wishing to model or analyze plant disease epidemics should be aware of the assumptions underlying the statistical procedures used as well as the implications of using various procedures. If care is taken, comparison of disease progress curves will be highly productive.

ACKNOWLEDGMENTS

I thank C. L. Campbell, W. M. Thal, and an anonymous reviewer for critically reviewing this manuscript.

Salaries and research support were provided by state and federal funds (especially a U.S. Department of Agriculture Crop Loss Grant) appropriated to the Ohio Agricultural Research and Development Center (OARDC), the Ohio State University (OARDC Journal Article 73-85).

REFERENCES

Berkey, C. S. 1982. Comparison of two longitudinal growth models for preschool children. *Biometrics* 38:221–234.

Bjornsson, H. 1978. Analysis of a series of long-term grassland experiments with autocorrelated errors. *Biometrics* 34:645–651.

Box, G. E. P. 1954. Some theorems on quadratic forms applied in the study of analysis of variance problems, II. Effects of inequality of variance and of correlation between errors in the two-way classification. *Ann. Math. Statist.* 25:484–498.

Box, G. E. P., and Jenkins, G. M. 1970. *Time Series Analysis, Forecasting and Control.* Holden-Day, Inc., San Francisco.

Butt, D. G., and Royle, D. J. 1974. Multiple regression analysis in the epidemiology of plant diseases, in: *Epidemics of Plant Disease: Mathematical Analysis and Modeling* (J. Kranz, ed.), pp. 78–114. Springer-Verlag, Berlin.

Campbell, C. L., Madden, L. V., and Pennypacker, S. P. 1980. Structural characterization of bean root rot epidemics. *Phytopathology* 70:152–155.

Carmer, S. G., and Walker, W. M. 1982. Baby Bear's dilemma: A statistical tale. *Agron. J.* 74:122–124.

Chew, V. 1976. Comparing treatment means: a compendium. *HortScience* 11:348–356.

Draper, N. R., and Smith, H. 1981. *Applied Regression Analysis,* 2d ed. John Wiley and Sons, New York.

Durbin, J., and Watson, G. S. 1951. Testing for serial correlation in least squares regression II. *Biometrika* 38:159–178.

Edminster, T. W. 1978. Concepts for using modeling as a research tool. *Tech. Manual* 520, U.S. Dep. Agr. Agric. Res. Serv. 18.

Fulton, W. C. 1979. On comparing values of Van der Plank's r. *Phytopathology* 69:1162–1164.

Gallant, A. R., and Goebel, J. J. 1976. Nonlinear regression with autocorrelated errors, *J. Am. Stat. Assoc.* 71:961–967.

Geisser, S., and Greenhouse, S. W. 1958. An extension of Box's results on the use of the F distribution in multivariate analysis. *Ann. Math. Statist.* 29:885–891.

Gill, J. L. 1978. *Design and Analysis of Experiments in the Animal and Medical Sciences,* vol. 2. Iowa State University Press, Ames.

Glasbey, C. A. 1979. Correlated residuals in non-linear regression applied to growth data, *Appl. Statist.* 28:251–259.

Gregory, L. V., Ayers, J. E., and Nelson, R. R. 1981. Reliability of apparent infection rates in epidemiological research. *Phytopathol. Z.* 100:135–142.

Griggs, N. M., Nance, W. L., and Dinus, R. J. 1978. Analysis and comparison of fusiform rust disease progress curves for five slash pine families. *Phytopathology* 68:1631–1636.

Grizzle, J. E., and Allen, D. M. 1969. Analysis of growth and dose response curves. *Biometrics* 25:357–381.

Hau, B., and Kranz, J. 1977. Ein Vergleich verschiedener Transformationen von Befallskurven. *Phytopathol. Z.* 88:53–68.

Hearne, E. M., Clark, G. M., and Hatch, J. P. 1983. A test for serial correlation in univariate repeated-measures analysis. *Biometrics* 39:237–243.

Hicks, C. R. 1973. *Fundamental Concepts in the Design of Experiments* 2d ed. Holt, Rinehart, and Winston, New York.

Jeger, M. J. 1982. Using growth curve relative rates to model disease progress of apple powdery mildew. *Prot. Ecol.* 4:49–58.

Jowett, D., Browning, J. A., and Haning, B. C. 1974. Nonlinear disease progress curves, in: *Epidemics of Plant Disease: Mathematical Analysis and Modeling* (J. Kranz, ed.), pp. 115–136. Springer-Verlag, Berlin.

Khatri, C. G. 1966. A note on a MANOVA model applied to problems in growth curves. *Ann. Inst. Statist. Math.* 18:75–86.

Koch, G. G., Amara, J. A., Stokes, M. E., and Gillings, D. B. 1980. Some views on parametric and non-parametric analysis for repeated measurements and selected bibliography. *Int. Stat. Rev.* 48:249–265.

Kranz, J. 1978. Comparative anatomy of epidemics, in: *Plant Disease, An Advanced Treatise,* vol. 2 (J. G. Horsfall and E. B. Cowling, eds.), pp. 33-62. Academic Press, New York.

Madden, L. V. 1980. Quantification of disease progression. *Prot. Ecol.* 2:159–176.

Madden, L. V., and Campbell, C. L. 1985. Description of virus disease epidemics in time and space, in: *Plant Virus Epidemiology: Monitoring, Modeling, and Predicting Outbreaks* (G. D. McLean, R. G. Garrett, and W. G. Ruesink, eds.), Academic Press, Australia.

Madden, L., and Pennypacker, S. P. 1979. Principal component analysis of tomato early blight epidemics. *Phytopathol. Z.* 95:364–369.

Mandel, J. 1969. A method for fitting empirical surfaces to physical or chemical data. *Technometrics* 11:411–429.

Milliken, G. A., and Johnson, D. E. 1984. *Analysis of Messy Data.* vol. 1. Lifetime Learning Publications, Belmont, California.

Monlezun, C. J., Blouin, D. C., and Malone, L. C. 1984. Contrasting split plot and repeated measures experiments and analyses. *Ann. Stat.* 38:21–27.

Morrison, D. F. 1976. *Multivariate Statistical Methods.* McGraw-Hill, New York.

Murray, B. G. 1979. *Population Dynamics, Alternative Models.* Academic Press, New York.

Neter, J., Wasserman, W., and Kutner, M. H. 1983. *Applied Linear Regression Models.* Richard D. Irwin, Inc., Homewood, Illinois.

Potthoff, J. H., and Roy, S. N. 1964. A generalized multivariate analysis of variance model useful especially for growth curve problems. *Biometrika* 51:313–326.

Rao, C. R. 1965. The theory of least squares when the parameters are stochastic and its application to the analysis of growth curves. *Biometrika* 52:447–458.

Richards, F. J. 1969. The quantitative analysis of growth, in: *Plant Physiology*, vol. VA (F. C. Steward, ed.), pp. 3–76. Academic Press, New York.

Rouse, D. I., MacKenzie, D. R., and Nelson, R. R. 1981. A relationship between initial inoculum and apparent infection rate in a set of disease progress data for powdery mildew on wheat. *Phytopathol. Z.* 100:143–149.

Sandland, R. L., and McGilchrist, C. A. 1979. Stochastic growth curve analysis. *Biometrica* 35:255–271.

Starkey, T. E. 1977. Analysis of plant disease epidemics. Ph.D. thesis, Pennsylvania State University, Univ. Park.

Statistical Analysis System, Inc. 1982. *SAS User's Guide: Statistics*. SAS Institute, Inc., Cary, North Carolina.

Thal, W. M., Campbell, C. L., and Madden, L. V. 1984. Sensitivity of Weibull model parameter estimates to variation in simulated disease progression data. *Phytopathology* 74:1425–1430.

Thiel, H., and Nagar, A. L., 1961. Testing the independence of regression disturbances. *J. Am. Stat. Assoc.* 56:793–806.

Van der Plank, J. E. 1963. *Plant Disease: Epidemics and Control*. Academic Press, New York.

White, G. C., and Brisbin, I. L. 1980. Estimation and comparison of parameters in stochastic growth models for barn owls. *Growth* 44:97–111.

Wishart, J. 1938. Growth rate determinations in nutrition studies with the bacon pig, and their analysis. *Biometrika* 30:16–28.

Zadoks, J. C. 1978. Methodology of epidemiological research, in: *Plant Disease, An Advanced Treatise*, Vol. 2 (J. G. Horsfall and E. B. Cowling, eds.), pp. 63–96. Academic Press, New York.

Part Two

The Influence of
Environment

Chapter 4

Role of Wet Periods in Predicting Foliar Diseases

A. L. Jones

The need for a susceptible host, a virulent pathogen, and favorable weather for development of an epidemic is well-known in plant pathology. Once researchers determine the contribution and influence of these various factors on disease development, it is often possible to identify key factors that are useful in developing practical disease management systems. The ideal system includes factors relating to the host, pathogen, and environment.

For diseases commonly controlled by routine spraying of fungicides, it is often reasonable to assume that a virulent pathogen and a susceptible host are present. For these diseases, weather is the main factor that influences disease severity. Among the weather factors, moisture, particularly the duration of wetness, stands out as a dominant factor, followed by temperature. In this chapter, I will discuss the nature, measurement, and study of leaf wetness and give some examples of how wetness duration is used as a basis for disease prediction.

THE NATURE OF WET PERIODS

Wetting of plants may be the result of natural or artificial events, may last for a few minutes or several days, and may be continuous or discontinuous. All of these factors are important in determining the significance of wet periods to disease development.

Natural Causes of Wet Periods

Rain can trigger spore discharge in many fungi and is a common cause of leaf wetness in all but arid and semiarid areas of the world. The duration of wetness from rain is often much longer than that caused by dew or fog. Dew and fog differ from rain because moisture develops on plant surfaces through condensation of water in warmer air onto the cooler surface of the plant. Dew formation generally occurs at night and may vary in different layers of the canopy. Leaf wetness is not caused directly by high relative humidity and may also result from guttation.

Artificial Causes of Wet Periods

Overhead irrigation of plants for increasing soil moisture, for frost protection, or for cooling crops during hot weather creates periods of surface moisture. Also, short periods of surface wetness are created when plants are sprayed with pesticides or artificially inoculated with spores suspended in water. The length of the wetness period is dependent not only on the duration of the irrigation cycle but also on the conditions of drying that occur after irrigation. Often wet periods from irrigation are much longer than anticipated due to poor drying conditions after the irrigation has ended.

Wet periods can also be caused by subjecting fruit and other plant parts to various kinds of dip treatments prior to storage. For example, fungal decay can be a problem on cranberries harvested with motorized rotating water reels in flooded bogs. The incidence of decay due to black rot, caused by *Strasseria oxycocci* and *Ceuthospora lunata,* increases with time as cranberry fruits are left in the bog water, and the problem becomes increasingly apparent with time as the fruits are held in cold storage (Stretch and Ceponis, 1983).

When dip or drench treatments with scald-inhibiting chemicals were first introduced in Michigan, blue mold and gray mold, caused by the fungi *Penicillium* spp. and *Botrytis* spp., respectively, become important storage problems on apples. Scald is a physiological disorder that results in a brown discoloration of the skin on susceptible apple cultivars during storage and marketing. When the fruits are dipped or drenched, fungal spores build up through repeated use of the treatment solution. Although the fruits are only dipped or drenched for a few minutes, the wet fruits are put into cold storage where the diseases develop.

Discontinuous Wet Periods

Wet periods are not always continuous. The host may be dry for a few minutes, several hours, or a day or more before being rewet by rain, dew, or fog. Attempts to define the minimum dry interval needed to stop infection by the pathogens that cause apple scab and cherry leaf spot have been made under controlled conditions. With scab, studies by Moore (1964) indicated that a dry period of

48 h but not of 24 h reduced scab infection significantly. Schwabe (1980) found that dry intervals of 16 h were required by ascospores and 32 h by conidia before two wet periods could be treated separately. However, studies by Roosje (1959) and field experience (Jones et al., 1980) indicate 8-hour dry periods are adequate to stop infection of apple leaves. With cherry leaf spot, interrupted wet periods resulted in less infection of sour cherry leaves than continuous wet periods (Eisensmith et al., 1982). Also, the level of infection decreased with increasing length of dry interruption.

In the field, the relationship of wetness duration to spore survival and infection is very difficult to interpret because of the uncertainty of knowing when the inoculum arrived at the infection site. Interrupted wetting may accelerate the death of partially germinated spores, but the spores may survive if wetting is too short to initiate germination. Thus, to answer the question of how much each wet period contributes to the final incidence of infection, knowledge of when the spores arrive and an understanding of the principal factors affecting their survival is needed.

Interaction with Temperature

The duration of surface wetness needed for germination and host penetration by each fungus depends on temperature. Therefore, in working out the hours of wetness needed for infection, the relationship to temperature must be established.

MEASURING WET PERIODS

Compared to the measurement of environmental parameters such as temperature, rain, and wind speed, leaf wetness duration is much harder to measure. In fact, a recent chapter on instrumentation for use in epidemiology research omits the problem of measuring leaf wetness duration (Pennypacker, 1978). Wetness is relative and standards for judging the degree of wetness are lacking. Without standard reference points it is difficult to compare the various sensors that have been developed for measuring wetness.

The calibration of sensors for measuring wetness is also difficult because data on their sensitivity to different intensities of surface water are rarely reported. A noteworthy exception is the study of Melching (1974). He measured the quantity of water deposited on a sensor by weighing it before and after wetting. Corresponding changes in resistance due to wetting were then measured with a galvanometer. Only when such data are available is it possible to properly calibrate and standardize sensors.

Two general approaches have been used to measure leaf wetness. The first is a mechanical method in which the contraction or expansion of a string or membrane element in response to wetting is measured. The instruments that use this method are similar in many ways to hygrometers that use hair or string elements to measure relative humidity. The second method is electronic and depends on measuring the change in resistance across an electronic circuit in response to wetness. In addition to making direct measurements, there have been attempts to predict wetness durations from other environmental factors.

Mechanical Instruments for Direct Assessment

The deWit leaf wetness meter has been in use for several years by plant pathologists to measure the duration of leaf wetness (Post, 1959). A hemp string, which expands when wet and contracts when dry, is the sensing element. Changes in the element are recorded on a revolving 7-day chart. A leaf wetness-temperature recorder was developed by Weltzien and Studt (1974) by replacing the hair element for humidity on a hygrothermograph with a hemp string element for wetness. Additional modifications of this system are described by MacHardy (1979) and by Zuck and MacHardy (1981). An advantage of using modified hygrothermographs is that wetness and temperature are recorded simultaneously.

When rain saturates the sensing element quickly, resulting in full deflection of the pen arm on the recorder, the duration of wetness can be measured relatively accurately with these instruments. However, it is difficult to estimate the duration of wet periods from a fine misty rain or from dew because the traces are intermediate between complete dryness and wetness. The amplitude of response can vary greatly between recorders. Quantification of wet periods from dew, fog, and drizzle is relatively poor with these instruments.

Approaches used by plant pathologists to measure the duration of wetness from dew include the Wallin-Polhemus and the Taylor recorders (Taylor, 1956; Wallin, 1963; Wallin and Polhemus, 1954). The former unit uses a lamb-gut sensor, while the latter uses an indelible pencil that makes a line on a revolving glass disk when the disk becomes wet from dew. Today, these mechanical units are rapidly being replaced by electronic devices. However, their use in early field studies demonstrated the importance of being able to monitor dew periods. See van der Wal (1978) for additional information on dew measurement.

Electronic Instruments for Direct Assessment

Electronic instruments for detecting and recording wetness periods were initially developed using radio tube circuits (Winters and Small, 1934). As transistor technology became available, the instruments were modified and improved. Many variations of sensors, circuits, and recording devices are now available for detecting leaf wetness duration (Davis and Hughes, 1970; Gillespie and Kidd, 1978; Lomas and Shashoua, 1970; Melching, 1974; Schurer and van der Wal, 1972; Small, 1978; Smith and Gilpatrick, 1980). These units are also used to control other kinds of epidemiology equipment such as spore traps. In addition to research, units have been used in integrated pest management programs where leaf wetness duration can be used to predict infection.

Most transistor-based electronic units use some type of artificial leaf sensor made up of two wires or two metal strips mounted in parallel. Current flows through the circuit only when moisture is present. The sensitivity of the unit can be adjusted by varying the level of resistance in the circuit (Melching, 1974).

Other units use wire clips that are attached directly onto the leaf to be monitored. Clips that attach directly to the leaf are difficult to use over long periods in the field. This is because the sensors are easily detached from leaves by wind. With many electronic units it is possible with slight adjustments to interchange the two types of sensors. Where temperature data are also required, the recording mechanism is commonly mounted on a standard thermograph to give the simultaneous recording of both wetness duration and temperature.

Rapid advances in the field of microelectronics led to the development of more sophisticated units for recording weather data in the field. In addition to a number of data loggers developed by commercial firms, units were developed that also incorporated a disease prediction function. An electronic fungus warning system was developed in Germany that uses a series of microchips (Richter and Hausserman, 1975). This unit monitors leaf wetness duration, relative humidity, and temperature. Warning lights indicate when wet periods were sufficiently long for infection of apple by the scab pathogen.

More sophisticated units became available with the development of microprocessors. The first was an on-site microcomputer specific for BLITECAST, a weather driven model for potato late blight (Krause et al., 1975). The unit collected temperature, humidity, and rainfall data in the potato field and displayed a numerical recommendation for the appropriate fungicide application schedule (MacKenzie and Schimmelpfennig, 1978). An on-site microcomputer for predicting primary apple scab infection periods was built with sensors for leaf wetness, temperature, and relative humidity (Jones et al., 1980). This unit differed significantly from the unit developed in Germany for apple scab because the predictions and data were retained in memory for later reference and changes in the predictive model could be made by reprogramming rather than modifying the construction of the unit.

Leaf wetness sensors that utilize a microprocessor are particularly suited for monitoring leaf wetness. This is because the instructions for interpreting the degree of wetness can be programmed into the sensor circuitry. For example, if rain is required for spore discharge from dry leaves, the sensor can be set to require a high level of moisture before indicating wetness. The degree of wetness can be set to the approximate level that is needed to initiate spore discharge. With appropriate adjustment, wetting from light dews or mist does not trigger the sensor. However, once the sensor is triggered by significant moisture, the sensitivity of the sensor is increased so that it detects low levels of moisture. This means that it must be fairly dry before the circuit will indicate dryness. After a specified dry period the sensitivity of the sensor is returned to the less sensitive state. The sensor remains sensitive for a while after drying in order to detect dew periods or light showers that may extend the infection process. Thus, an advantage of microprocessor circuits for disease prediction is that they can be programmed to detect specific kinds of wet periods that are important in the epidemiology of the disease.

The on-site microcomputer developed by Jones et al. (1984) was soon programmed to predict other diseases because of its flexibility. Based on rules developed by Spotts (1977a), units were programmed to predict black rot of grapes and to time fungicides for black rot control (Ellis et al., 1984b). A model to predict the risk of fire blight infection to apple blossoms and a second to predict cherry leaf spot are currently being evaluated in Michigan by the author. Models to predict anthracnose (Danneberger et al., 1984) and Pythium blight (Nutter et al., 1983) on turf are being validated. A disease prediction system for Botrytis leaf blight of onion developed by Lacy and Pontius (1983) has been installed and field tested in the microcomputer (Lacy, 1985). Models for forecasting early blight (Pennypacker et al., 1983) and late blight (Krause et al., 1975) have been integrated and are being tested for scheduling fungicide sprays on tomatoes and potatoes in the North Central region of the United States. Beside disease prediction, units were programmed to simultaneously provide information on daily temperature and rainfall and make degree-day accumulations from selected dates and for different base temperatures (Jones et al., 1984). Degree-day accumulations are used for estimating stages of insect development.

Methods of Indirect Assessment

Early workers frequently used hours of high humidity as an indication of wet period duration. To predict apple scab infection periods, Preece and Smith (1961) substituted the hours of ≥ 90% relative humidity after rain for hours of leaf wetness in Mills' table to obtain "Smith periods." Predictions based on hours of high humidity often indicate a higher infection risk than warranted (Jones et al., 1980). In practice, it is difficult to separate periods of high humidity initiated by rain or dew from normal diurnal fluctuations in humidity. Because of the poor correlation between periods of high humidity and duration of leaf wetness (Lomas and Shashoura, 1970; Wallin, 1963), humidity is only a rough predictor of wetness and its use may result in inaccurate predictions. Therefore, direct recording of wet periods is preferred to using humidity as an indication of wetness in predictive models.

USE OF WET PERIOD DURATION IN PREDICTIVE MODELS

Moisture, particularly the duration of leaf wetness from rain, is a key factor in many predictive systems. Initially, these systems were derived empirically from the results of greenhouse and field experiments. More recently, multiple regression and other curve fitting techniques have been used to relate the duration of leaf wetness and temperature to infection or to increase in disease severity. In some systems, wetness duration is assessed directly, while in others, it is assessed indirectly from humidity and rain.

Empirical Models Based on Leaf Wetness Duration

Apple Scab   Scab is important in most apple producing areas of the world. The disease requires extensive and costly control measures except in semiarid regions where the disease is sporadic. In most apple growing regions the fungus overwinters only in infected leaves on the orchard floor. In spring, about the time buds emerge, mature ascospores can be found in perithecia within the overwintered leaves. When the overwintered leaves become wet, the ascospores are released and carried to the new growth by wind. Ascospores require free water on the surface of the leaves and fruit for germination and infection, and the duration of the wet period required for infection is temperature dependent.

Control of apple scab is accomplished primarily with fungicide applied in a series of approximately 12 sprays in wet climates, 3 to 5 sprays in drier climates. Weekly to biweekly sprays are made for a period that extends from the first appearance of green apple tissue in spring to a few weeks before harvest. Many scab control fungicides are effective only when applied before infection occurs, but some are also effective for a few hours up to about 3 days after the beginning of infection periods. These latter fungicides are referred to as curative or after-infection type fungicides and are very useful in instances where the grower can predict when infection occurred.

Mills (1944) published a series of curves indicating the "approximate hours of wetting necessary for primary scab leaf infection in an orchard containing an abundance of inoculum." Mills estimated it took about 1.5 times longer for light infection in the field than required in moist chamber experiments conducted in Wisconsin by Keitt and Jones (1926) (Lewis, 1978). Using these curves, it was possible to predict whether apple scab infection was likely or not depending on how long the leaves were wet and what the temperature was during the wet period. The predictive system of Mills was unique because it indicated different levels of risk ranging from light to severe. Mills' system has been used worldwide for timing scab control sprays with curative action.

After Mills introduced his predictive system, considerable effort was made to adapt it for use in other apple growing areas of the world. An international symposium was held by the World Meteorological Organization in 1963 specifically to exchange information on the implementation of the Mills system. In some regions, scab prediction was centralized and a few trained personnel provided summaries by telephone, radio, or television to all the growers in the region. The second approach was to develop various types of monitoring equipment and interpretive aids for on-the-farm prediction. Today, small microcomputers are available that monitor the environment and provide specific on-site predictions to growers with specific scab predictions (Jones et al., 1984). Microcomputers are as effective as the Mills system for predicting scab infection periods (Jones et al., 1980; Ellis et al., 1984a).

The apple scab predictive system identifies wet periods that favor infection to leaves. Early in the growing season wetting and temperature requirements for infection of young leaves and fruit of apple are considered to be identical, but just prior to harvest much longer wet periods are necessary for infection of fruit than of leaves (Schwabe, 1982; Tomerlin and Jones, 1983). Recently, Schwabe et al. (1984) established that the wetting requirements for infection to fruit increase exponentially with increasing fruit age starting about 1 week after full bloom. This information should make it possible to control apple scab with fewer fungicide applications than is currently being practiced during the latter part of the growing season.

Black Rot of Grapes   Although it might be expected that the success of Mills' system for apple scab would stimulate the development of predictive systems for other diseases, only a few systems have been developed using similar criteria. A recent example is a system for black rot of grapes. Black rot, caused by *Guignardia bidwellii,* is one of the most destructive diseases of grapes in the northeastern United States. The relation of duration of wetness to infection was established in growth chamber experiments and verified in the field (Spotts, 1977a). A table, similar to that of Mills' in 1944 was constructed giving the minimum number of hours of leaf wetness necessary for infection at various temperatures (Mills and La Plante, 1951). Although this research made it possible to predict black rot infection periods, such information was of limited value to grape growers because the only registered grape fungicides were protectants. They would not control the infections when applied 2 to 3 days after an infection period. However, two experimental fungicides, fenarimol and triadimefon gave good postinfection control in greenhouse trials (Spotts, 1977b) and triadimefon was found effective in field trials (Spotts, 1979). Recently, the epidemiology information on black rot infection was incorporated into a disease predictor for use in identifying black rot infection periods and timing of applications of triadimefon (Ellis et al., 1984b).

Multiple Regression Models

Cedar Apple Rust   Cedar apple rust is the most important of three apple rust diseases that occur east of the Rocky Mountains. It is caused by the fungus *Gymnosporangium juniperi-virginianae* which infects red cedar and apple. Wind-blown basidiospores, produced in telial horns in galls on cedar trees, infect apple from the pink stage of bud development to about 6 weeks after the petal fall stage of bud development. Fungicides are used for control on susceptible apple cultivars in areas where red cedars are abundant and their destruction is not practical. Timing of the fungicide sprays is on a calendar basis with treatments being applied every 7 to 10 days. Four to six sprays of a fungicide are applied each year to control this disease on susceptible cultivars.

Recently, researchers in New York state developed a system for predicting infection of apple by basidiospores of *G. juniperi-virginianae* and scheduling fungicide applications to control cedar apple rust (Aldwinckle et al., 1980). In this weather-based prediction system, the duration of the wet period is used to determine (1) when basidiospores are formed on the telial horns of the over-wintered galls on cedar, and (2) the favorability of the weather for infection of apple leaves by the basidiospores. Thus, this system assesses the presence of initial inoculum and then predicts the efficacy of the initial inoculum.

Multiple regression was used to establish the period of wetness required for basidiospore formation (Pearson et al., 1977). Environmental factors necessary for the discharge of basidiospores were established under field conditions (Pearson et al., 1980). Then, using data from infection studies conducted in controlled-temperature incubators, regression equations were developed to predict the minimum number of hours of leaf wetness duration required for light and for severe infection (Aldwinckle et al., 1980). This information was summarized in a table (Pearson et al., 1981) and in diagrams (Seem and Russo, 1984) for grower use in conjunction with wetness and temperature monitoring equipment. Automation of this system using current computer and electronic sensing technology should be easy.

Cherry Leaf Spot   Cherry leaf spot, caused by the fungus *Coccomyces hiemalis,* is a common disease on sour and sweet cherries in the Great Lakes area and mid-Atlantic states. Ascospores from apothecia in leaves that overwinter on the orchard floor initiate primary infections in the spring. Conidia from acervuli on infected, living leaves initiate secondary infections. The duration of wetness needed for infection of leaves from conidia was established in moist chamber experiments in Wisconsin (Keitt et al., 1937).

An equation relating duration of leaf wetness at various temperatures to disease incidence in greenhouse trials was constructed using multiple regression techniques (Eisensmith and Jones, 1981a). The relation of wetness and temperature to infection was presented in graphic form by plotting predictions made by the regression model on a three-dimensional response surface. When the regression model was tested in the field, it was reasonably accurate in predicting whether observed wet periods would result in infection.

To use the model in the field, an index value that reflected the relative favorability of the weather for infection was computed, and threshold values corresponding to light, moderate, or severe levels of infection were established empirically. Fungicides with curative properties gave good leaf spot control when applied after predicted infection periods (Eisensmith and Jones, 1981b). The epidemiology information on leaf spot infection was incorporated into a nomogram that allows growers to determine index values based on leaf wetness duration and temperature information. The model was also incorporated into a microcomputer, but commercial use of the microcomputer for leaf spot prediction must await registration of curative fungicides.

A weather-based predictive model for anthracnose (*Colletotrichum graminicola*) in annual bluegrass (Danneberger et al., 1984) and a model for predicting the wetting requirements for scab infection of apple fruit with increasing age (Schwabe et al., 1984) were developed following the approach used with leaf spot.

Models Involving the Indirect Assessment of Wetness

The association of periods of high moisture with observed increases in disease has been used to develop a number of disease warning systems. Late blight of potato, early blight of tomatoes, and peanut leaf spot are reviewed here in some detail. Other forecasting techniques that are based in part on moisture have been reported by Backman et al. (1984), Dainello and Jones (1984), Gillespie and Sutton (1979), Kemheller and Diercks (1983), Nutter et al. (1983), and Thomas (1983).

Potato Late Blight   Potato is an important food crop in many regions of the world. Because potato late blight is also severe in most of these areas, many attempts have been made to forecast late blight and schedule control treatments with fungicides (Fry, 1982). Late blight is favored by cool, wet weather. BLITE-CAST is a computer model for forecasting late blight and scheduling fungicide applications on potato for control (Krause et al., 1975; MacKenzie, 1981). The hours of high relative humidity and the temperature during those periods are used in a portion of the model to calculate severity values as an index of potential pathogen growth. The frequency at which fungicide sprays are needed depends on how rapidly the severity values accumulate.

Potato late blight forecast models are fairly accurate in the region where they were developed, but they must usually be modified for use elsewhere. Although a region-to-region modification of predictive models is not unusual, late blight appears to be an extreme example. Unlike the models described in the previous section, the effects of moisture duration and temperature on infection are known only on a relative basis. However, moisture duration is very important in the epidemiology of potato late blight (Rotem and Reickert, 1964). The potential of increasing the exactness of late blight models by quantification of the wetness time for infection, and by using leaf wetness measurement rather than estimates, should be investigated. Also, BLITECAST may be more effective for timing after-infection or curative fungicides than for timing protective fungicides (Fohner et al., 1984).

Early Blight of Tomatoes   Early blight, caused by *Alternaria solani,* is favored by warm, wet weather. FAST is a computer model for forecasting early blight and scheduling fungicide applications for its control (Madden et al., 1978; Pennypacker et al., 1983). The model interprets the epidemiological consequence of leaf wetness time, hours of high relative humidity, amount of rainfall, and the temperatures during those periods to develop specified critical levels established for the initiation of spray applications and for adjusting the interval of the spray schedule once spraying is initiated. Specific infection periods are not predicted. Rather, the model indicates the relative favorability of the weather, particularly moisture, for disease development.

Peanut Leafspot Two fungi, *Cercospora arachidicola* and *Cercosporidium personatum*, cause early and late leafspot, respectively, of peanuts. Leafspot development is favored by humid weather. Jensen and Boyle (1966) developed a graph for predicting the rate of increase in leafspot infection based on the hours of continuous relative humidity at or above 95% and the minimum temperature during the humid period. Humidity is used in this system as an indirect measurement of leaf wetness. A computer program was developed in Georgia to deliver worded advisories (Parvin et al., 1974), and electronic environmental monitoring stations were used in Virginia to collect the data needed to make daily advisories (Phipps and Powell, 1984). A recent evaluation of the predictive model for scheduling fungicide applications indicated greater disease in plots sprayed according to advisories but no differences in peanut yield (Phipps and Powell, 1984). Because an average of 4.25 fewer sprays were required with the advisory system, it was economically more efficient.

FUTURE RESEARCH NEEDS AND CONCLUSIONS

The recent increase in the number of forecasting and predictive systems based on moisture duration and the temperature during those periods underscores the importance of these variables in predictive models. Although moisture is difficult to measure, recent advances in the field of microelectronics have given plant pathologists the ability to monitor and record wetness periods with more precision than with earlier methods. As predictions become more accurate and the number of diseases that can be predicted increases, pest management activities for enhancing the efficiency of control programs should increase on a number of crops.

REFERENCES

Aldwinckle, H. S., Pearson, R. C., and Seem, R. C. 1980. Infection periods of *Gymnosporangium juniperi-virginianae* on apple. *Phytopathology* 70:1070–1073.

Backman, P. A., Crawford, M. A., and Hammond, J. M. 1984. Comparison of meteorological and standardized timings of fungicide applications for soybean disease control. *Plant Dis.* 68:44–46.

Dainello, F. J., and Jones, R. K. 1984. Continuous hours of leaf wetness as a parameter for scheduling fungicide applications to control white rust in spinach. *Plant Dis.* 68:1069–1072.

Danneberger, T. K., Vargas, Jr., J. M., and Jones, A. L. 1984. A model for weather-based forecasting of anthracnose on annual bluegrass. *Phytopathology* 74:448–451.

Davis, D. R., and Hughes, J. E. 1970. A new approach to recording the wetting parameter by the use of electronic resistance sensors. *Plant Dis. Rep.* 54:474–479.

Eisensmith, S. P., and Jones, A. L. 1981a. A model for detecting infection periods of *Coccomyces hiemalis* on sour cherry. *Phytopathology* 71:728–832.

Eisensmith, S. P., and Jones, A. L. 1981b. Infection model for timing fungicide applications to control cherry leaf spot. *Plant Dis.* 65:955–958.

Eisensmith, S. P., Jones, A. L., and Cress, C. E. 1982. Effects of interrupted wet periods on infection of sour cherry by *Coccomyces hiemalis*. *Phytopathology* 72:680–682.

Ellis, M. A., Madden, L. V., and Wilson, L. L. 1984a. Evaluation of an electronic apple scab predictor for scheduling fungicides with curative activity. *Plant Dis.* 68:1055–1057.

Ellis, M. A., Madden, L. V., and Wilson, L. L. 1984b. A microcomputer for predicting grape black rot infection periods. (Abstr.) *Phytopathology* 74:1269.

Fohner, G. R., Fry, W. E., and White, G. B. 1984. Computer simulation raises questions about timing protectant fungicide application frequency according to a potato late blight forecast. *Phytopathology* 74:1145–1147.

Fry, W. F. 1982. Disease forecasting: Epidemiological considerations, in: *Principles of Plant Disease Management*, pp. 105–126. Academic Press, New York, 378 pp.

Gillespie, T. J., and Kidd, G. E. 1978. Sensing duration of leaf moisture using electrical impedance grids. *Can. J. Plant Sci.* 58:179–187.

Gillespie, T. J., and Sutton, J. C. 1979. A predictive scheme for timing fungicide applications to control alternaria leaf blight to carrots. *Can. J. Plant Pathol.* 1:95–99.

Jensen, R. E., and Boyle, L. W. 1966. A technique for forecasting leafspot on peanuts. *Plant Dis. Rep.* 50:810–814.

Jones, A. L., Fisher, P. D., Seem, R. C., Kroon, J. C., and van deMotter, P. J. 1984. Development and commercialization of an in-field microcomputer delivery system for weather-driven predictive models. *Plant Dis.* 68:458–463.

Jones, A. L., Lillevik, S. L., Fisher, P. D., and Stebbins, T. C. 1980. A microcomputer-based instrument to predict primary apple scab infection periods. *Plant Dis.* 64:69–72.

Keitt, G. W., and Jones, L. K. 1926. Studies of the epidemiology and control of apple scab. *Wis. Agr. Exp. Sta. Res. Bull.* 73, 104 pp.

Keitt, G. W., Blodgett, E. C., Wilson, E. E., and Magie, R. O. 1937. The epidemiology and control of cherry leaf spot. *Wis. Agr. Exp. Sta. Res. Bull.* 132, 117 pp.

Krause, R. A., Massie, L. B., and Hyre, R. A. 1975. BLITECAST: A computerized forecast of potato late blight. *Plant Dis. Rep.* 59:95–98.

Kremheller, H. T., and Diercks, R. 1983. Epidemiologie und Prognose des Falshcen Mehtaues (*Pseudoperonospora humuli*) an Hopfen. *Zeitschrift für Pflanzenkrankheiten und Pflanzenschutz* 90:599–616.

Lacy, M. L. 1985. Timing onion leaf blight sprays with a dedicated disease predictor in Michigan. (Abstr.) *Phytopathology* 75: (in press).

Lacy, M. L., and Pontius, G. A. 1983. Prediction of weather-mediated release of conidia of *Botrytis squamosa* from onion leaves in the field. *Phytopathology* 73:670–676.

Lewis, F. H. 1978. Dr. W. D. Mills and his system of predicting apple scab infection. *N. Y. State Agr. Exp. Sta. Special Report*, 28, pp. 37–38.

Lomas, J., and Shashoua, Y. 1970. The performance of three types of leaf-wetness recorders, *Agr. Meteorol.* 7:159–166.

MacHardy, W. E. 1979. A simple, quick technique for determining apple scab infection periods. *Plant Dis. Rep.* 63:199–204.

MacKenzie, D. R. 1981. Scheduling fungicide applications for potato late blight with BLITECAST. *Plant Dis.* 65:394–399.

MacKenzie, D. R., and Schimmelpfennig, H. 1978. Development of a microcomputer unit for forecasting potato late blight using the Pennsylvania State University BLITECAST system. (Abstr.) *Am. Potato J.* 55:384.

Madden, L., Pennypacker, S. P., and MacNab, A. A. 1978. FAST, a forecast system for *Alternaria solani* on tomato. *Phytopathology* 68:1354–1358.

Melching, J. S. 1974. A portable self-contained system for the continuous electronic recording of moisture conditions on the surface of living plants. *U.S. Dep. Agr. Serv.* ARS-NE-42, 13 pp.

Mills, W. D. 1944. Efficient use of sulfur dusts and sprays during rain to control apple scab. *N.Y. Agr. Exp. Sta. Ithaca Ext. Bull.* 630, 4 pp.

Mills, W. D., and LaPlante, A. A. 1951. Diseases and insects in the orchard. *N.Y. State Agr. Exp. Sta. Ithaca Ext. Bull.* pp. 18–21.

Moore, M. H. 1964. Glasshouse experiments on apple scab I. Foliage infection in relation to wet and dry periods. *Ann. Appl. Biol.* 53:423–435.

Nutter, F. W., Cole, H., Jr., and Schein, R. D. 1983. Disease forecasting system for warm weather Pythium blight of turfgrass. *Plant Dis.* 67:1126–1128.

Parvin, D. W., Jr., Smith, D. H., and Crosby, F. L. 1974. Development and evaluation of a computerized forecasting method for Cercospora leafspot of peanuts. *Phytopathology* 64:385–388.

Pearson, R. C., Aldwinckle, H. S., and Seem, R. C. 1977. Teliospore germination and basidiospore formation in *Gymnosporangium juniperi-virginianae:* A regression model of temperature and time effects. *Can. J. Bot.* 55:2832–2837.

————. 1981. Cedar apple rust. *N. Y. Agr. Exp. Sta. Tree Fruit IPM Dis. Information Sheet* 5, 2 pp.

Pearson, R. C., Seem, R. C., and Meyer, F. W. 1980. Environmental factors influencing the discharge of basidiospores of *Gymnosporangium juniperi-virginianae.* *Photopathology* 70:262–266.

Pennypacker, S. P. 1978. Instrumentation for epidemiology, in: *Plant Disease: An Advanced Treatise,* vol. 2: How disease develops in populations (J. G. Horsfall and E. B. Cowling, eds.), pp. 97–118. Academic Press, New York.

Pennypacker, S. P., Madden, L. V., and MacNab, A. A. 1983. Validation of an early blight forecasting system for tomatoes. *Plant Dis.* 67:287–289.

Phipps, P. M., and Powell, N. L. 1984. Evaluation of criteria for the utilization of peanut leafspot advisories in Virginia. *Phytopathology* 74:1189–1193.

Post, J. J. 1959. Het instrumentarium voor het bepalim van infecteperioden. *Meded. Dir. Tuinbouw Neth.* 22:365–371.

Preece, T. F., and Smith, L. P. 1961. Apple scab infection weather in England and Wales, 1956–60. *Plant Pathol.* 10:43–51.

Richter, V. J., and Haussermann, R. 1975. Ein elektronisches Schorfwarngerät. *Anz. Schädlingskde., Pflanzenschutz, Umweltschutz.* 48:107–109.

Roosje, G. S. 1959. Laboratoriumonderzoek van infectias door ascosporen en conidien. *Meded. Dir. Tuinbouw Neth.* 22:15–23.

Rotem, J., and Reickert, I. 1964. Dew: A principal moisture factor enabling early blight epidemics in a semiarid region of Israel. *Plant Dis. Rep.* 48:211–215.

Schurer, K., and van der Wal, A. F. 1972. An electronic leaf wetness recorder. *Neth. J. Plant Pathol.* 78:29–32.

Schwabe, W. F. S. 1980. Wetting and temperature requirements for apple leaf infection by *Venturia inaequalis* in South Africa. *Phytophylactica* 12:69–80.

————. 1982. Wetting and temperature requirements for infection of mature apples by *Venturia inaequalis* in South Africa. *Ann. Appl. Biol.* 100:415–423.

Schwabe, W. F. S., Jones, A. L., and Jonker, J. P. 1984. Changes in the susceptibility of developing apple fruit to *Venturia inaequalis*. *Phytopathology* 74:118–121.

Seem, R. C., and Russo, J. M. 1984. Simple decision aids for practical control of pests. *Plant Dis.* 68:656–660.

Small, C. G. 1978. A moisture-activated electronic instrument for use in field studies of plant diseases. *Plant Dis. Rep.* 62:1039–1043.

Smith, C. A., and Gilpatrick, J. D. 1980. Geneva leaf-wetness detector. *Plant Dis.* 64:286–288.

Spotts, R. A. 1977a. Effect of leaf wetness duration and temperature on the infectivity of *Guignardia bidwellii* on grape leaves. *Phytopathology* 67:1378–1381.

———. Chemical eradication of grape black rot caused by *Guignardia bidwellii*. *Plant Dis. Rep.* 61:125–128.

———. 1979. Use of Bay Meb 6447 for eradication of grape black rot caused by *Guignardia bidwellii*. *Plant Dis. Rep.* 63:967–969.

Stretch, A. W., and Ceponis, M. J. 1983. Influence of water immersion time and storage period on black rot development in cold-stored, water-harvested cranberries. *Plant Dis.* 67:21–23.

Taylor, C. F. 1956. A device for recording the duration of dew deposits. *Plant Dis. Rep.* 40:1025–1028.

Thomas, C. E. 1983. Fungicide applications based on duration of leaf wetness periods to control Alternaria leaf blight of cantaloup in south Texas. *Plant Dis.* 67:145–147.

Tomerlin, J. R., and Jones, A. L. 1983. Development of apple scab on fruit in the orchard and during cold storage. *Plant Dis.* 67:147–150.

Van der Wal, A. F. 1978. Moisture as a factor in epidemiology and forecasting, in: *Water Deficits and Plant Growth,* vol. 5: Water and Plant Disease (T. T. Kozlowski, ed.), pp. 253–295. Academic Press, New York.

Wallin, J. R. 1963. Dew: Its significance and measurement in phytopathology. *Phytopathology* 53:1210–1216.

Wallin, J. R., and Polhemus, D. N. 1954. A dew recorder. *Science* 119:194–295.

Weltzien, H. C., and Studt, H. G. 1974. A combination temperature-leaf wetness recorder for improved apple scab control in Lebanon. *Plant Dis. Rep.* 58:133–135.

Winters, R., and Small, C. G. 1934. An automatic moisture-recording device. *Phytopathology* 24:284–288.

Zuck, M. G., and MacHardy, W. E. 1981. Recent experiences in timing sprays for control of apple scab: Equipment and test results. *Plant Dis.* 65:995–998.

Chapter 5

Role of Soil Water in Population Dynamics of Nematodes

T. C. Vrain

With water I made all living things.
The Koran

SOIL WATER AND NEMATODE HABITAT

Soil nematodes are aquatic, aerobic organisms that inhabit the system of inter-connected spaces between soil particles, a habitat they share with all other soil organisms and plant roots. Populations of nematodes vary in size in response to pressures from the edaphic environment that affect regulatory control of such factors as sex ratio and rates of reproduction and death. The availability and quality of food and the capacity of the nematodes to migrate and reach it are among the main determinants of population fluctuations.

The proportion of soil space occupied by water and by air influences the activities of ectoparasites and those stages of endoparasites exposed to the soil environment. Once sedentary endoparasites are in the roots, they obtain a steady supply of water, food, and oxygen from the plants and are not influenced by moisture to the same extent as ectoparasites. Variations in the amount of water in the soil are accompanied by variations in the amount of air and of all the chemicals dissolved in the water: oxygen and carbon dioxide, salts, root exudates, pheromones, and microbial metabolites. Soil water variations also affect activities of nematodes directly exposed in the soil: egg development, hatching, migration, penetration or feeding, development, and reproduction. In a draining soil, certain proportions of the spaces are occupied by different amounts of moisture and are differentially favorable to nematode activity. Nematodes near the soil surface are subjected to extremes of temperature and moisture variations, during which they are either inactive or die, whereas those inhabiting the deeper layers are protected from these extremes but are subject to periods of anoxia.

The principles of nematode movement over the surfaces of soil particles in thin films of water were established by Wallace (1959b). When soil dries, pores are drained and the thickness of the film of water covering soil particles diminishes. Nematode movement decreases until the film of water disappears and the nematodes are immobilized in spaces filled with moisture-saturated air. As the air becomes drier, species that have developed mechanisms to survive this dry environment coil and become quiescent.

Many studies consider only the quantity, not the quality, of water in the soil and the picture drawn from published results is not quite coherent. Most reports on the effect of soil moisture, or the influence of rainfall or irrigation, on the distribution or the population dynamics of a nematode species give moisture measurements in absolute quantities, relating them only to the particular soil studied. In greenhouse studies, moisture levels are reported as percent of soil (either as weight by weight or weight by volume), percent of water-holding capacity, or percent of field capacity. In field studies, weekly or monthly rainfall and amount of water added by irrigation are reported. Sometimes the amount of moisture at different depths is measured along with rainfall. This type of information is useful in clarifying the trends of population dynamics in a particular soil at different levels of soil wetness but does not provide a means of comparison with other soils. We now know that nematode populations increase most rapidly at intermediate moisture levels close to field capacity. Field studies show correlations between population trends and accumulated rainfall, and that different species of nematodes respond differently to rainfall patterns in different soils.

The significance and relevance of the correlations between amount of rain and nematode activity are difficult to determine due to the lag time imposed by soil drainage and the many interactions between soil moisture content and other factors affecting population dynamics. It is easier to relate population fluctuations to amounts of rainfall, or irrigation, than it is to show precisely the biological and physical processes involved. If we are to quantify the relationship between nematode activity and amount of water in the soil we need to determine what proportion of soil spaces are occupied by water and how fast nematodes hatch, move, reproduce, and die in soil spaces full or partially full of water.

Attempts to utilize published results in a unifying model are impeded by the lack of a common unit of measurement (e.g., water potential). Although soil water potential is related to water content of the soil, one cannot be inferred from the other. Therefore, results obtained experimentally on the effect of soil moisture on nematode population dynamics reported on a water content basis are difficult to interpret in terms of water potential because of the complex nature of this relationship. This chapter outlines techniques to determine moisture potentials in soils, reviews the direct and indirect effects of soil moisture on nematode movement and development, and stresses the necessity of using quantitative approaches in soil moisture studies.

CHARACTERIZATION AND MEASUREMENT OF SOIL WATER

Ecological studies of nematodes require characterization of soil in terms of texture, structure, and chemical or mineralogical composition. We need to characterize the space available to nematodes in different kinds of soil in order to evaluate the constraints placed on nematode movements and to develop conceptual models including the most relevant effects of soil moisture. Since nematodes live in the system of interconnected pores between the inert supporting skeleton of soil particles, the descriptive measurements of texture can be disregarded once we know the physical parameters of that living space. These parameters are mainly the available space between pore necks relative to nematode movement, the moisture potential determining which pores are filled with water and which are partially or totally empty of water, and the concentration of oxygen and carbon dioxide in the water and the air contained by these spaces.

Soil Water Content

The mass of water per volume of soil characterizes the soil water content. Although there are several ways to measure soil water, there is as yet no universally recognized standard method.

The gravimetric method uses samples of soil collected with a probe or an auger. The wet weight is measured at the time of sampling. The samples are then dried in an oven at 105° C for at least 24 h and the water content is usually reported as the ratio of the weight loss in drying relative to the dry weight of the samples.

With moisture blocks, electrodes embedded in blocks of gypsum, Mylar, or fiberglass are placed in the soil. The electrical resistance inside the porous blocks in equilibrium with soil moisture is measured and a calibration curve gives the soil moisture content. Moisture blocks are preferable to the gravimetric method because, once installed and calibrated, they can be connected to a recorder to give continuous measurements without disturbing the soil.

With a neutron probe, a source of fast neutrons (2 to 4 megaelectron volts, MeV) is placed into the soil. Neutrons are deflected and lose their kinetic energy to approximately 0.03 eV by repeated collisions with different atomic nuclei in the soil. The hydrogen nuclei of water molecules are the most effective at slowing down and scattering these neutrons. As water increases in the soil, so does the number of slowed neutrons. Thus a cloud of slow neutrons forms instantly around the probe, which contains a detector, and can be measured on a monitor. When properly calibrated the neutron probe provides measurements of soil moisture content over the entire range of variation of soil water. This method is accurate, nondestructive, and more rapid than the gravimetric method (Greacen, 1981).

Energy Potential of Soil Water

Measuring soil water content provides little information about the bioavailability of water in a particular soil. For example, a sandy soil may be saturated when it contains 10% water while a much finer soil would be practically dry. The structure, texture, and the manner in which water is retained in a soil are important factors that determine the availability of water. Soil particles are irregularly shaped and leave irregular spaces between them. These particles aggregate in crumbs separated from each other by wide spaces, the macropores, and there are also long channels made by roots, earthworms, or insects. The water in these spaces has potential energy, and when different levels of potential energy of water exist between two points, the water flows. The water is held in the interstices and by adsorption on the solid surfaces. In unsaturated soils, water covers solid particles with thin films of variable shapes and thicknesses. Since water is held, it requires energy to be made available; we think of its energy potential as being negative. The concept of water potential replaces the arbitrary classification of gravitational, capillary, and hygroscopic water, providing a unified measure by which the state of water in different soils can be compared.

The energy potential, or suction, is a measure of the intensity with which water is held in a soil. If water, held in a capillary tube standing in water, rises to a height of h cm, it would take a certain pressure applied at the top of the tube to push the water down to the same level as the water outside the capillary tube. Think of the air-water meniscus in that capillary as exerting the suction upward and the height of the column of water as a measure of that suction, in centimeters of water. When a soil holds water at a suction of h cm, a suction greater than h needs to be applied to be able to use that water or remove it from the soil. The suction draining a pore must overcome the surface tension forces acting around the perimeter of the necks. As the diameter of pore necks decreases, the suction required to empty the pores increases.

The intensity with which water is held in a capillary tube bounded by an air-water meniscus can be defined in terms of the reduction of its energy below that of free water. Soil water can also have its energy reduced when salts that increase its osmotic pressure are dissolved in it. The energy reduction due to the menisci in soil is called matric potential and that due to dissolved salts is the osmotic potential. The words suction, tension, and negative pressure are all equivalent and allow expression of the osmotic and matric potential in positive terms. Suction can be measured and expressed as the pressure exerted by a column of water h cm high, or in bars, or in atmospheres. However, the SI unit for pressure or suction is the Pascal (Pa), with 1 kPa equivalent to 10 cm. If the suction s is exerted by the meniscus in a capillary tube, then it is a function of the diameter of the tube d and the relation is approximately $d = 300/s$, with d in micrometers (μm) and s in kPa. For example, in a soil where suction is 10 kPa, the pores with necks of a diameter greater than 30 μm are emptied of water.

Measurement of Soil Suction

Total soil moisture potential or suction, the sum of matric and osmotic potentials, can be measured with a tensiometer or a thermocouple psychrometer. A tensiometer consists of an airtight tube full of water with a ceramic cup at one end and a manometer at the other end. The solution in the soil comes in contact and equilibrates with the water in the tube through the capillary pores of the ceramic cup. The soil solution, with its suction, draws out some water from the airtight tube causing a drop of pressure inside which is measured by the manometer. Tensiometers can be left in place to follow the variations of the soil suction, as soil water content increases with rainfall and irrigation and decreases with drainage, evaporation, and plant uptake. The range of suctions measureable by tensiometry is 0 to 80 kPa, which encompasses most of the soil wetness range in agricultural soils.

The thermocouple psychrometer is very useful beyond the range of potentials measured by tensiometry but is not practical or accurate for measurement of low soil potentials. The psychrometer measures the relative humidity (vapor pressure) of the atmosphere in unsaturated soil, where the vapor potential is equal to the suction, since air only allows water molecules to pass from the soil to the thermocouple wires doing the actual measurement. Vapor pressure is very dependent on temperature, which must be measured very accurately.

The Soil Moisture Characteristic Curve

The relationship between the configuration of the solid matrix and the suction forces withdrawing the water from between particles determines the amount of water held on the surfaces of that matrix. When a slight suction is applied to a wet soil, water is drawn from the largest pores first. A gradual increase in suction results in the emptying of progressively smaller pores and a decrease in the thickness of the water film covering the large soil particles. The amount of water retained at values of matric suction between 0 and 100 kPa depends on the distribution of pore size and is therefore dependent on the structure of each soil.

The moisture characteristic curve is determined experimentally and is obtained by plotting the moisture content of a soil against the suction applied using a tension plate or a pressure plate. Increasing suction values are applied in succession, and each time the soil moisture content at equilibrium is determined. The tension plate can be used only in the low suction range (below 100 kPa), whereas the pressure plate can usually hold pressures up to 2000 kPa. If the ceramic plate is replaced by a cellulose acetate membrane, pressures exceeding 10,000 kPa can be applied.

The curve obtained allows calculation of the proportion of different classes of pore sizes in the soil. A relatively flat moisture characteristic curve indicates a soil with a small proportion of macropores, due to greater bulk density or finer texture. Since the amount of water retained at low values of suction depends on soil structure, determination of soil moisture characteristic curves should be made with undisturbed samples. We can measure soil water content in situ with a neutron probe and measure soil moisture potential using tensiometers. When a soil sample is taken, the soil structure in the sample is disturbed and soil particles and aggregates are perturbed as are the macropores between them. If the macropores are perturbed and partially filled with small particles, the moisture characteristic of that sample is flattened and the measurement of available space to the nematode is underestimated.

The soil moisture characteristic curves can be obtained by applying progressively higher suctions to initially saturated soils (desorption), and/or by sorption with initially dry soils that are progressively wetted. However, the water content of the soils for a particular value of the matric potential is higher with desorption than with sorption and the relation between matric potential and water content depends on whether the curves are obtained by drying or wetting soils. For most soils, especially those of fine texture, the moisture characteristic curves depend on the past history of wetting and drying. These curves are called hysteresis curves (Fig. 1).

A soil is at field capacity when water in the macropores has drained by gravity and the only water remaining is held by capillarity and adsorption. For most soil, it is difficult to determine when drainage is completed. The definition of the water potential at field capacity is somewhat arbitrary, depending on what

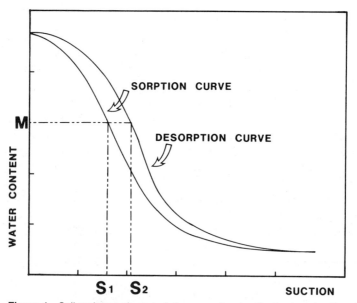

Figure 1 Soil moisture characteristic curve showing the hysteresis effect. For a moisture content M the suction can have any value between S_1 and S_2.

is considered to be negligible drainage. In agricultural soils, the water potential at field capacity is between 3 kPa for sandy soils and 30 kPa for loams and silt loams. These values vary with temperature, which affects the rate of drainage, and with soil management practices that alter soil structure and pore size distribution. Field capacity is not a well-defined soil property and, although it is a useful concept to design irrigation systems, it is of limited value in studies of the response of nematodes to water.

The Soil Atmosphere

As they empty of water, the soil pores fill with air, which constitutes the soil atmosphere. The rate of movement of air from the atmosphere to the sites of respiration in the soil greatly influences the activity of soil organisms, including roots, and that rate depends on soil moisture content. The respiration of plant roots and microorganisms enriches the soil atmosphere with carbon dioxide and depletes the oxygen. When soil pores become filled with water after a heavy rain, or if the soil pores are so fine that they are not emptied rapidly by drainage, the movement of air is very limited, and the soil is quickly depleted of oxygen and becomes anaerobic. Moreover, the soil is made up of large and small aggregates, with large pores and cracks that can drain quickly so that the superficial layers of these aggregates become aerobic while their cores remain anaerobic.

Only when the appropriate physical parameters that describe the living space of nematodes are obtained for a particular soil can we extrapolate from published results on the rates of activity of nematodes at different values of these parameters.

DIRECT EFFECTS OF WATER POTENTIAL

Egg Development, Hatching, and Emergence

Eggs of ectoparasitic species are laid in the soil, whereas eggs of endoparasitic species are somewhat protected either in plant roots or in an egg mass. The egg-mass matrix of *Meloidogyne javanica* (an endoparasite) shrinks as it loses water when suction is increased from 0 to 20 kPa. Wallace (1968a) speculated that the shrinkage causes mechanical pressure on the eggs, thus inhibiting hatching, while eggs laid singly in the soil may hatch at higher suctions. The gelatinous matrix is also an efficient obstacle to water loss. Embryonic and juvenile development in the matrix are not affected in an atmosphere at 98% relative humidity (2900 kPa suction) but are damaged when exposed to that suction for 0.5 hours without the protection of the matrix (Wallace, 1968a). The rate of free egg hatch of *M. javanica* in nonelectrolyte solutions is optimal between 0 and 250 kPa (Wallace, 1966) and decreases as suction increases to 1550 kPa. Egg hatch and emergence from cysts of *Heterodera schachtii* or from egg sacs of *M. javanica* peak at suctions corresponding to the point of inflection of the moisture characteristic curve of the soil they are in (Wallace, 1955; 1966). On dialysis membranes over polyethylene glycol solutions at 10 to 1600 kPa, percentage hatch of these free eggs decreases rapidly at suctions above 400 kPa (Baxter and Blake, 1969a). The volume of eggs decreases as extracellular fluid is lost; this is taken as an indication that the water potential of these eggs is close to 400 kPa.

Eggs at different stages of development survive high tensions differently and it seems that juvenile stages in eggs survive the best. When soil is placed in polyethylene glycol for 7 days, an increasing proportion of *M. incognita* eggs die at 1600 to 5000 kPa, and hatch of surviving eggs is very rapid when the egg masses are transferred to water (De Guiran, 1979). *Meloidogyne javanica* juveniles survive better than eggs in soils at 16, 30, and 110 kPa but no differences in survival were seen at 1500 and 9200 kPa (Towson and Apt, 1983).

Movement

Nematodes move by undulatory propulsion and their speed and efficiency in soil are determined by many interrelated soil characteristics, including size, arrangement, and rugosity of soil particles; size of pore necks; amount of water held in the pores; suction; friction forces; temperature; oxygen content; and nematode characteristics such as body length, wave frequency of the undulations, and muscular power. Nematode movement in films of water of variable thicknesses can be observed and the speed and distance traveled by a particular species can be estimated.

The speed of undulatory progression through water over a rigid substrate depends on forces that press the nematode onto the substrate. Their speed increases as lateral slip of movement becomes less in thinner films of water. *Heterodera schachtii* juveniles placed on alginate jelly in petri dishes, where the thickness of water films can be controlled, increase their speed up to 30 cm/h as the thickness of the film is decreased from 50 μm to approximately 2 μm. However, their movement is almost totally inhibited in layers of water less than 1 μm thick (Wallace, 1958a). Using cinematography of nematodes moving in films of water, Wallace (1959b) recorded speeds of 2.4 mm/min for *H. schachtii* juveniles and 21.6 mm/min for *Aphelenchoides ritzemabozi*. These speeds would allow migrations of 3.5 and 31 m/day, respectively. These *H. schachtii* juveniles were not migrating in a gradient but, since their tracks were plotted and the totality of their migration recorded, the estimates of speed are probably valid. If nematodes can move several meters per day in optimal conditions, there must be some severe limitations to their migration in soils, because many experiments record movements of only a few centimeters per day to a few centimeters per month.

Obviously, soil moisture can be suboptimal but several other soil factors interacting directly or indirectly with moisture content can also reduce the rate of movement of nematodes. Soil particles are not uniform but contain a wide range of particle and aggregate sizes. With a sandy loam fractionated by sieving into five classes of particle sizes, Wallace (1958a) showed an interaction effect of soil moisture and particle size on migration. The movements of *H. schachtii* and *Ditylenchus dipsaci* in single layers of soil particles can be observed directly under the microscope. When placed at different suctions in soil fractions with different particle sizes, juveniles of *H. schachtii* or *D. dipsaci* moved farthest in their respective optimum particle size fractions when suction was close to the point of inflection of the moisture characteristic (Wallace, 1958b; 1961) (Fig. 2). However, *D. dipsaci* moved faster in coarser soil with large pore spaces and *H. schachtii* moved faster in finer soil.

Meloidogyne hapla (Wong and Mai, 1973), *Pratylenchus penetrans* (Kable and Mai, 1968), and *Tylenchorhynchus icarus* (Wallace and Greet, 1964) move faster when the majority of the pores are draining rather than when they are full of water or totally empty at high suctions. Kable and Mai (1968) found that penetration of alfalfa roots by *P. penetrans* occurred at 100 kPa in a sandy loam for which the accessible pore space was theoretically completely drained. Townshend and Weber (1971) examined the migration of *P. penetrans* in a loamy sand and two finer loams at two bulk densities, each held at eight different moisture tensions. Migrations in 7 days plotted along the moisture characteristics show that the nematodes moved over a very broad range of 1 to 100 kPa of moisture tension. At the higher bulk density, movement was restricted between 1 and 31 kPa in the loamy sand, and between 3 and 100 kPa in the loams. As shown by the moisture characteristics of these soils, the nematodes apparently

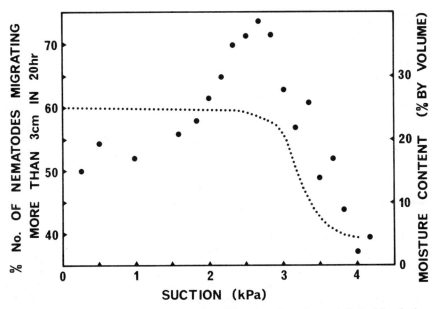

Figure 2 Mobility of *Ditylenchus dipsaci* at different suctions in sand. Dotted line is the moisture characteristic. *(Adapted from Wallace, 1961.)*

moved in pore spaces (10 to 0.3 μm) smaller than their body width. Either nematodes move in the very thin film of water covering soil particles of drained pore spaces, or they are able to squeeze through pore necks smaller than their own diameter.

High expenditures of energy may cause mortality at moisture tensions favorable to migration of *P. penetrans* (Kable and Mai, 1968), whereas *Tylenchorhynchus dubius* and *Rotylenchus robustus* survived with little change in the size of their populations for 3 months in soils at moisture tensions varying from 0.3 to 1550 kPa (Simons, 1973). This suggests that some nematodes move little when not attracted by food, thereby increasing their rate of survival.

The soil is constantly subjected to drying or wetting through drainage, evaporation, transpiration through plants, rainfall, or irrigation, and moisture gradients are constantly being formed, disturbed, and reestablished. When there are no gradients or stimuli to direct nematode migrations, the distance they travel becomes very short but is still a function of soil moisture potential and its many interactions with other soil and nematode characteristics. Luc (1961) showed that *Hemicycliophora paradoxa* migration in a vertical column of clay was very limited when the moisture potential was constant, but was more extensive when there was a moisture gradient. *Ditylenchus dipsaci* moves toward the higher moisture content of a gradient (Wallace, 1961). The mechanism of orientation is so sensitive that this nematode responds to a 1% moisture content difference per 10 cm of sand with particle size 250 to 500 μm. Indeed, Wallace (1960) showed that *Globodera rostochiensis* juveniles redistribute themselves according to moisture gradient in a sand of uniform particle size. Using a 5-cm cube placed on a sintered glass plate, he established a slight differential in suction between the top and the bottom of the cube of sand. The nematodes introduced in the center of the cube redistributed themselves in a few hours and moved toward the wet end of the gradient, regardless of its position at the bottom or at the top of the cube. The 5-cm difference (0.5 kPa) between top and bottom represented a more or less important moisture gradient in proportion to the suction applied. The percentage of larvae moving to the wet end of the gradient increased with the steepness of that gradient and, as shown in other experiments (Wallace 1958b), speed increased until a maximum was reached at a suction of 10 kPa. Prot (1979) suggested that juveniles of *M. javanica* migrating toward the wet end of a gradient when placed in a horizontal column of sand may be responding to gradients other than the moisture gradient, such as pH, aeration, osmotic pressure, or soluble salt concentrations.

Where there is a moisture gradient there is a flow of water, however small the flow. The flow of water through pore spaces interferes only slightly with nematode direction and progression. Peters (1953) found that percolating water tends to carry juveniles of *G. rostochiensis* downward in soils. Wallace (1959a) indicated that downward velocity of nematodes is increased when water percolates slowly through a column of sand, but that the relationship between nematode speed and rate of flow is independent of nematode efforts.

Penetration and Feeding

There is no information on the effects of soil water on penetration and feeding because it is difficult to design experiments to separate them from effects on movement to the roots. Increasing number of *M. hapla* juveniles invade lettuce roots as suction increases from 2 to 12 kPa in an organic soil but there is no penetration at 0 kPa. In the same soil, however, movement of juveniles is influenced by moisture potential in exactly the same way (Wong and Mai, 1973). Invasion of corn roots by *P. penetrans* and *P. minyus* (Townshend, 1972) is optimal between 1 and 10 kPa, although juveniles and adult nematodes can penetrate roots at moisture tensions of 0 to 100 kPa. Again, the nematodes must migrate a short distance before reaching the roots so the measured effect is not on penetration per se.

Nematodes move along roots to find the best feeding or penetration site. Feeding of ectoparasites is presumably independent of soil moisture potential since they are in contact with the water film of the root. Infectivity of nematodes depends on the amount of energy they have expended before coming in contact with a root because most of their energy is used in migrating to the root and locating a site for penetration. Considerable strength and energy are required for penetration, so nematodes that have exhausted their energy reserves may be unable to invade the roots. Van Gundy et al. (1967) found significantly greater lipid and protein reserves in bodies of *M. javanica* juveniles held in a moist soil (3.4 kPa) than in those in a dryer soil (5.0 kPa) where the nematodes utilized their energy reserves faster to maintain the speed equivalent to that in the moist soil (Fig. 3).

Development and Reproduction

Development and reproduction depend on feeding and movement. When nematodes are in roots or in the rhizoplane, their development is presumably not affected by soil moisture tension as long as the plant maintains adequate rates of absorption and translocation. We have no information on a direct effect of low moisture potentials on development and reproduction in soils but there is an abundance of research on the effect of very high moisture potentials.

Anhydrobiosis, the ability to live in a dry environment, was once thought to be a characteristic that only a few species of nematodes shared with other soil organisms like protozoans, tardigrades, and rotifers. Only recently have we realized that many species of Tylenchida are able to enter the anhydrobiotic state. Susceptibility to high moisture tension varies among different species of nematodes. Species of *Trichodorus* are notorious for their lack of resistance to desiccation (Rossner, 1971). *Rotylenchus robustus* is very susceptible when compared to *Tylenchorhynchus dubius,* and species of *Tylenchorhynchus* and *Paratylenchus* are less susceptible than species of *Helicotylenchus* (Simons, 1973). There also appears to be distinction at the level of trophic groups. In a forest soil, Arpin (1975) found that after suction had reached 9800 kPa, 93 to 100% of plant parasitic nematodes survived while only 10 to 15% of microbi-

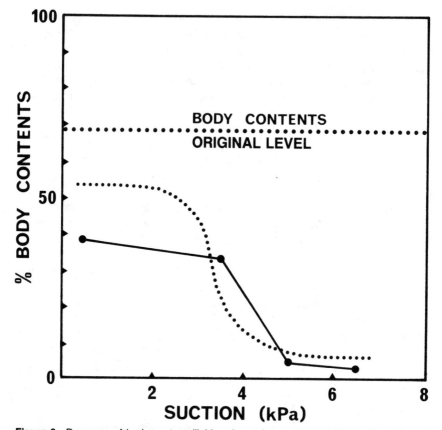

Figure 3 Decrease of body content (lipid and protein reserves) of *M. javanica* juveniles after 10 days at 22° C in soils at four different moisture tensions. Dotted line indicates moisture characteristic of the soil. *(Adapted from Van Gundy et al., 1967.)*

vorous and none of the predatory nematodes survived. The impacts of anhydrobiosis on population dynamics of nematodes are manifold: energy reserves are used, development and reproduction are delayed, and a proportion of the population does not survive the process. However, anhydrobiotic nematodes are more resistant to freezing temperatures (Bosher and McKeen, 1954; Townshend, 1984) and to nematicides (Freckman et al., 1980), and some species are more resistant to high temperature (Demeure, 1978), while others are not (Townshend, 1984).

For most nematodes, each life stage resists anhydrobiosis to a different degree. Therefore, the composition of the population at the onset of the drying period determines the proportion of surviving nematodes. The eggs of ectoparasites are usually more susceptible to desiccation than eggs of endoparasites that are protected by a gelatinous matrix, an impermeable cyst wall, or by the surrounding plant tissue. Second-stage juveniles of endoparasites can be quite resistant (Towson and Apt, 1983) when they are the only soilborne stage of the species, while second-stage juveniles of ectoparasites appear very susceptible (Simons, 1973) and the third-stage and fourth-stage juveniles are often more resistant.

A proportion of a nematode population is anhydrobiotic any time the soil is not very wet. Townshend (1984) reported that 3 and 13% of *P. penetrans* in soils at 1.9 and 19 kPa were anhydrobiotic. The standard extraction techniques do not allow for the examination of the anhydrobiotic segment of a nematode community. When field soil is stored in plastic bags, anhydrobiotic nematodes may be rehydrated by water vapor diffusion in a few hours so that in slightly dry soils the active nematode population is overestimated.

Drying of soil is a relatively slow process as water vapor moves rapidly by diffusion in drained pores. Nematode species that survive dehydration have the capacity to slow their rate of water loss while undergoing physiological changes. The amount of water inside a nematode depends on the osmotic and matric potential of the soil solution, with the rate of water exchange regulated by internal mechanisms in order to maintain turgor pressure. To survive dehydration, *Aphelenchus avenae* must be exposed to moist air at least 99.2% relative humidity (RH) (900 kPa) for 72 h (Crowe and Madin, 1975). Its water content falls from 75 to 80% when active to 2% or less when in equilibrium with dry air. *Ditylenchus dipsaci* and *Aphelenchoides besseyi* (Huang and Huang 1974) survive dehydration better in aggregates. *Aphelenchus avenae* in pellets of 100 μg take only 50 to 60 h to reach the anhydrobiotic state when exposed to 97% RH (3900 kPa) (Crowe and Madin, 1975). The nematodes in the center of these aggregates are protected by the outer layers and have time to adapt while slowly losing water. After physiological preparation, these nematodes can survive 0% RH (greater than 10^6 kPa) for a very long time. When the period of preparation is shorter than 60 h, or if they are exposed to air drier than 97% RH, the nematodes lose water at a much faster rate and a large proportion (99%) do not survive subsequent exposure to dry air. It is doubtful whether nematodes can aggregate in drying soil as they do in drying plant parts or in vitro, so results of experiments done with single nematodes are probably more representative of what happens in the soil.

Rossner (1971) indicated that *P. penetrans* can survive in air-dried soil for at least 11 months, while Townshend (1984) showed that anhydrobiotic survival of that species in field soil is greatly increased when soil is air dried slowly. When the moisture suction of a silt loam was increased from 19 kPa to more than 150 kPa in 18 days, coiled nematodes increased from 3 to 74%, and 23% of the population survived 60 days in the dry soil, but when the drying time was shortened to 3 days, only 31% of the nematodes were coiled and only 3% of the original population survived 60 days.

The requirement for slow dehydration suggests that nematodes prepare for the extensive water loss they experience. Madin and Crowe (1975) examined the carbohydrate and lipid content of *A. avenae* during slow dehydration and showed that glycerol and trehalose were synthesized in large quantities while glycogen and lipid content decreased by 95% and 35%. Huang and Huang (1974) showed that *Aphelenchoides besseyi* starved for 2 weeks did not withstand dehydration as well as better fed nematodes. *Scutellonema cavenessi* with adequate intestinal reserves could withstand desiccation while starved specimens with little reserve did not survive (Demeure et al., 1978). The lipid content of female *Pratylenchus thornei* decreased by 30% during 5 months of anhydrobiosis in dried field soil (Storey et al., 1982) while, in water, the rate of utilization of lipids was 1% per day. The net result is that by quickly expending approximately one-third of their food reserves at the induction of anhydrobiosis, a large proportion of these nematodes were able to survive in dry soil for several months.

The return to activity, when the soil becomes moist again, is also a critical stage. The morphological and metabolic changes that prepare nematodes for dehydration are probably reversed during rehydration. Respiration of *Anguina tritici* from dried wheat galls increased for several hours when the nematodes were placed in air at 95% RH (6200 kPa) (Bhatt and Rohde, 1970). Increase in oxygen uptake during the rehydration of *A. avenae* (Madin and Crowe, 1975) was associated with a 40% loss of lipid and an increase in glycogen content. *Pratylenchus thornei* has large intestinal vacuoles after rehydration; these vacuoles are not found in fresh or anhydrobiotic nematodes. Glazer and Orion (1983) suggested that large amounts of lipids must be used during rehydration. Even in the best of adaptive conditions most nematodes cannot repeat the process several times without losing their viability (Crowe and Madin, 1975; Womersley, 1980). However *D. dipsaci* and *A. tritici* synthesize trehalose at the expense of glycogen, not lipids (Womersley et al., 1982), and they can survive repeated dehydration-rehydration processes.

INDIRECT EFFECTS OF WATER POTENTIAL

Population dynamics of nematodes are affected by several components in soil water, such as oxygen and other chemicals that vary with the quantity of water held in the soil.

Available Oxygen

Wet, well-drained soils consist of a mosaic of anaerobic volumes embedded in an aerobic matrix. The composition of the air in the surface layers of an arable soil is usually similar to that of the atmosphere: oxygen content is around 20%, nitrogen 79%, and carbon dioxide between 0.2 and 0.7%. The diffusion of oxygen through water is very much slower than through air and is effectively blocked in saturated soils or during drainage, when air-filled spaces are not connected.

Oxygen usually moves from the soil atmosphere to the soil water. The rate of this movement is determined by the difference in oxygen concentration between soil atmosphere and soil water and by the surface of exchange. Nematodes are better supplied with oxygen in thin layers of water than in thick layers of water, where their body is far from the air-water interface. When oxygen diffusion to the nematodes decreases, their activities are slowed. Below a certain level their metabolism is not provided with enough oxygen and nematode activity stops. Measurements of the available oxygen are made in different ways and the published data are still too sparse to provide a coherent picture of the relations among moisture potential, aeration, and nematode activities.

Van Gundy and Stolzy (1963a, 1963b) measured oxygen diffusion with a thin platinum electrode, maintained at a constant electric potential, that measures the chemical reduction of elemental oxygen. The resulting current measures the flux of oxygen to the electrode acting as a sink, thus simulating a respiring nematode. The flux of oxygen around a probe inserted into moist soil represents the oxygen supplying power of that soil, which can be used to estimate how much oxygen is available to the nematode body and how fast its physiological systems are functioning. This technique does not work in dry soils but these soils usually have adequate aeration.

Plant roots and soil microorganisms require considerable oxygen in relation to the oxygen reserve in the soil, which is usually enough to sustain life for only 2 to 3 days. When the soil becomes water saturated, most nematodes slow their metabolism, decrease their oxygen demand to a minimum, and cease development and reproduction. If the demand for oxygen by nematodes and other soil organisms exceeds that which can be supplied by diffusion, their activities decrease and anaerobic conditions develop. As soon as anaerobic conditions occur, bacteria with a metabolism based on anaerobic respiration or fermentation processes start to multiply. Their metabolic wastes are reduced compounds, many of which are toxic to nematodes.

Depletion of oxygen in soil water around nematodes is more detrimental to ectoparasitic than to endoparasitic forms. Different nematode species and stages are affected differently by suboptimal oxygen and excess carbon dioxide concentrations. *Xiphinema americanum* and *Paratrichodorus christiei* do not survive in low oxygen tensions as well as *Tylenchulus semipenetrans* or *M. javanica*, which are less sensitive once they are partially or totally embedded in roots (Van Gundy and Stolzy, 1963a; Van Gundy et al., 1962). Nematode species with a high requirement for oxygen might fare better in lighter sandy soils where the films of water on soil particles may be very thin for long periods of time. Species with a low requirement for oxygen would do better in heavier fine soils where films of water are thicker longer and the oxygen diffusion rates are lower than in coarser soils. Perhaps this why *P. penetrans* and *P. crenatus* are found in light soils, and *P. thornei* and *P. minyus* inhabit heavier soils.

Some soil nematodes remain relatively active for a short time in microaerobic and anaerobic soils by fermentative catabolism of glycogen (Cooper and Van Gundy, 1970). The intestinal reserves cannot be used for extensive physiological adaptation to anoxia, as is done in anhydrobiosis, because oxygen is needed for lipid oxidation.

Low oxygen tensions do not affect all activities of nematodes equally but the effects of anoxia can be seen in population dynamics. Embryonated eggs do not survive 1 day without oxygen, but second stage larvae in eggs can survive up to 6 days. Hatch does not occur in nonaerated water (Wallace, 1968a; Baxter and Blake, 1969b), and the rate of development of *M. javanica* embryos approaches a maximum at an oxygen concentration of 15% (Baxter and Blake, 1969b) while the percentage hatch of eggs increases as oxygen concentration increases from 0.2 to 21%. Wong and Mai (1973) reached similar conclusions and their results with *M. hapla* also suggest that hatching is tolerant of carbon dioxide concentrations up to 10%.

Hatching of eggs of *Hemicycliophora arenaria* is almost completely inhibited at oxygen diffusion rates between 0 and 6×10^{-8} g cm^{-2} min^{-1} but is normal at 11×10^{-8} g cm^{-2} min^{-1}, whereas the hatching of eggs of *T. semipenetrans* is still inhibited at 30×10^{-8} g cm^{-2} min^{-1} (Van Gundy and Stolzy, 1963b; Van Gundy et al., 1962). Molting of *H. arenaria* is increasingly and irreversibly inhibited at oxygen diffusion rates below 45×10^{-8} g cm^{-2} min^{-1} for 7 days (Van Gundy and Stolzy, 1963b), whereas its migration is not affected by low oxygen diffusion rates above 1×10^{-8} g cm^{-2} min^{-1}. Migration of *M. javanica,* however, is related linearly to oxygen diffusion rates between 0 and 50×10^{-8} g cm^{-2} min^{-1} at 5 kPa of suction. After 18 to 24 days in soil with only 2 to 3 or 10% oxygen, the juveniles had significantly higher body contents, motility, and infectivity than those in well-aerated soil (Van Gundy and Stolzy, 1963a; Van Gundy et al., 1967). The conservation of energy reserves in poorly aerated soil increases the life span and maintains the infectivity of this nematode. Like hatching, migration and invasion of *M. hapla* are lowered by oxygen concentration below 21%, but unlike hatching these two processes are reduced by carbon dioxide concentrations above atmospheric level (Wong and Mai, 1973).

The linear relationship of population density of *H. arenaria* to oxygen diffusion rate in soil (Fig. 4) demonstrated in soil tests over 1 month by Van Gundy and Stolzy (1963b) suggests that oxygen diffusion is a determining factor in survival and reproduction. The number of eggs deposited by females of *H. arenaria* decreases with oxygen diffusion rates from 45×10^{-8} to 1×10^{-8} g cm^{-2} min^{-1} because of arrested development of the oocytes. *Meloidogyne incognita* reproduction decreases with oxygen concentrations between 21 and 3.5% (Van Gundy and Stolzy, 1961, 1963a). Cooper et al. (1970) showed that reproduction of *A. avenae* is unaffected at 10% oxygen, even with 5% carbon dioxide, but is completely inhibited at 4% oxygen.

Frequently repeated anoxia (short interval fluctuations between high and low oxygen concentration) seem to be very disruptive of processes requiring high oxygen tension. The rate of reproduction of *H. arenaria* over a 30-day period in soil decreased by 80 and 94% when oxygen was replaced with nitrogen for 12 or 24 h every 3 days (Van Gundy et al., 1968).

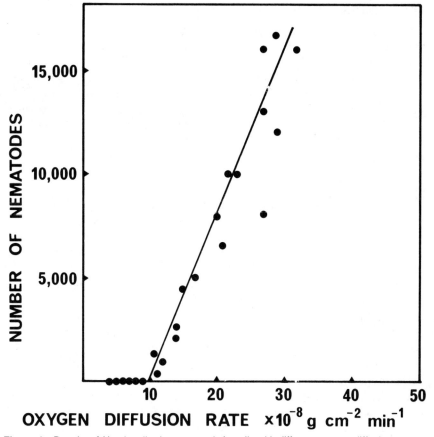

Figure 4 Density of *Hemicycliophora arenaria* in soils with different oxygen diffusion rates. *(Adapted from Van Gundy and Stolzy, 1963b.)*

Interaction of Water Potential with Temperature

Temperature affects the suction at which nematode mobility is maximal. For *D. dipsaci* in sand, mobility is reduced rapidly as the suction increases above the point of inflexion of the moisture characteristic at high (25 to 30° C) or low (5 to 10° C) temperatures, whereas between 10 and 20° C mobility is high over a wide range of suction (Webster, 1964). Somewhat different results were obtained by Townshend (1972) and Santos (1973). Penetration of corn roots by *P. penetrans* in several soils increased, and migration of males of *Meloidogyne* spp. in sand increased with increased temperature at high moisture tensions (i.e., the nematodes migrated through smaller pores as temperature increased, or they migrated faster in the very thin film of water covering soil particles of drained pores).

Chemicals in Soil Water

The accepted credo is that nematodes are probably not affected by osmotic pressure in agricultural soils, because around field capacity it rarely exceeds 200 kPa and the mobility of most nematodes tested is not affected up to 1000 kPa. However, consider that nematodes must experience ample variations in the osmotic pressure of their aquatic habitat, when the soil pores are filled and the solution becomes diluted, or when the soil dries out and the soil solution becomes very concentrated. Furthermore, there is no direct evidence for osmotic or ionic regulation in nematodes (Wright and Newall, 1976). Our knowledge is based primarily on their capability to survive in hyposmotic or hyperosmotic solutions. Wallace and Greet (1964) noted that increasing concentrations of urea from 10^{-3} to 10^{-1} M (2.2 to 220 kPa) significantly decreased the movement of *Tylenchorhynchus icarus* and increased their oxygen consumption (Fig. 5). That the metabolism of the nematode increased while movement decreased suggests an active process of physiological adaptation. A mechanism by which nematodes may survive changes in osmotic pressure in soil is an ability to change their internal concentration of carbohydrates into compounds less active osmotically.

Soil salinity in arid regions may have a direct effect on nematodes but it certainly has an indirect effect through host plants. As salinity increases, nematode damage to plants increases and nematode densities decrease. High salt concentrations inhibit and delay hatching of *Heterodera, Meloidogyne,* and *Ditylenchus* species (Dropkin et al., 1958). Gradients of salts at concentrations close to those observed in moist soils cause juveniles of *M. javanica* and *M. incognita* to move toward the region of lowest concentration. Prot (1979) suggested that what appears to be hydrotaxis, an orientation in a moisture gradient, may actually be chemorepulsion. Almost all attempts to measure the effect of different ion concentrations on nematode physiology or long-term survival have involved the use of distilled water or single-salt solutions. Tolerance and survival of nematodes is probably better in soil than in vitro, because changes in soil take place gradually, giving nematodes time to adapt. An exception may occur when large amounts of soluble fertilizer are applied.

If indeed there is some physiological process by which nematodes actively exchange water, ions, and molecules to and from the soil solution, then they must expend energy to maintain this process (Reversat, 1981). The further the soil solution is from optimal composition and osmotic pressure, the more energy nematodes need to expend in order to maintain their internal equilibrium and the greater the impact on their population dynamics.

No effect of the pH of the soil solution in agricultural soils on the physiology of nematodes has been demonstrated. However, since plant growth can be considerably affected by soil pH, the indirect effect on nematodes through the host plants is important (Burns, 1971).

Chemicals are carried from the roots to the nematodes over great distances in the soil water or soil atmosphere. Prot and Netscher (1979) noted that root-

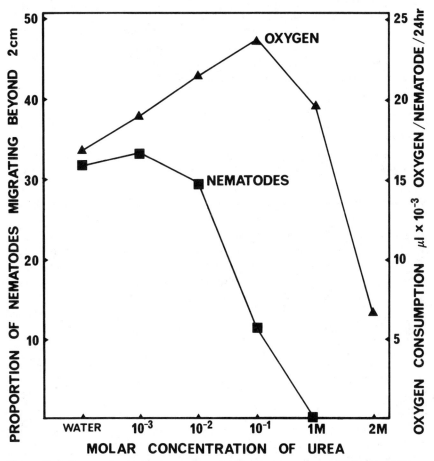

Figure 5 Influence of osmotic pressure on movement and oxygen consumption of *Tylenchorhynchus icarus*. *(Adapted from Wallace and Greet, 1964.)*

knot juveniles are attracted from as far as 75 cm to tomato seedlings in the field. Nematodes have sex pheromones to aid in locating their mates. Pheromones released by either males or females have been found for every nematode species but one, for which studies have been reported (Bone and Shorey, 1978). If nematodes must expend energy to locate mates it is likely they would do so only when the supply of food is secure, in the vicinity of a root, and not in the large spaces of soil where there are no roots. The pheromones used for recognition would need to diffuse in the water only over very short distances.

Nematode primer pheromones, if they exist, would be transported in the soil water, and their role in regulating nematode physiology and behavior needs to be demonstrated. These pheromones would have a role in regulating the onset of reproductive maturity, or sexual responsiveness, and in regulating the population fluctuations by appropriately changing the capacity for development, reproduction, or survival. Temperature can greatly influence the pheromone system (Bone and Shorey, 1978), as can low oxygen tensions (Rende et al., 1982), and very low concentrations of certain nematicides (Hough and Thomason, 1975). We lack information on the establishment of diffusion gradients of pheromones in soil water, no doubt because of the difficulties encountered in conducting experiments in natural soils on pheromone communication.

Metabolic By-products

Nematodes feeding at the same site in large numbers must produce large amounts of excretory products with direct effects on their physiological and reproductive rates, or with indirect effects if the excretions are utilized by other organisms whose respiration or metabolic products may affect the nematodes. The decomposition of organic matter results in production and release into the soil solution of a large array of metabolic by-products, some of which, especially those of fermentative metabolism, are toxic to nematodes. The impact of these processes may not be important to nematode population dynamics since the amount of fresh organic matter in soil is usually small. When a large quantity of undecomposed organic matter is incorporated into any moist soil, however, a succession of microorganisms utilize the available nutrients and the soluble carbohydrates allow a very rapid build-up of bacteria, many of them anaerobes. In a few hours to a few days, these sugars are utilized and aerobic bacteria deplete soil oxygen so much that even nonsaturated soils become anaerobic. Additionally, large quantities of fatty acids are produced and there is a decrease in the soil solution pH. Johnston (1959), Sayre et al. (1965), and Hollis and Rodriguez-Kabana (1966) have shown the toxic effects of large amounts of butyric acid produced by *Clostridium butyricum* from a source of rapidly decomposable organic matter such as sucrose, cornmeal, or rye. Saprophagous nematodes are not affected to the same extent as plant parasites. Even though butyric acid is very nematicidal in vitro, it is inactivated in soil by dissociation, precipitation with calcium and magnesium ions, adsorption, and metabolization by other microorganisms.

Phenolic compounds are also produced during the decomposition of organic matter. Hollis and Rodriguez-Kabana (1966) showed that nematicidal concentrations are rarely reached under field conditions except in the close vicinity of the decomposing matter. The decline of nematode populations can also be correlated with the increase in concentration of hydrogen sulfide in anaerobic soils (Rodriguez-Kabana et al., 1965).

When other bacteria utilize the protidic part of the soil amendment, ammonia is produced and the soil solution becomes more alkaline. Ammonia is nematicidal but the concentrations reached in the field are infrequently nematicidal, although there are many reports of decline of nematode populations attributed to decomposition of nitrogenous substances (Singh and Sitaramaiah, 1973). Nitrites are more nematicidal than nitrates or ammonium salts; since nitrite oxidation is faster than ammonium oxidation, the level of nitrites in soil water is usually low. However nitrites could become harmful to nematodes when heavy doses of ammonia or ammonium fertilizers are applied (Vassallo, 1967).

Predation and Parasitism

Moisture potential influences the host-parasite interactions of nematodes and their parasites and predators. *Nematophthora gynophila,* a parasite of *H. avenae* (Kerry et al., 1980), is active from May to July, when female nematodes are exposed on roots and when there is ample moisture in the soil, but is less active during dry summers in well-drained coarse soils. The decreased degree of parasitism has been attributed to the possible inhibition of zoospore movement in the drier soils. Two parasites of *Heterodera schachtii* eggs have widely different requirements for soil moisture (Nigh et al., 1980). *Acremonium strictum* parasitizes eggs only in moist soils, whereas *Fusarium oxysporum* grows and parasitizes large proportions of eggs in dry soils. When the soil dries, some fungi grow long after nematodes are immobilized, at matric tensions well above the minimum for movement of nematodes. However, as the macropores empty of water, waterborne spores of parasitic fungi probably cannot reach nematodes nor can the fungal mycelium track or recognize its prey since nematodes do not move in dry soil or they are not caught in the mycelial webs of predatory species.

Jaffee and Zehr (1983) observed that *Criconemella xenoplax* is more susceptible to infection by *Hirsutella rhossiliensis* in solutions of certain salts at tensions of 30 to 300 kPa than in distilled or tap water, or solutions of polyethylene glycol at the same moisture potentials. If the degree of parasitism in soil is influenced by osmotic tension, parasitism might be increased by application of fertilizers at appropriate times.

Nematode attraction to one another includes that of predacious nematodes to their prey. There must be some recognition processes (kairomones) transmitted through the water whereby predators can track other nematodes. No such recognition or transmission have yet been shown; most studies impose a foreign in vitro environment that probably disturbs normal production, diffusion, and recognition of the attractants in the water.

Interactions with Host Plant

Nematodes move through the soil in their quest for a gradient of root exudates they can follow in order to reach a host root. For many organisms, reduced activity is an adaptation to environmental stress. Some nematode species may respond in this manner to lack of food or of a chemical track leading to it. Except for a few species with an extremely limited host range and quiescent survival stages, most nematodes, whether they are active or not, do not remain infective for more than a few weeks to a few months without food.

Soil moisture controls many physiological processes in plant roots and affects nematode population dynamics through plant growth. In roots growing under moisture tensions that restrict growth, the number of nematodes present is often lower than that supported by roots growing at optimal moisture tensions. In birdsfoot trefoil a soil moisture tension of 98 kPa resulted in lower overall forage yield and decreased populations of *P. penetrans,* whereas at lower tensions (16 to 1.6 kPa) yield was increased and so was the concentration of nematodes in the roots (Willis and Thompson, 1969). However, the more rapid development of nematodes at high moisture levels can retard plant growth, whereas the low moisture levels that reduce the reproduction of nematodes may allow the plants to grow back as much as those in uninfested soil. The density of *P. penetrans* in roots of apple seedlings grown at 10 kPa was higher than in roots at 0.6 kPa (Kable and Mai, 1968), but density of *P. minyus* in roots of wheat increased with plant weight and soil moisture potential (Kimpinski et al., 1976). Mathur et al. (1981) showed that populations of *H. avenae* increase considerably on wheat and barley when soil moisture tension is low and plant growth is adequate. Populations decline at high moisture tensions when plant growth is decreased (Van Gundy et al., 1968); however, densities of *H. arenaria* decline in a citrus orchard soil when irrigation and soil moisture increase. These studies also indicate that, at low moisture tensions, the rate of nematode population increase depends not only on root growth but also on soil aeration.

The sex ratio of *Pratylenchus alleni* in soybeans is affected by soil moisture potential (Norton and Burns, 1971). The ratio of males to females was 1.2 when plants were grown in dry soil at tensions of 62 to 1550 kPa, but was 2.8 in moist soil at 31 to 62 kPa. Growth of soybean plants in the dry soil was significantly less than in the moist soil. These results agree with other studies in which nematodes produced more males when the plants were stressed.

Roots are not passive to nematode feeding and burrowing. They have the ability to compensate for and repair injuries but, if adequate quantities of water or oxygen are not available to the roots, the tissues are not regenerated and the local food source of nematodes is depleted. When this happens, the development and reproduction of sedentary endoparasites ceases and they may die. Migratory endoparasites can either migrate to another part of the root where there is healthy tissue or, if conditions permit, migrate to another root as all ectoparasites do. If the conditions do not favor migration, the nematodes quickly use their energy reserves and die.

Heavily parasitized plants exploit a smaller volume of soil than do unparasitized plants and are more stressed by extremes of moisture variations. Their root systems are smaller, with altered physiological functions, and they absorb less water and nutrients. Tomato plants growing in soil adequately supplied with water are less susceptible to damage and support a large number of *M. incognita.* When growing in soils with less than optimal water content, the plants are stressed, display considerable root injury and support fewer nematodes (Barker et al., 1976). Soybean roots growing in coarse sand favorable to movement of *Belonolaimus longicaudatus* are more damaged in a relatively wet soil, decreasing the available food source and nematode reproduction, than in a dry or very wet sand where the nematodes are either immobilized or cannot move efficiently (Robbins and Barker, 1974).

Interaction of Moisture with Soil Type

Most species of plant parasitic nematodes are more successful in soils with high content of sand particles and little clay. In soils of different textures, conditions are favorable for movement for different lengths of time depending on the rate of drainage after rainfall or irrigation. Under the same climatic influences, clay loam soils dry less rapidly than sandy soils since the hydraulic conductivity is lower at low moisture tensions. When moisture tensions favorable to nematode activities are reached in a fine soil, the favorable period of time should be shorter than in sandy soils because, as drying continues, the hydraulic conductivity of clay soils exceeds that of sandy soils and the former dry faster.

Kable and Mai (1968) showed that the range of moisture favorable to population growth varies with soil texture: finer textured soils favor nematode activities at higher moisture tensions than coarser soils. There is presumably an effect of poor diffusion of oxygen in small pores of finer soils still full of water at high moisture tensions, while pores of coarser soils, empty of water, allow a better diffusion of oxygen at lower moisture tensions.

Dispersal

Nematodes are spread in the anhydrobiotic state in dry topsoil carried by strong winds. Orr and Newton (1971) reported 28 genera including *Meloidogyne* collected in dust traps placed 2 m above the ground and suggested that several hundred thousand acres of sandy land in west Texas, infested with root-knot nematode, may have been contaminated in this manner.

Floodwaters from rain or irrigation are effective means of spreading *D. dipsaci* and probably other species, as nematode-infested areas often enlarge in the direction of natural drainage or of irrigation. Faulkner and Bolander (1970) observed the contamination of soils by several nematode species including *Meloidogyne* sp. when irrigation water was taken from canals rather than from wells in the Columbia basin of northwestern United States. A few years later, *M. chitwoodii* was described from this region and its distribution suggests that it spread through irrigation canals.

CONCLUSIONS

The behavior of many of the essential components of water in soils and their interactions have been studied, but they form such a complex system that many nematologists are discouraged from attempting to abstract and reduce it. Many effects of soil water potential on nematode activities that are now understood and quantified can be incorporated into simulation models. Ferris and Duncan

(1980) incorporated the influence of soil moisture, oxygen, and porosity into a simulation model by weighting the favorability of different soil textures to nematode movement.

The microenvironment of nematodes varies considerably both vertically and horizontally with the soil profile and its interactions with moisture potential and oxygen diffusion rates. Soil water effects have a dynamic nature and their variability within fields require numerous measurements to characterize them. Fortunately, automated systems for on-site measurements of moisture potential are becoming increasingly available. Long (1984) described a field system with frequent automatic readings and tape recordings of data from 56 tensiometers placed at various depths, with vacuum transducers whose output voltage varied with the soil suction.

Simulation models of nematode population dynamics use rates of development of nematodes and these rates are affected by environmental factors. Jones (1975b) suggested that the usable pore space of a soil be defined as the proportion of soil spaces that can accommodate nematode migration. Many models use effective heat units, measurement of cumulative heat above a threshold, expressed in degree days or degree hours. We could also use the effective moisture potential (EMP), as the cumulative time when the thickness of water films in macropores in soil permits a certain rate of hatching, movement, and development of nematodes. Variables that govern EMP are the soil water tension, or the amount of water received by the soil, soil texture and porosity, rate of drainage, and a few other factors modifying soil water such as temperature, osmotic pressure, pesticides, and nematode pheromones.

The summation of effective temperature time is interrupted for those days when moisture content of the soil does not allow nematodes to move. Nielsen (1961) assumed that during droughts nematode activity and respiration is zero. Rapoport and Tschapek (1967) suggested that many soil organisms are active for only short periods when soil is draining after rain or irrigation. Jones (1975a) suggested that variations in soil moisture are most likely to affect nematode activity during the growing season when soil pores are practically empty of water. After each rainfall during the growing season, there is a certain period of time, while the soil drains, when the pores are filled with acceptable levels of water and nematodes are active. This is supported by the findings of Wallace (1958b, 1968b) that indicate that nematodes are most active at moisture potentials in the range of 10 to 25 kPa, when the soil macropores are drained. Jones (1975a) used accumulated rainfall and effective temperature to deduce nematode activity and anticipate the extent of damage to the sugar beet crop in England.

If rates of activity of nematodes are decreased when soils are saturated, different modes of watering, such as frequent light rains vs. heavy infrequent rains, would have an effect. Perhaps for that reason, drip irrigation should prove more favorable to nematode activity than flood irrigation.

Many principles of the effects of moisture potential on nematode activities are now established and the knowledge gathered using one soil can be extended to other soils only by relating moisture content to moisture potential for each soil. Therefore, any study attempting to relate the variations of soil moisture content to any nematode activity should include an accurately measured moisture characteristic curve or give measurements obtained by tensiometry.

REFERENCES

Arpin, P. 1975. Sur quelques aspects des interactions sol-nématodes dans des biocénoses forestières ou herbacées. *Rev. Ecol. Biol. Sol* 12:57–67.

Barker, K. R., Shoemaker, P. B., and Nelson, L. A. 1976. Relationships of initial population densities of *Meloidogyne incognita* and *M. hapla* to yield of tomato. *J. Nematol.* 8:232–239.

Baxter, R. I., and Blake, C. D. 1969a. Some effects of suction on the hatching eggs of *Meloidogyne javanica*. *Ann. Appl. Biol.* 63:183–190.

———. 1969b. Oxygen and the hatch of eggs and migration of larvae of *Meloidogyne javanica*. *Ann. Appl. Biol.* 63:191–203.

Bhatt, B. D., and Rohde, R. A. 1970. The influence of environmental factors on the respiration of plant parasitic nematodes. *J. Nematol.* 2:277–285.

Bone, L. W. and Shorey, M. H. 1978. Nematode sex pheromones. *J. Chem. Ecol.* 4:595–612.

Bosher, J. E., and McKeen, W. E. 1954. Lyophilization and low temperature studies with the bulb and stem nematode *Ditylenchus dipsaci* (Kuhn 1858) Filipjev. *Proc. Helminth. Soc. Wash.* 21:113–117.

Burns, N. C. 1971. Soil pH effects on nematode populations associated with soybeans. *J. Nematol.* 3:238–245.

Cooper, A. F., and Van Gundy, S. D. 1970. Metabolism of glycogen and neutral lipids by *Aphelenchus avenae* and *Caenorhabditis* sp. in aerobic, microaerobic and anaerobic environments. *J. Nematol.* 2:305–315.

Cooper, A. F., Van Gundy, S. D., and Stolzy, L. H. 1970. Nematode reproduction in environments of fluctuating aeration. *J. Nematol.* 2:182–188.

Crowe, J. H., and Madin, K. A. 1975. Anhydrobiosis in nematodes: Evaporative water loss and survival. *J. Exp. Zool.* 193:323–334.

De Guiran, G. 1979. Survie des nématodes dans les sols secs et saturés d'eau: Oeufs et larves de *Meloidogyne incognita*. *Rev. Nématol.* 2:65–67.

Demeure, Y. 1978. Influence des températures élevées sur les états actifs et anhydrobiotiques du nématode *Scutellonema cavenessi*. *Rev. Nématol.* 1:13–19.

Demeure, Y., Reversat, G., Van Gundy, S. D., and Freckman, D. W. 1978. The relationship between nematode reserves and their survival to desiccation. *Nematropica* 8:7–8.

Dropkin, V. H., Martin, G. C., and Johnson, R. W. 1958. Effect of osmotic concentration on hatching of some plant parasitic nematodes. *Nematologica* 3:115–126.

Faulkner, L. R., and Bolander, W. J. 1970. Acquisition and distribution of nematodes in irrigation waterways of the Columbia Basin in eastern Washington. *J. Nematol.* 2:362–367.

Ferris, H., and Duncan, L. 1980. Consideration of edaphic factors in quantifying nematode stress on plant growth. *J. Nematol.* 12:220.

Freckman, D. W., Demeure, Y., Munnecke, D., and Van Gundy, S. D. 1980. Resistance of anhydrobiotic *Aphelenchus avenae* to methyl bromide fumigation. *J. Nematol.* 12:19–22.

Glazer, I., and Orion, D. 1983. Studies on anhydrobiosis of *Pratylenchus thornei*. *J. Nematol.* 15:333–338.

Greacen, E. L. 1981. Soil water assessment by the neutron method. (CSIRO Australia, ed.). Division of soils, CSIRO, Adelaide, Australia, 140 pp.

Hollis, J. P., and Rodriguez-Kabana, R. 1966. Rapid kill of nematodes in flooded soil. *Phytopathology* 56:1015–1019.

Hough, A., and Thomason, I. J. 1975. Effects of aldicarb on the behavior of *Heterodera schachtii* and *Meloidogyne javanica*. *J. Nematol.* 7:221–228.

Huang, C. S., and Huang, S. P. 1974. Dehydration and the survival of rice white tip nematode *Aphelenchoides besseyi*. *Nematologica* 20:9–18.

Jaffee, B. A., and Zehr, E. I. 1983. Effects of certain solutes, osmotic potential, and soil solutions on parasitism of *Criconemella xenoplax* by *Hirsutella rhossiliensis*. *Phytopathology* 73:544–546.

Johnston, T. M. 1959. Antibiosis of *Clostridium butyricum* Praxmowski on *Tylenchorhynchus martini* Fielding 1956, (Nematoda, Phasmidia) in submerged rice soils. Ph.D. thesis, Louisiana State Univ., 62 pp.

Jones, F. G. W. 1975a. Accumulated temperature and rainfall as measures of nematode development and activity. *Nematologica* 21:62–70.

——. 1975b. The soil as an environment for plant parasitic nematodes. *Ann. Appl. Biol.* 79:113–139.

Kable, P. F., and Mai, W. F. 1968. Influence of soil moisture on *Pratylenchus penetrans*. *Nematologica* 14:101–122.

Kerry, B. R., Crump, D. H., and Mullen, L. A. 1980. Parasitic fungi (*Nematophthora gynophila*), soil moisture and multiplication of the cereal cyst nematode, *Heterodera avenae*. *Nematologica* 26:57–68.

Kimpinski, J., Wallace, H. R., and Cunningham, R. B. 1976. Influence of some environmental factors on populations of *Pratylenchus minyus* in wheat. *J. Nematol.* 8:310–314.

Long, F. L. 1984. A field system for automatically measuring soil water potential. *Soil Sci.* 137:227–230.

Luc, M. 1961. Note préliminaire sur le déplacement de *Hemicycliophora paradoxa* Luc (Nematoda–Criconematidae) dans le sol. *Nematologica* 6:95–106.

Madin, K. A. C., and Crowe, J. H. 1975. Anhydrobiosis in nematodes: Carbohydrate and lipid metabolism during dehydration. *J. Exp. Zool.* 193:335–342.

Mathur, B. N., Arya, H. C., Handa, D. K., and Mathur, R. L. 1981. Biology of cereal cyst nematode *Heterodera avenae* in India: III. Factors affecting population level and damage to host crops. *Indian J. Mycol. Plant Pathol.* 11:5–13.

Nielsen, O. 1961. Respiratory metabolism of some populations of enchytraid worms and free-living nematodes. *Oikos* 12:17–35.

Nigh, E. A., Thomason, I. J., and Van Gundy, S. D. 1980. Effect of temperature and moisture on parasitization of *Heterodera schachtii* eggs by *Acremonium strictum* and *Fusarium oxysporum*. *Phytopathology* 70:889–891.

Norton, D. C., and Burns, N. 1971. Colonization and sex ratios of *Pratylenchus alleni* in soybean roots under two soil moisture regimes. *J. Nematol.* 3:374–377.

Orr, C. C., and Newton, O. H. 1971. Distribution of nematodes by wind. *Plant Dis. Rep.* 55:61–63.

Peters, B. G. 1953. Vertical migration of potato root eelworm. *J. Helminthol.* 27:107–112.

Prot, J. C. 1979. Horizontal migrations of second-stage juveniles of *Meloidogyne javanica* in sand in concentration gradients of salts and in a moisture gradient. *Rev. Nematol.* 2:17–21.

Prot, J. C., and Netscher, C. 1979. Influence of movements of juveniles on detection of fields infested with *Meloidogyne*, in: *Root-knot Nematodes (Meloidogyne species) Systematics, Biology and Control* (F. Lamberti and C. E. Taylor, eds.), pp. 193–203, Academic Press, New York.

Rapoport, E. H., and Tschapek, M. 1967. Soil water and soil fauna. *Rev. Ecol. Biol. Sol* 4:1–58.

Rende, J. F., Teft, P. M., and Bone, L. W. 1982. Pheromone attraction in the soybean cyst nematode *Heterodera glycines* race 3. *J. Chem. Ecol.* 8:981–991.

Reversat, G. 1981. Consumption of food reserves by starved second-stage juveniles of *Meloidogyne javanica* under conditions including osmobiosis. *Nematologica* 27:207–214.

Robbins, R. T., and Barker, K. R. 1974. The effect of soil type, particle size, temperature, and moisture on reproduction of *Belonolaimus longicaudatus*. *J. Nematol.* 6:1–6.

Rodriquez-Kabana, R., Jordan, J. W., and Hollis, J. P. 1965. Nematodes: Biological control in rice fields: Role of hydrogen sulfide. *Science* 148:524–526.

Rossner, J. 1971. Einfluss der Austrocknung des Bodens auf Wandernde Wurzel-nematoden. *Nematologica* 17:127–144.

Santos, M. S. 1973. Mobility of males of *Meloidogyne* spp. and their responses to females. *Nematologica* 19:521–527.

Sayre, R. M., Patrick, Z. A., and Thorpe, H. J. 1965. Identification of a selective nematicidal component in extracts of plant residues decomposing in soil. *Nematologica* 11:263–268.

Simons, W. R. 1973. Nematode survival in relation to soil moisture. Dept. of Nematology Agr. Univ., Wageningen, The Netherlands, 85 pp.

Singh, R. S., and Sitaramaiah, K. 1973. Control of plant parasitic nematodes with organic amendments of soil. *Exp. Sta. Bull.* 6, G. B. Pant Univ., Pantnagar, India, 289 pp.

Storey, R. M. J., Glazer, I., and Orion, D. 1982. Lipid utilization by starved and anhydrobiotic individuals of *Pratylenchus thornei*. *Nematologica* 28:373–378.

Townshend, J. L. 1972. Influence of edaphic factors on penetration of corn roots by *Pratylenchus penetrans* and *P. minyus* in 3 Ontario soils. *Nematologica* 18:201–212.

———. 1984. Anhydrobiosis in *Pratylenchus penetrans*. *J. Nematol.* 16:282–289.

Townshend, J. L., and Webber, L. R. 1971. Movement of *Pratylenchus penetrans* and the moisture characteristics of three Ontario soils. *Nematologica* 17:47–57.

Towson, A. J., and Apt, W. J. 1983. Effect of soil water potential on survival of *Meloidogyne javanica* in fallow soil. *J. Nematol.* 15:110–115.

Van Gundy, S. D., Bird, A. F., and Wallace, H. R. 1967. Aging and starvation in larvae of *Meloidogyne javanica* and *Tylenchulus semipenetrans*. *Phytopathology* 57:559–571.

Van Gundy, S. D., McElroy, F. D., Cooper, A. F., Stolzy, L. H. 1968. Influence of soil temperature, irrigation and aeration on *Hemicycliophora arenaria*. *Soil Sci.* 106:270–274.

Van Gundy, S. D., and Stolzy, L. H. 1961. Influence of soil oxygen concentrations on the development of *Meloidogyne javanica*. *Science* 134:665–666.

———. 1963a. Oxygen diffusion rates and nematode movement in cellulose sponges. *Nature London* 200:1187–1189.

————. 1963b. The relationship of oxygen diffusion rates to the survival, movement, and reproduction of *Hemicycliophora arenaria*. *Nematologica* 9:605–612.

Van Gundy, S. D., Stolzy, L. H., Szuszkiewicz, T. E., and Rackham, R. L. 1962. Influence of oxygen supply on survival of plant parasitic nematodes in soil. *Phytopathology* 52:628–632.

Vassallo, M. A. 1967. The nematicidal power of ammonia. *Nematologica* 13:155.

Wallace, H. R. 1955. The influence of soil moisture on the emergence of larvae from cysts of the beet eelworm *Heterodera schachtii* Schmidt. *Ann. Appl. Biol.* 43:477–484.

————. 1958a. Movement of eelworms, I. The influence of pore size and moisture content of the soil on the migration of larvae of the beet eelworm *Heterodera schachtii* Schmidt. Ann. Appl. Biol. 46:74–85.

————. 1958b. Movement of eelworms, II. A comparative study of the movement in soil of *Heterodera schachtii* Schmidt and of *Ditylenchus dipsaci* (Kuhn) Filipjev. *Ann. Appl. Biol.* 46:86–94.

————. 1959a. Movement of eelworms. IV. The influence of water percolation. *Ann. Appl. Biol.* 47:131–139.

————. 1959b. The movement of eelworms in water films. *Ann. Appl. Biol.* 47:366–370.

————. 1960. Movement of eelworms. VI. The influence of soil type, moisture gradients and host plant roots on the migration of the potato-root eelworm *Heterodera rostochiensis* Wollenweber. Ann. Appl. Biol. 48:107–120.

————. 1961. The orientation of *Ditylenchus dipsaci* to physical stimuli. *Nematologica* 6:222–236.

————. 1966. The influence of soil moisture stress on the development, hatch and survival of eggs of *Meloidogyne javanica*. *Nematologica* 12:57–69.

————. 1968a. The influence of soil moisture on survival and hatch of *Meloidogyne javanica*. *Nematologica* 14:231–242.

————. 1968b. Undulatory locomotion of the plant parasitic nematode *Meloidogyne javanica*. *Parasitology* 58:377–391.

Wallace, H. R., and Greet, D. N. 1964. Observations on the taxonomy and biology of *Tylenchorhynchus macrurus* (Goodey, 1932) Filipjev, 1936 and *Tylenchorhynchus icarus* sp. nov. *Parasitology* 54:129–144.

Webster, J. M. 1964. Interaction of temperature and suction in relation to movement of eelworms. *Nature London* 202:574–575.

Willis, C. B., and Thompson, L. S. 1969. The influence of soil moisture and cutting management on *Pratylenchus penetrans*. Reproduction in birdsfoot trefoil and the relationship of inoculum levels to yields. *Phytopathology* 59:1872–1875.

Womersley, C. 1980. The effect of different periods of dehydration/rehydration upon the ability of second-stage larvae of *Anguina tritici* to survive a desiccation at 0% relative humidity. *Ann. Appl. Biol.* 95:221–224.

Womersley, C., Thompson, S. N., and Smith, L. 1982. Anhydrobiosis in nematodes. II: Carbohydrate and lipid analysis in undesiccated and desiccated nematodes. *J. Nematol.* 14:145–153.

Wong, T. K., and Mai, W. F. 1973. *Meloidogyne hapla* in organic soil: Effects of environment on hatch, movement and root invasion. *J. Nematol.* 5:130–138.

Wright, D. J., and Newall, D. R. 1976. Nitrogen excretion, osmotic and ionic regulation in nematodes, in: *The Organization of Nematodes* (N. A. Croll, ed.), pp. 163–210. Academic Press, New York.

Chapter 6

Development and Impact of Regional Cereal Rust Epidemics

A. P. Roelfs

The rusts are important diseases of many of the cereal crops that comprise a major part of the world's food supply. Historically, epidemics of stem, leaf, and stripe rust of wheat have been among the most devastating of the plant diseases. The pathogens *Puccinia graminis* Pers. f. sp. *tritici*, *P. recondita* Rob. ex Desm. f. sp. *tritici*, and *P. striiformis* Westend. f. sp. *tritici* reproduce asexually and produce large numbers of uredospores that are frequently transported by wind to distances up to thousands of kilometers within a few days. The hosts, various *Triticum* spp., are widely grown. Wheat is self-pollinated, and large areas are often sown to a cultivar or a group of cultivars that may be genetically similar in resistance to one or more of the wheat rusts. When environmental conditions are favorable for disease development and virulent, aggressive genotypes of the pathogen are present, the stage is set for a regional rust epidemic.

Rust epidemics have sometimes been local, affecting only a single field, cultivar, or geograpical entity such as a valley or low-lying area. Failure of local epidemics to become more widespread results from limitations in at least one of the factors of disease (host, pathogen, environment, or time). For this chapter I have arbitrarily defined regional epidemics as those that cover at least 50% of the area occupied by the crop on a continental basis and that cause an average loss of 5% or more.

HOST FACTORS IN REGIONAL EPIDEMICS

Distribution of Susceptible Phenotypes

Regional epidemics are most apt to develop when highly susceptible cultivars are grown over an extensive area. The cultivars need not be of identical, or even similar, genotypes except for the lack of resistance factors effective against the predominant pathogen genotype. In the United States, over 350 different cultivars of wheat are grown annually. Within regions, however, the soils, climate, pests, markets, economic incentives, and cultural practices are similar, so that farmers tend to select from a limited group of cultivars. There are approximately 50 cultivars grown per region annually but 20 cultivars make up roughly 30 to 50% of the national production in any year (Roelfs, 1985b). The cultivars differ widely in some characters and are very similar in others. Diversity for resistance cannot be determined except for individual pathogen phenotypes. For example, Centurk (a cultivar of *Triticum aestivum*) has genes *Sr5, 6, 8, 9a,* and *17* for stem rust resistance, and Lee has *Sr9g, 11,* and *16* but *Puccinia graminis* f. sp. *tritici* race 113-RTQ virulent on *Sr17* is virulent on both cultivars, race 113-RKQ virulent on *Sr17* is virulent on Centurk but not Lee. Race 113-RKQ avirulent on *Sr17* is avirulent on Centurk and Lee, while race 113-RKQ virulent on *SR17* is avirulent on Lee but virulent on Centurk.

Additionally, some hosts seem to be more susceptible than other susceptible hosts. For example, when the susceptible cultivars Baart and Prelude were crossed, progeny were obtained that were more susceptible than either parent (Skovmand et al., 1978). Both of these cultivars are themselves so susceptible to stem rust that they are frequently killed. Some pathogen cultures are more aggressive than others (see Katsuya and Green, 1967). In our greenhouse seedling tests race 15 and 56 produce larger uredia than other cultures from North America.

Spacing of Plants and Fields

Cereal crops are generally grown in rows. Row spacing for wheat varies from about 15 to 30 cm with about 400 tillers per square meter. When farmers increase row spacing they generally increase the seeding rate per linear unit and vice versa. In areas where high numbers of tillers per plant are formed the seeding rate per square meter is normally decreased. Thus, there is little space between adjacent tillers to impede the spread of disease. Multilines (mixtures of phenotypically similar lines with different resistance genes) are being widely promoted and tested but currently make up a minor portion of the world wheat production. Multilines of oats have been developed for crown rust control (Browning and Frey, 1969) as has a multiline of wheat for stripe rust control (Allan et al., 1983). In the People's Republic of China wheat is interplanted with other crops, but this practice is not common over entire regions nor is it used to control rust diseases.

Although relatively little research has been done on the effect of spacing between fields of wheat, the distances between fields must have a significant effect on rust development. Two large wheat growing areas in the United States are Kansas and North Dakota with approximately 5.3 and 4.2 million hectares (ha) of wheat, respectively (Fig. 1). Yet, wheat is grown on less than 20% of the areas of these states.

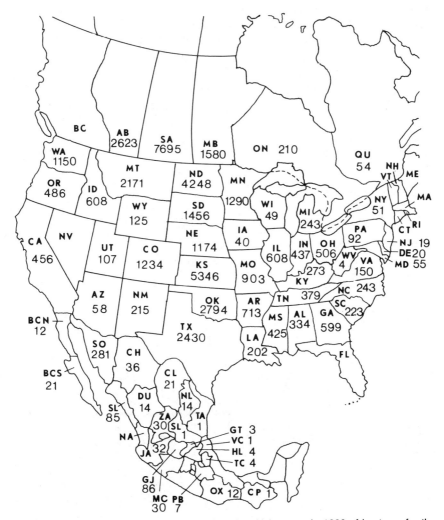

Figure 1 Distribution of wheat in North America. Values are in 1000 of hectares for the period 1978 to 1981.

When the pathogen race involved is virulent on only some of the cultivars, the relevant measure of spacing must be calculated for distance between fields of susceptible cultivars. If two or more pathogen races are involved, then the spacing may vary for the different pathogen races. Likewise, when a host resistance functions only under some environmental conditions, then spacing changes as conditions change.

Life Cycle of the Host

Wheat is an annual crop with a growing period of 3 to 11 months. Thus, the cereal rusts cannot survive from one year to another on the same plant, as occurs with perennial crops. In areas with mild winters wheat is planted in the fall and harvested in late spring or summer. Where winters are severe wheat is seeded in the spring (Figs. 2 and 3). Thus, in most areas of the world there is a period when wheat is not grown. Exceptions exist where the date of sowing of winter wheat overlaps the harvesting of the previous crop (e.g., a small area of the Pacific Northwest of the United States) or where elevation changes greatly over short distances (e.g., Kenya). Here rust may spread directly from the mature to the newly sown crop with favorable environmental conditions. Where self-sown wheat occurs between cropping seasons a "green bridge" of host material is provided on which the pathogen can survive between seasons.

Alternate Hosts

The alternate hosts for *P. graminis* are several species of *Berberis* and *Mahonia* (Roelfs, 1985a). For *P. recondita* alternate hosts are species of *Thalictrum, Isopyrum, Anchusa,* and *Clematis* (Samborski, 1985), while for *P. striiformis* there is no known alternate host (Stubbs, 1985). The alternate host and the

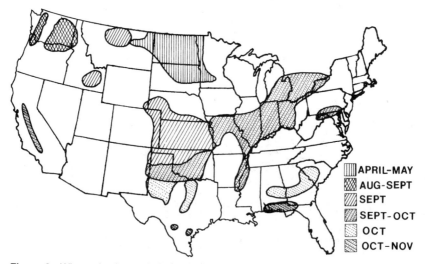

Figure 2 Wheat planting periods for the various areas of the United States.

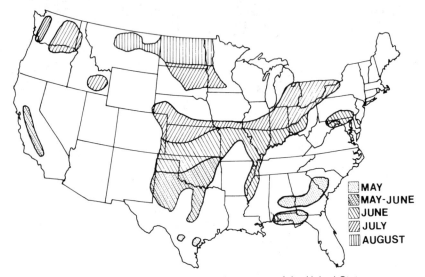

Figure 3   Wheat harvesting periods for the various areas of the United States.

teliospores provide a mechanism for the pathogen to survive the winter and reinfect wheat in the spring. The sexual cycle occurs during the transition from teliospores to aeciospores on the alternate hosts, so they provide not only a local source of inoculum at the early stages of the wheat growth, but also an opportunity for genetic recombination of virulence types.

The major stem rust epidemics in North America (which occurred in the years 1878, 1906, 1916, 1935, 1937, 1953, and 1954) show no evidence of being the result of aeciospores spread from barberry. They all seem to have originated from air-borne uredospores from the south (Craigie, 1945; Roelfs, 1982). Many local epidemics have resulted from aeciospores spread from barberry in Minnesota, North Dakota, and Wisconsin (Roelfs, 1982). A local epidemic and its associated loss is shown in Fig. 4. Such epidemics were eliminated in Denmark by barberry eradication (Hermansen, 1968).

The only area where sexual reproduction occurs regularly in *P. recondita* is in eastern Siberia (Chester, 1946, pp. 62-67; Saari and Prescott, 1985). The role of the alternate host in epidemics and the effects of the alternate host on diversity in virulence are discussed in Roelfs and Groth (1980) and Groth and Roelfs (1982).

Accessory Host

An accessory host is a noncereal grass, either native or introduced, on which the asexual stage of the pathogen survives. The role of accessory hosts is not

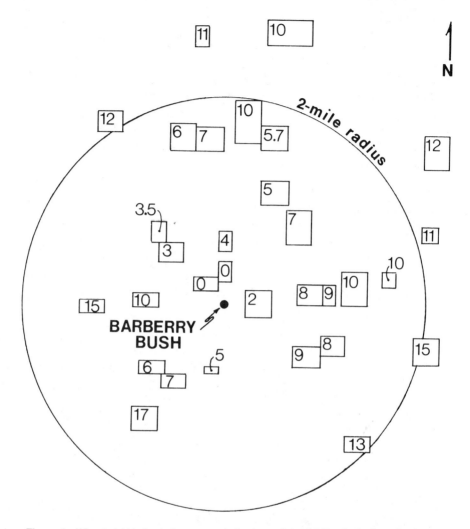

Figure 4 Wheat yield in bushels per acre in the immediate vicinity of a barberry bush near Alert in Decatur County, Indiana. Yields outside the rusted area were 15 to 25 bushels per acre. *(Adapted from Beeson, 1923.)*

well understood in cereal rust epidemiology in most areas of the world. Interesting reviews are available for Israel which is part of the area of origin of wheat (Anikster and Wahl, 1979; Wahl et al., 1984). In North America, the wild, noncultivated *Hordeum jubatum* L. is often infected with *P. graminis* but most inoculum is from uredospores produced on wheat rather than on *H. jubatum*. After wheat harvest in Nebraska, stem rust is often common on *H. jubatum*; however, it is not known whether this forms a green bridge to the fall sown wheat. An accessory host such as *H. jubatum* may also be infected with *P.*

graminis f. sp. *secalis,* which is avirulent on wheat. Stripe rust is common on wild *Hordeum* spp. in Chile, but these cultures are generally avirulent on wheat. Often where accessory hosts are infected, self-sown plants of the crop are also infected and occur in greater numbers nearer to newly sown fields. The pathogen races present on accessory hosts may differ in relative frequency from their occurrence on the cereal crop, as with *P. graminis* Pers. f. sp. *avenae* on wild oats, *Avena fatua* L., in the United States.

In North America, there is no available evidence to indicate that regional epidemics arise from initial inoculum generated on accessory hosts. However, accessory hosts can maintain a green bridge between crop seasons, and in the areas with mild climates they may provide a continuous source of inoculum. In studies of virulence changes and their causes in the cereal rusts one should be aware of the potential effects of accessory host, especially during periods of low pathogen population levels.

PATHOGEN FACTORS IN REGIONAL EPIDEMICS

Virulence Combinations

Resistance in cereals and virulence in rust fungi interact in a typical gene-for-gene relation. Most of the current wheat cultivars have at least one gene for stem rust resistance even though it may be ineffective. In areas where stem rust has historically been severe, most cultivars have three to six genes for resistance. These combinations of resistance genes normally consist of some that are ineffective against all local pathogen strains, some that are effective against different portions of the pathogen population, and often one resistance gene that is effective against the entire pathogen population. Thus, in the study of regional epidemics it is essential to know the specific virulence combinations in the pathogen, their frequency, and their distribution. A regional epidemic is normally caused by a single pathogen genotype that makes up most of the pathogen population. Local spread from a single focus also generally involves a single pathogen genotype; however, when environmental conditions are favorable over a large area and many foci overlap, then several pathogen phenotypes occur together. During an epidemic in the southeastern United States in 1972 nearly every overwintering location sampled had a different pathogen race (Roelfs and McVey, 1973).

Aggressiveness

Little is known about variation in aggressiveness among field cultures of rust fungi. However, in an F_2 population of *P. graminis* there was great variation in latent period length (7 to 14 days), number of spores produced, and number of uredia per unit of inoculum. In greenhouse tests, cultures of race 15 and 56 can be distinguished from cultures of other races on a susceptible host by the numerous large, erumpent uredia. In several field tests in the northern Great Plains with above-normal temperatures, race 151-QSH of *P. graminis* created a more severe epidemic than race 15-TNM which normally produces the more severe epidemics. Greater aggressiveness of race 15B-1 than race 56 at prevailing temperatures was in part responsible for the prevalance of race 15B-1 between 1950 and 1961 (Katsuya and Green, 1967). *Puccinia striiformis* genotypes that were studied in Europe from less polluted areas have been reported to be more sensitive to air pollution than the native European cultures (Stubbs, 1985).

These mostly inadequately tested observations lead me to believe that cultures differ in aggressiveness, and that the differences may be sufficient to affect development of regional epidemics. This may be why stem rust epidemics in North America have been caused by a few of the many races that existed, and why the epidemic causing races were not always those with the greatest number of virulence genes.

ENVIRONMENTAL CONDITIONS IN REGIONAL EPIDEMICS

Favorable environmental conditions are necessary for disease development, but they may differ for different diseases. For instance, optimal temperatures for infection are 20, 20, and 11° C for leaf, stem, and stripe rust of wheat (Hogg et al., 1969). Stripe rust tends to be more important in the cooler periods of the year and in wheat grown at high elevations. Leaf rust is more important where the fall and/or spring and early summer are warm and moist. Stem rust is important in warmer areas when frequent free water (usually dew) is available on the wheat leaves for spore germination and infection. The rust pathogens generally have a wide range on both sides of the optimal temperature at which near maximum development occurs, with a rapid reduction in development to the minimum and maximum limits (see Fig. 2 in Roelfs, 1985b).

Environmental conditions directly affect host resistance, infection frequency, length of the latent period, and the rate and duration of fungal sporulation. Free water, most frequently as dew, is required for spore germination. Dew formation is usually heaviest on clear, still nights when the leaf temperature decreases considerably from heat radiation to the sky. Rain and, near large bodies of water, fog may also be frequent sources of free water. *Puccinia recondita* can infect in 3 to 4 h, whereas *P. graminis* requires 6 h under optimal conditions. For germination of uredospores of *P. graminis* free water must be present for 6 h at air temperatures around 18° C, and this must be followed by 3 to 4 h of continued free water combined with a light intensity of more than 10,000 lux and a gradual temperature rise to 26° C (Rowell, 1984).

For wheat rusts the latent period between inoculation and sporulation by the pathogen is 7 to 10 days under optimal environmental conditions. However, latent periods can be as long as 30 days for stem rust when night temperatures are cool (1 to 5° C) and daytime temperatures are less than 10 to 15° C. The latent period for leaf and stripe rust may be even longer when plants are infected late in the fall and sporulation does not occur until spring.

Cereal rust fungi produce large numbers of uredospores. *Puccinia graminis* can produce 5000 uredospores per uredium per day during active sporulation, resulting in up to 5 million uredospores per day from a severely infected tiller. Rust uredia commonly sporulate for 21 days or more under favorable conditions. Thus, a hectare of heavily infected wheat can produce 1.5 trillion uredospores per day for several weeks. Senescence of leaves as plants mature can reduce the length of the sporulation period. Early senescence can be induced by severe rust infection or by environmental stress. Conversely, under cool conditions the accumulation of nutrient reserves around the uredium can prolong the life of that portion of the leaf and the uredium for several weeks.

Strong winds at 1 to 3 km altitude blowing from the disease source toward less mature crop are conducive to long-distance spore transport. Ideal conditions for deposition occur when rain falls through spore-laden air scrubbing out the uredospores. Following such a rain, conditions are often favorable for infection (Roelfs et al., 1970). The terminal velocity of a uredospore is approximately 1 cm/s in still air. Thus, sedimentation from 1 to 3 km in still air requires 24 to 36 h. Of course, downdrafts could bring spores down faster.

Viable uredospores are often transported regionwide but fail to initiate epidemics. The host cultivars in the region may be resistant to the predominant pathogen genotypes, the environmental conditions following dissemination may be unfavorable, or the spores may arrive too late in the growing season to allow an epidemic to develop before the host matures. Wheat stem rust epidemics often fail to develop in some areas because the temperatures are too cool. Conversely, stripe rust development in the Great Plains of North America is retarded by the hot summer weather. Wheat is grown in areas that are usually too arid for rust development. The line of demarcation between areas conducive to rust and those not conducive varies from year to year depending upon rainfall patterns. Regional epidemics often have one or more of their boundaries along such environmental limits.

Environmental conditions also have dramatic effects on certain types of resistance. For instance, the resistance conditioned by the host genes $Sr6$ and $Sr15$ becomes ineffective at temperatures above 22 and 20° C. Resistance conditioned by $Sr13$ functions best above 26° C, especially with intense light (Roelfs and McVey, 1979).

The development of regional epidemics of rust requires that environmental conditions be favorable for disease development over large areas. It is not essential that the environmental conditions are always optimal; an epidemic can develop if the conditions are frequently favorable. A 10-fold increase in disease in 5 days is common, and with ideal conditions wheat stem rust can increase 100-fold in 3 days (Rowell, 1968). Thus, in 9 days of ideal conditions, stem rust may increase from an almost undetectable level of one uredium per 100 tillers to 100% severity (1000 pustules per tiller). Yield losses of 50% commonly occur if the disease severity reaches 100% before the mid-dough stage of host development. Fortunately, environmental conditions this favorable rarely occur daily, but they can occur several times per growing season in the northern Great

Plains. In general, for regional epidemics to develop, spore transport must occur early enough so that the crop throughout the region receives initial inoculum 30 or 40 days before normal leaf senescence for leaf rust or before plant maturity for stem rust. Earlier spore transport is required for stripe rust, because it develops best at cooler temperatures.

TIME IN REGIONAL EPIDEMICS

Much more time is required for development of regional epidemics than for local epidemics. The pathogen must spread to the top of the crop canopy in the source area. In south Texas, this may require more than 2 months. Inoculum must be transported throughout the region early enough for serious disease to develop before crop maturity in the target area. An entire region may be infected from a single episode of spore transportation and deposition or from consecutive episodes. When an entire region receives inoculum from a single source, then the time required for multiple episodes is only the length of time between the first and last transportation-deposition episode. If the first effective dispersal affects only a portion of the region and must provide the inoculum for a second effective dispersal to the rest of the region, then time is also generally required for disease increase after the first episode. Long-distance dispersal generally results in infection densities of about one uredium per 10 m of row. Generally, infections that occur during active growth of the host plant are within the canopy when they sporulate, restricting spore escape.

The wheat stem rust fungus begins to increase from overwintering uredia in central and northern Texas in April. By early May in favorable years adequate inoculum is available at the top of the canopy for effective long-distance transport. Spread of the epidemic to a distance of about 1500 km can occur in 55 days (Fig. 5).

CULTURAL PRACTICES AFFECTING REGIONAL EPIDEMICS

Green Bridges between Seasons

A number of cultural practices can enhance epidemics if they result in self-sown plants growing during the period between normal wheat crops. In Parana, a state of Brazil, self-sown wheat is a common weed in the summer soybean crop. Rust inoculum moves from late maturing wheat fields to self-sown plants in adjacent fields that were harvested a month or more earlier and planted to soybeans. After the late wheat field is tilled and planted to soybeans, self-sown wheat plants appear. Inoculum then moves from self-seeded wheat plants in the early soybean fields to those in late seeded fields. Rust then spreads back to the early seeded wheat fields following the early soybean crop. Weather conditions during the summers are favorable for rust, and the distance the spores must travel varies from a few meters to a kilometer, which eliminates the need for long-distance transport. A green bridge also exists in many areas of the world with self-sown

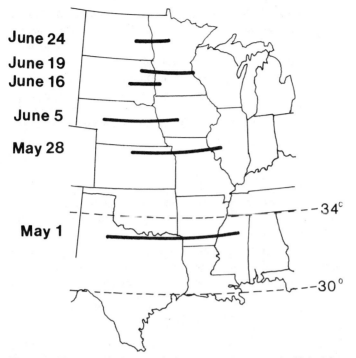

June 24
June 19
June 16
June 5
May 28
May 1

34°
30°

Figure 5 The normal advance of wheat stem rust across the United States. Overwintering may occur north to the 34th parallel and almost always in trace amounts to the 30th parallel. *(Data, in part, from Hamilton and Stakman, 1967.)*

plants along roadsides, irrigation ditches, threshing floors, fences and hedge rows, and in fields not in production for a season.

Irrigation Practices

Irrigation can enhance epidemic development by direct addition of free water on the foliage and by maintaining soil moisture and promoting dense foliage which is favorable for dew formation. There are large areas of irrigated wheat production, however, that have not been the source of inoculum for a regional epidemic. This may be partly because of the generally low relative humidity in areas where irrigation is practiced. Sprinkler irrigation during the day may provide spores with enough moisture to germinate but inadequate time for infection. Also large water drops may eliminate inoculum by dislodging spores and washing them to the ground. One example in which irrigation aided in the development of a local leaf rust epidemic was in the Mayo and Yaqui valleys of Mexico (Dubin and Torres, 1981). In the north China Plain where irrigation is extenisve, it is felt that irrigation helps the pathogen to survive between seasons on the self-sown plants. Irrigation canals and ditches often provide ideal sites for survival.

A possible source of stem rust inoculum for the United States and Canada is northeastern Mexico, but only 50,000 ha of wheat is grown during the dry winter period. The wheat cultivars currently grown are highly resistant to stem rust and only traces of stem rust have been observed. In a limited area small fields, primarily of barley, are grown under irrigation for forage. These fields have occasionally been found to be infected with traces of wheat stem rust. Although the Mexican crop in the spring currently has little role in epidemiology of rust epidemics in the rest of North America, barley could serve as a green bridge through the summer for reintroduction of stem rust into the southern United States in the fall.

Variations in Planting and Harvesting Dates

Earlier planting dates may advance the harvest time enough to escape epidemics when favorable conditions for disease or arrival of inoculum normally occur late in the plant's growth cycle. Mexican farmers had successfully avoided rust epidemics by early planting prior to the introduction of resistant cultivars (Borlaug, 1954). Adjustment of seeding date may be used to weaken the green bridge between crops, by timing the transfer of inoculum from crop to self-sown plants or vice versa so that it occurs during periods unfavorable for pathogen spread or increase. In Brazil the multiple green bridge would be weakened by reducing the time between the earliest and latest planting dates for both soybeans and wheat in an area. In Kansas winter wheats that matured 2 weeks earlier than the original Turkey types were developed to reduce the likelihood of hot dry winds at flowering. Additionally, the earlier wheats eliminated the last cycle of rust increase, greatly reducing local disease severities as well as the amount of inoculum that might be transported northward.

DISEASE FACTORS IN REGIONAL EPIDEMICS

Critical Month

The critical month was described by Chester (1946) for wheat leaf rust in Oklahoma. Wheat is planted in Oklahoma in September and October and by the end of fall it is often infected with leaf rust. Fall severities vary greatly among locations and years. During early winter the disease level generally remains constant, or in severe winters, decreases. By the beginning of March, temperatures increase and conditions again become favorable for disease development. April starts the period in Oklahoma when weather is nearly always within the range favorable for disease increase. Thus, the weather in March determines if the pathogen can move successfully from the surviving overwintering uredia on the lowest leaves to the newly produced leaves. Temperatures are marginal for infection during most dew periods and freezing temperatures may occur. The plant grows during warm periods and the old infected leaves senesce. Chester (1946) described the "critical month" as the first 30-day period in the spring for which the average daily temperature exceeds 10° C. This system of estimating disease severity of wheat leaf rust in Oklahoma was used for many years to predict losses. If such a system was developed for stem rust the "critical month" would be the 30-day period in which the mean daily temeperature for the preceding month was 15° C. A similar system has been recently developed for stripe rust in the Pacific Northwest area of the United States (Coakley and Line, 1981).

Detection Thresholds

The difficulty in detecting single uredia on widely scattered plants in the approximately 1 million ha area where stem rust on wheat has overwintered at least once in the past 15 years has forced us to develop various techniques for detecting overwintering foci. A highly susceptible cultivar is planted in dispersed sites. Plot shape has not been important, but at least 10 m of row has been necessary. Overwintering disease usually is reported only from commercial fields in areas where we detected overwintering in our plots. Plots are inspected during April and early May when the disease has had adequate time to develop a focus up to 1 m in diameter.

Outside the stem rust overwintering areas, the arrival of inoculum and the detection of disease is monitored annually. Trap plots of highly susceptible cultivars are planted. In areas where the disease may appear rarely or at very low intensity, large stips 1 m wide and 100 m long are most effective. In areas where stem rust has historically been important, 10 m of linear row is generally adequate. Surveys are made of these plots and commercial fields with stops every 32 km. Crop growth stage, disease severity and incidence (percent of tillers infected) are recorded. Experience has shown that in the case of stem rust we can detect an infection frequency from exogenous inoculum at a frequency of one uredium per 1000 tillers. Detection of single uredia in overwintering areas is more difficult because the uredia occur low in the canopy and often on dying leaves or those infected with other diseases.

Inoculum has also been detected by trapping uredospores on 5-mm-diameter glass rods used as impaction traps placed about 15 cm above the crop canopy. Rod traps are more efficient than glass microscope slides (Roelfs et al., 1968) and will detect locally produced inoculum from disease levels of approximately one uredium per 1000 tillers of either stem or leaf rust. Impaction traps are not efficient enough for detecting exogenous inoculum, which is best detected in rain samplers (Roelfs et al., 1970). Because most exogenous inoculum is rain deposited (see the earlier section in this chapter, Environmental Conditions in Regional Epidemics), rain samples are very useful in predicting the effective establishment of wheat rusts after long-distance transport (Roelfs et al., 1970; Nagarajan and Joshi, 1985). Data from spore traps must be interpreted with caution, especially for exogenous inoculum, since not all spores trapped are viable and uredospores of different rust pathogens may be very similar morphologically. Spores exposed to rain, sunlight, and varying humidities, and mixed with other spores, pollen grains, and a range of other organic and inorganic materials are especially difficult to identify. Spore identification is normally done at 100 × magnification to permit evaluation of samples from large numbers of locations, replicates, and treatments which increases the problems of identification.

Exogenous inoculum can arrive in amounts below the threshold of detection. The resultant infections are apparent only later as foci develop. This can lead to an error of one generation time (14 to 21 days) or more in estimating the initial infection date, which in turn can cause either the amount of exogenous inoculum or the disease increase rate to be greatly overestimated. Additionally, errors in estimating the arrival date of exogenous inoculum can affect conclusions about the apparent source of the inoculum.

Onset Time

Regional epidemics of rust develop only when the initial infections occur at or before flowering. Under ideal conditions disease onset of a week earlier can double the severity of wheat stem rust. Historically the date of the first observation of wheat stem rust in the United States has been recorded (Hamilton and Stakman, 1967) (Fig. 5). A more meaningful observation would include the stage of host development at disease onset. In Minnesota, about 40% of the variation in loss of spring sown oats (*Avena sativa* L.) due to stem rust was explained by date of onset (Roelfs and Long, 1980). Had these disease onset data been improved by including host growth stage they would surely have explained more of the variation in disease loss. The most severe losses were always in fields planted later than the average planting date.

Rate of Disease Increase

Because the latent period varies with temperature, a single event of spore deposition and infection occurring over a large area can become apparent as sporulation initiated over a period of days. The initial sporulation occurs in the warmer areas first and then progresses to the cooler areas over the period of a few days to a few weeks. To the untrained observer this process appears to be a continuous wave of disease spread with daily infections.

Disease increase is normally expressed as the change in percent disease (using a modified Cobb scale, Peterson et al., 1948) over time in days. When rust epidemics are to be compared where the host growth cycle varies greatly in length there is an advantage to using a host growth stage scale as the measurement of time (Calpouzos et al., 1976).

Initial disease severities are generally lower than the 1% on the modified Cobb scale. They can be quantified by converting the percent scale to a number of uredia per tiller. For wheat stem rust 1% severity is equivalent to 10 pustules per tiller. The conversion factor for leaf rust is: 18 pustules per tiller equal 1%. Estimating severity of stripe rust due to its semisystemic nature must include some estimation of lesion size.

Inoculum Density

Although each uredospore produced by the cereal rust fungi has the potential to cause an infection, not all do. When single uredospores of *P. graminis* with a germination rate near 100% are placed on susceptible host tissue under ideal conditions for spore germination and infection only 1 of 10 will result in an infection (see Rowell 1984, for details of the infection process). *Puccinia recondita* uredospores under similar conditions may produce up to one uredium per three uredospores.

Additionally, the effectiveness of host resistance may depend in part on the amount and quality of inoculum produced on surrounding plants (Roelfs et al., 1972). The stem rust resistance of the cultivar Thatcher is illustrative. In plots with heavy infection (high inoculum level) terminal severities of Thatcher averaged 86%, with a susceptible host response. Thatcher had a mean terminal severity of 2% with a moderately susceptible host response a few kilometers away in a similar geographic area (low inoculum level). In this experiment, adequate spores were present so that the susceptible check cultivar Baart had a terminal severity of 90% in both areas (Nazareno and Roelfs, 1981). Thatcher was damaged by stem rust during the epidemic years 1953 and 1954 when most of the wheat area was planted to cultivars more susceptible than it. In the late 1960s and early 1970s Thatcher was undamaged when most of the cultivars grown were resistant.

IMPACT OF REGIONAL EPIDEMICS

Loss in Yield

Rust can greatly reduce the amount of grain produced by a plant. A severe early natural infection of stem rust or leaf rust at St. Paul, Minnesota can result in a 50% loss compared to a rust-free check in 2 out of 3 years. It appears that this is the maximum loss possible under current levels of inoculum in this area with spring wheat. The major portion of this loss is due to a reduction in grain size. The number of tillers and florets per spike are generally determined before the rust becomes severe, particulary for spring wheats. Occasionally, however, any of the three wheat rusts can become severe enough to kill the entire field prior to grain formation. A reduction in the number of grains can occur, when severe rust kills tillers that would normally survive to produce seed or when lodging and stem breakage occurs and the grain, although formed, cannot be harvested mechanically.

The history of rust epidemics worldwide was reviewed by Chester (1946) and Hogg et al. (1969). The regional epidemics since 1918 in the United States are shown in Table 1. Although the total is very variable, about 29 million ha of wheat are grown annually in the United States. Losses greater than 50% on a statewide basis are infrequent but have occurred. Since 1918, the greatest statewide losses due to stem rust were 56.5 and 51.6% in 1935 for North Dakota

Table 1 Losses resulting from regional epidemics of wheat stem and leaf rust in the United States since 1918

| | Losses due to stem rust | | | | | | | | | | Losses due to leaf rust, 1938 | |
| | 1923 | | 1935 | | 1937 | | 1953 | | 1954 | | | |
State	%	Tonnes	%	Tonnes	%	Tonnes	%	Tonnes	%	Tonnes	%	Tonnes
CO	10.0	55,576	6.0	12,079	Tr	Tr	0.6	7,211	Tr	Tr	2.0	10,710
IL	1.5	26,476	1.0	9,477	11.0	160,434	0.3	4,876	Tr	5,209	15.0	199,942
IN	Tr	Tr	Tr	Tr	12.0	142,849	1.0	12,712	0.5	5,801	3.0	23,858
IA	1.0	3,829	6.5	13,413	19.0	101,212	6.0	4,857	15.0	10,271	1.0	3,567
KS	1.0	23,321	12.0	252,570	6.6	305,904	1.5	60,082	3.0	152,561	12.0	585,867
MI	2.0	8,858	0.5	2,618	2.5	13,745	Tr	Tr	Tr	Tr	1.0	5,373
MN	15.0	114,851	51.6	607,042	10.0	107,454	13.4	68,045	18.0	62,324	8.0	90,158
MO	—	—	4.0	32,122	30.0	507,371	2.0	22,835	1.0	11,348	20.0	236,269
MT	18.0	293,986	0.6	5,624	0.0	0	3.0	96,352	2.0	46,121	Tr	Tr
NE	4.0	36,938	15.0	185,351	6.0	81,870	2.0	48,712	5.0	85,650	20.0	404,144
ND	12.0	258,262	56.5	1,939,554	25.0	535,554	37.7	1,605,878	42.9	1,417,012	2.0	43,110
OH	2.0	22,740	1.5	19,876	2.0	27,046	1.0	18,717	2.0	26,149	2.5	32,435
OK	—	—	Tr	Tr	1.0	18,714	1.0	19,497	2.0	39,390	27.0	622,145
PA	Tr	Tr	Tr	Tr	3.0	22,478	0.2	1,189	0.2	1,066	10.0	66,641
SD	10.0	83,744	29.0	306,049	12.0	58,974	35.2	484,961	21.0	190,972	—	—
TX	2.0	9,174	10.0	35,555	Tr	Tr	Tr	Tr	Tr	Tr	5.0	50,624
VA	0.2	5,264	3.0	7,432	2.0	5,438	2.0	4,301	3.0	5,967	5.0	12,513
WA	Tr	Tr	—	—	—	—	Tr	Tr	Tr	Tr	0.1	1,481
WV	Tr	Tr	Tr	Tr	2.0	1,380	2.0	706	2.5	8,045	—	—
WI	1.0	6,109	1.5	9,873	19.8	14,105	2.9	1,388	1.2	5,427	Tr	Tr
WY	Tr	Tr	10.0	6,112	Tr	Tr	3.2	6,150	Tr	Tr	0.5	4,691
U.S.	4.6	20,713,145	16.7	17,133,463	8.1	23,834,018	7.2	31,895,018	7.3	26,448,572	8.7	25,088,536

Tr = trace.

and Minnesota, respectively. Durum wheat losses due to stem rust were 66% in 1953 for the 746,000 ha and 79% in 1954 for 523,000 ha grown in Minnesota, North Dakota, and South Dakota. Leaf rust resulted in a 50% loss in Georgia in 1972 (Roelfs, 1978).

Rust losses in Australia have been substantial, but loss estimates are not available. Scattered rust oversummers in the moister areas of Australia. Epidemics have occurred when new combinations of pathogen virulence were introduced or developed through parasexualism or mutation (Luig, 1985).

Epidemics have occurred in India and Pakistan (Nagarajan and Joshi, 1985), but the extent of annual losses is unknown. Rust survives the hot dry summers in the southern hills of India and in the foothills of the Himalayas in the northern areas of both countries. Losses occur when cultivars are susceptible, the inoculum arrives early in the season, and environment is especially favorable.

Losses in China were fairly common in the periods of political turmoil during the 1950s through the mid-1970s. Control of volunteer plants, timely planting, and use of resistant cultivars have all resulted in reduced losses in recent years (Roelfs, 1977; Hu and Roelfs, 1985).

Losses due to rust in Europe are usually local and erratic. Epidemics are associated with early arrival and/or increase of inoculum in the spring. Stripe rust is an important disease in winter cereal areas, with Great Britain and The Netherlands particularly threatened (Zadoks and Bouwmann, 1985). Hogg et al. (1969) reported no epidemics of stem rust after 1958, and leaf rust resulted in fewer epidemics and less loss. No major leaf rust losses have occurred since 1961. Fungicides are currently used along with resistant cultivars to control stripe rust in northwestern Europe (Stubbs, 1985).

In South America, the cereal rusts occur but seldom result in more than local or areawide epidemics. Losses are difficult to estimate in some of the areas because of the severity of many other foliar diseases. Epidemics are associated with high initial levels of inoculum and with years in which the winters are mild and moist.

Loss in Grain Quality

Nearly all crops have some quality standards by which they are judged and which affect their market value. Test weight (weight per volume) of wheat is such a quality standard in much of the world. Shriveled grain, a common result of severe rust epidemics, often reduces test weight enough to result in a much lower price. Grain produced during a severe rust epidemic can be of such poor quality that it is useful only for animal feed, which means its price is determined in the United States by the price of corn and sorghum, the predominant feed grains. Although shriveled grain tends to have higher protein content, a desirable trait, shriveled grain also has a lower flour extraction, an undesirable trait.

Rust epidemics can also indirectly lower the grain quality by forcing growers to use a lower quality resistant wheat in place of a high quality susceptible wheat. The 1953 and 1954 rust epidemics destroyed the United States durum crop.

Growers turned to the resistant cultivars Wells and Lakota that had small seed (undesirable durum quality) because the large-seeded Langdon and the Yuma durums were too susceptible to grow economically. It required 14 years to develop the large-seeded resistant cultivar Leeds (Lebsock et al., 1966).

Losses in Forage

In some areas of the world, including the southern Great Plains of the United States, fall planted wheat is important as fall and winter forage for cattle and sheep. The value of forage in this area is often the profit margin for growing wheat. Winter sown oats in the Gulf Coast states is often planted only for forage. Rust decreases the amount of forage directly by killing the leaf tissue, and indirectly by weakening the plant and reducing its growth. Severe epidemics can result in losses of 100%. Additionally, the uredospores of the rusts can be a physical irritant to the grazing animals, resulting in watery eyes and runny noses.

Economic Impact

When disease destroys a portion of the crop, the price for the remaining crop increases. Thus, growers that produce some wheat may receive nearly a normal income. Growers with severe disease loss suffer while other growers with little or no disease loss may profit from the higher prices. The total supply of grain may be adequate following a rust epidemic, but the amount of grain in the high-quality grades may be in very short supply. This usually results in an unusually large price differential between grades. Currently with a surplus of wheat in the United States it would take a very serious epidemic to produce a major change in price. In areas of the world where the population consumes the entire harvest to meet just the basic food needs, even local epidemics can have a significant impact on both prices and survival.

Rust epidemics have changed grain exporting and grain self-sufficient countries into importers, for example, Pakistan (Hassan, 1978) and Mexico (Dubin and Torres, 1981). As a result of the serious stem rust epidemics of North America in the 1950s pasta products were made from bread wheat rather than from the high-quality durum wheats.

Prevention of losses due to rust is costly. Plant breeders must make crosses and select for resistant lines, reducing the available gene pool for commercial production. Good lines otherwise may have to be reentered in the crossing block to obtain rust resistance. Time and space are required for evaluation of rust resistance. The cost of evaluation varies with the level, type, and genetic complexity of the resistance required and the severity of the disease in the area. Many breeding programs spend from 10 to 20% of their time on rust resistance. The breeding program usually involves local, national, and international pathologists. Field surveys of disease intensity and severity, physiological race surveys, basic studies on the pathogen, basic and applied studies on the disease, searches for resistant germplasm, basic and applied studies on the nature of resistance, basic and applied studies on genetics of resistance, germplasm classification, improvement, and storage all are necessary to provide the inputs that result in a constant flow of improved resistant cultivars to the grower. As these items are funded by various groups, the entire cost to be assigned to each cultivar is not readily available. But in 1984, perhaps 250,000 dollars per rust-resistant cultivar would be a reasonable estimate in the United States.

The following example illustrates the time and problems in developing a rust-resistant cultivar. Thatcher was developed by a cross made in 1914 between an adapted cultivar of a hard red spring wheat (Marquis) and a stem rust resistant durum wheat cultivar (Immillo). A line selected in 1918 was released to growers in 1928 as Marquillo. This cultivar never was important; however, in 1921 a sib line of Marquillo had been crossed by a Marquis/Kanred derivative. Kanred/Marquis cross had been made in 1918 to obtain the immunity of Kanred to certain *P. graminis* races in a hard red spring wheat. The line (later named Thatcher) was selected in 1925 and released to growers in 1934. Although it took 20 years to develop, it was among the most successful resistance sources ever developed (Schafer et al., 1984). It provided spectacular resistance to the epidemics of 1935 and 1937. Although it was damaged in the epidemics of 1953 and 1954, it still is widely used as a background for stem rust resistance in the hard red spring wheats.

Chemical sprays are used either alone or in combination with host resistance to control stripe rust in Europe (Stubbs, 1985) and in the Pacific Northwest of the United States (Line, 1982) and to control leaf rust in Brazil and Paraguay. Obvious costs are the price of the chemical and its application. The cost of the chemical is usually adjusted to include the cost of developing and testing the chemical. However, other costs are not as obvious. Ground-drawn rigs can reduce the yield by 2 to 4% by direct damage to the crop. Aircraft sprayers eliminate this loss but usually are more expensive. Monitoring systems are necessary to determine if and when spraying is required even in years when the disease is absent (Rowell, 1968). Most, like Epimul (Kampmeijer and Zadoks, 1977), have been developed at public expense.

Although I have avoided assigning a value to the cost of cereal rust epidemics, it is substantial. Worldwide, the equivalent of hundreds of years of scientific investigation are spent on rust control annually.

REFERENCES

Allan, R. E., Line, R. F., Peterson, Jr., C. J., Rubenthaler, G. L., Morrison, K. J., and Rohde, C. R. 1983. Crew, a multiline wheat cultivar. *Crop Sci.* 23:1015–1016.

Anikster, Y., and Wahl, I. 1979. Coevolution of the rust fungi on Gramineae and Liliaceae and their hosts. *Annu. Rev. Phytopathol.* 17:367–403.

Beeson, K. E. 1923. Common barberry and black stem rust in Indiana. *Purdue Univ. Dep. Agr. Exp. Bull.* 118, 8 pp.

Borlaug, N. E. 1954. Mexican wheat production and its role in the epidemiology of stem rust in North America. *Phytopathology* 44:398–404.

Browning, J. A., and Frey, K. J. 1969. Multiline cultivars as a means of disease control. *Annu. Rev. Phytopathol.* 7:355–382.

Calpouzos, L., Roelfs, A. P., Madson, M. E., Martin, F. B., Welsh, J. R., and Wilcoxson, R. D. 1976. A new model to measure yield losses caused by stem rust in spring wheat. *Univ. Minnesota Tech. Bull.* 307, 23 pp.

Chester, K. S. 1946. *The Nature and Prevention of the Cereal Rusts as Exemplified in the Leaf Rust of Wheat.* Chronica Botanica Co., Waltham, 269 pp.

Coakley, S. M. and Line, R. F. 1981. Quantitative relationships between climatic variables and stripe rust epidemics on winter wheat. *Phytopathology* 71:461–467.

Craigie, J. H. 1945. Epidemiology of stem rust in western Canada. *Sci. Agr.* 25:285–401.

Dubin, H. J., and Torres, E. 1981. Causes and consequences of the 1976–1977 wheat leaf rust epidemic in northwest Mexico. *Annu. Rev. Phytopathol.* 19:41–49.

Groth, J. V., and Roelfs, A. P. 1982. Effect of sexual and asexual reproduction on race abundance in cereal rust fungus populations. *Phytopathology* 75:1503–1507.

Hamilton, L. M., and Stakman, E. C. 1967. Time of stem rust appearance on wheat in the western Mississippi basin in relation to the development of epidemics from 1921 to 1962. *Phytopathology* 57:609–614.

Hassan, S. F. 1978. Rust problem in Pakistan, in: *Wheat Research and Production in Pakistan* vol. 1 (M. Tahir, ed.), pp. 90–93. PARC, Islamabad.

Hermansen, J. E. 1968. Studies on the spread and survival of cereal rust and mildew diseases in Denmark. *Friesia Nord. Mykol. Tidsskrift* 3:161–359.

Hogg, W. H., Hounam, C. E., Mallik, A. Ḳ., and Zadoks, J. C. 1969. Meteorological factors affecting the epidemiology of wheat rusts. *World Meteorol. Org. Tech. Note* 99, 143 pp.

Hu, C. C., and Roelfs, A. P. 1985. The wheat rusts in the People's Republic of China. *Cereal Rusts Bull.* 13:11–28.

Kampmeijer, P., and Zadoks, J. C. 1977. Epimul, a simulator of foci and epidemics in mixtures of resistant and susceptible plant, mosaics and multilines. *Cent. Agr. Publ. Doc. Wageningen,* 50 pp.

Katsuya, K., and Green, G. J. 1967. Reproductive potentials of races 15B and 56 of wheat stem rust. *Can. J. Bot.* 45:1077–1091.

Lebsock, K. L., Gough, F. J., and Sibbith, L. D. 1966. Leeds: A rust resistant large kerneled durum. *Farm Res.* 24(5):9–14.

Line, R. L. 1982. Chemical control of stripe rust and leaf rust on wheat in the pacific northwest. (Abstr.) *Phytopathology* 72:972.

Luig, N. H. 1985. Epidemiology in Australia and New Zealand, in: *The Cereal Rusts,* vol. 2, *The Diseases, Their Distribution, Epidemiology, and Control* (A. P. Roelfs and W. R. Bushnell, eds.), pp. 61–101. Academic Press, Orlando.

Nagarajan, S., and Joshi, L. M. 1985. Epidemiology in the Indian sub-continent, in: *The Cereal Rusts,* vol. 2, *The Diseases, Their Distribution, Epidemiology, and Control.* (A. P. Roelfs and W. R. Bushnell, eds.), pp. 371–402. Academic Press, Orlando.

Nazareno, N. R. X., and Roelfs, A. P. 1981. Adult plant resistance of Thatcher wheat to stem rust. *Phytopathology* 71:181–185.

Peterson, R. F., Campbell, A. B., and Hannah, A. E. 1948. A diagrammatic scale for estimating rust intensity on leaves and stems of cereals. *Can. J. Res. Sect. C.* 26:496–500.

Roelfs, A. P. 1977. Foliar fungal diseases of wheat in the People's Republic of China. *Plant Dis. Rep.* 61:836–841.

———. 1978. Estimated losses caused by rust in small grain cereals in the United States— 1918–76. *U. S. Dep. Agr. Misc. Publ.* 1363, 85 pp.

———. 1982. Effects of barberry eradication on stem rust in the United States. *Plant Dis.* 66:177–181.

———. 1985a. Wheat and rye stem rust, in: *The Cereal Rusts,* vol. 2, *The Diseases, Their Distribution, Epidemiology, and Control* (A. P. Roelfs and W. R. Bushnell, eds.), pp. 3–37. Academic Press, Orlando.

———. 1985b. Epidemiology in North America, in: *The Cereal Rusts,* vol. 2, *The Diseases, Their Distribution, Epidemiology, and Control* (A. P. Roelfs and W. R. Bushnell, eds.), pp. 403–434. Academic Press, Orlando.

Roelfs, A. P., Dirks, V. A., and Romig, R. W. 1968. A comparison of rod and slide samplers used in cereal rust epidemiology. *Phytopathology* 58:1150–1154.

Roelfs, A. P., and Groth, J. V. 1980. A comparison of virulence phenotypes in wheat stem rust populations reproducing sexually and asexually. *Phytopathology* 70:855–862.

Roelfs, A. P., and Long, D. L. 1980. Analysis of recent oat stem rust epidemics. *Phytopathology* 70:436–440.

Roelfs, A. P., and McVey, D. V. 1973. Races of *Puccinia graminis* f. sp. *tritici* in the USA during 1972. *Plant Dis. Rep.* 57:880–884.

———. 1979. Low infection types produced by *Puccinia graminis* f. sp. *tritici* and wheat lines with designated genes for resistance. *Phytopathology* 69:722–730.

Roelfs, A. P., McVey, D. M., Long, D. L., and Rowell, J. B. 1972. Natural rust epidemics in wheat nurseries as affected by inoculum density. *Plant Dis. Rep.* 56:410–414.

Roelfs, A. P., Rowell, J. B., and Romig, R. W. 1970. Sampler for monitoring cereal rust uredospores in rain. *Phytopathology* 60:187–188.

Rowell, J. B. 1968. Chemical control of the cereal rusts. *Annu. Rev. Phytopathol.* 6:243–262.

———. 1984. Controlled infection by *Puccinia graminis* f. sp. *tritici* under artificial conditions, in: *The Cereal Rusts,* vol. 1, *Origins, Specificity, Structure, and Physiology* (W. R. Bushnell and A. P. Roelfs, eds), pp. 291–332. Academic Press, Orlando.

Saari, E. E., and Prescott, J. M. 1985. World distribution in relation to economic losses, in: *The Cereal Rusts,* vol. 2, *The Diseases, Their Distribution, Epidemiology, and Control.* (A. P. Roelfs and W. R. Bushnell, eds.), pp. 259–298. Academic Press, Orlando.

Samborski, D. J. 1985. Wheat leaf rust, in: *The Cereal Rusts,* vol. 2, *The Diseases, Their Distribution, Epidemiology, and Control* (A. P. Roelfs and W. R. Busnell, eds.), pp. 39–59. Academic Press, Orlando.

Schafer, J. F., Roelfs, A. P., and Bushnell, W. R. 1984. Contributions of early scientists to knowledge of cereal rusts, in: *The Cereal Rusts,* vol. 1, *Origins, Specificity, Structure, and Physiology.* (W. R. Bushnell and A. P. Roelfs, eds.), pp. 3–38. Academic Press, Orlando.

Skovmand, B., Wilcoxson, R. D., Shearer, B. L., and Stucker, R. E. 1978. Inheritance of slow rusting to stem rust in wheat. *Euphytica* 27:95–107.

Stubbs, R. W. 1985. Stripe rust, in: *The Cereal Rusts,* vol. 2, *The Diseases, Their Distribution, Epidemiology, and Control* (A. P. Roelfs and W. R. Bushnell, eds.), pp. 61–101. Academic Press, Orlando.

Wahl, I., Anikster, Y., Manisterski, J., and Segal, A. 1984. Evolution at the center of origin, in: *The Cereal Rusts*, vol. 1, *Origins, Specificity, Structure, and Physiology*. (W. R. Bushnell, and A. P. Roelfs, eds.), pp. 39–77. Academic Press, Orlando.

Zadoks, J. C., and Bouwmann, J. J. 1985. Epidemiology in Europe, in: *The Cereal Rusts*, vol. 2, *The Diseases, Their Distribution, Epidemiology, and Control* (A. P. Roelfs, and W. R. Bushnell, eds.), pp. 329–369. Academic Press, Orlando.

Part Three

Disease Prediction and Management

Chapter 7

Disease and Pathogen Detection for Disease Management

J. O. Strandberg

Detection of plant pathogens and the diseases they incite is a key component of disease management. Detection is the process of establishing the presence of pathogens or disease through the observation of entities or observable processes of disease. Often, devices such as the microscope or even remote sensing equipment may assist in the observation, but in order to detect disease we must examine, inspect, and observe. Thus, observation is also a key component of disease management. Through detection and observation, we attempt to transform fact into useful information.

IMPLICATIONS OF DISEASE AND PATHOGEN DETECTION FOR DISEASE MANAGEMENT

Disease management decisions are concerned mostly with probabilities. Most often, they are concerned with what might happen, but when sufficient information is available, they may also consider what is likely to happen. Disease management, like much of integrated pest control, may be "doing more of what you would have done in the first place if you'd only thought of it." (Haskell,

1977). Many advances in plant disease research that benefit disease managers serve mostly to increase the number of reasonable options available to them. Correct and timely information is essential in order to exercise the best of these options. Observation and detection of disease and pathogens are informational components essential to this process.

There are numerous examples of the processes involved in disease and pathogen detection that could be cited, but most of these have not yet been formally or obviously applied to disease management in a systematic or direct manner. Nevertheless, a close and important relation exists between disease detection and disease management which is implied by the principles of epidemiology (Fry, 1982). Disease management decisions can significantly affect the future status of the crop system including subsequent disease damage, but information about the apparent or present state of the system (the true state is seldom known) greatly affects these decisions. Disease and pathogen detection are only two small but very important components of the information base potentially available to the manager. Other components may include disease incidence, damage, yield loss, crop phenology, and weather.

Although it is not immediately obvious, all of these informational components are only made useful through some form of inspection, observation, or detection. The processes of detection are fundamental in nature, but depend greatly on what is to be detected. An instrument may be employed to detect some physical process or entity. For example, information on temperature or weather is obtained through observation of instruments or their records which detect changes in these entities with time. However, crop phenology information must usually be obtained through periodic observations by people and it is an analysis or comparison of the acquired information that detects changes with time. Detection of pathogens or plant disease is similar in many respects, but often the detection of the pathogen itself is incredibly difficult. In other cases, we must conveniently rely on detection of disease to convince us that the pathogen is (or has been) present. In short, diseases or pathogens should be detected by processes that are appropriate for the purpose.

What Questions Can Be Answered?

Three basic questions about disease or pathogens that can be answered by detection and observation are:

1. Is it there or not?
2. How much is there?
3. Where is it?

Determining when these things happened, or more commonly, when we became aware of them is often implicit in the fact of detection. Answers to these questions greatly affect disease management decisions.

Often the processes employed in the detection of plant disease or pathogens are largely determined by complicating factors such as the distribution and abundance of the pathogen, its interaction with the host and environment, when infection of plants occurs, and the development and recognition of symptoms. Opportunities for detection depend on our knowledge of the biology of host and pathogen; available knowledge determines how and when detection may be accomplished. Although the study of pathogen biology has sometimes become separated from the study of epidemiology and disease development, it is often useful to consider reproduction, persistence, dispersal, host colonization, and disease development simply as some of the more observable life processes of the pathogen. Systematic observations of these processes can furnish new and useful information. In disease management, this systematic observation has been called scouting.

WHY AND WHEN DETECTION IS IMPORTANT

It is logical to assume that, given the opportunity of encountering a susceptible host at the right time and in a favorable environment, a pathogen can incite disease with some finite probability. Thus, detection of disease or gaining information supporting a very strong probability of disease (for example, detection of the pathogen) can lead to similar assumptions by the disease manager. One example is the detection of disease symptoms or the pathogen itself associated with seeds or propagation materials well in advance of planting. This might be considered as detection of disease before the fact. If cabbage seeds are tested and found to be infested with the incitant of black rot, *Xanthomonas campestris* pv. *campestris,* it may be assumed that planting them in most cabbage growing areas of the world will almost surely result in serious disease (Williams, 1980). Similarly, a high incidence of pea root rot caused by *Aphanomyces euteiches,* which is detected when a sample of soil is evaluated by bioassay with pea plants in a greenhouse, forecasts an unacceptably high probability of root damage if peas are planted on that site and moist soils prevail (Sherwood and Hagedorn, 1958). Alternatively, the significance of detection of disease or pathogens that are likely to be placed in an environment totally unfavorable for disease development must be evaluated in that light. Perhaps there is almost no probability of disease. Detection before the fact can result in unique opportunities for disease management.

Detection of disease or pathogens after the fact is also important, but is usually based on more immediately observable events and processes. As an example, detection of initial observable events and processes. As an example, detection of initial inoculum is sometimes possible. Spore trapping has sometimes been used to detect initial inoculum in or near crop sites. Useful management programs may also function with only the benefit of implied detection of initial inoculum. If the pathogen is known to be present and inoculum production is possible, weather can be evaluated to determine if new inoculum production is likely, and in the case of apple scab, whether or not this inoculum is likely to be successful (Jones, 1976). The BLITECAST program which forecasts favorable conditions for late blight of potato is based on similar assumptions (MacKenzie, 1981). Thus, it is possible and often appropriate to make disease management decisions based on the implied or assumed as well as the actual detection of initial inoculum.

Detection of disease or pathogens in an established crop may also be considered as detection after the fact. Here, several additional possibilities for information useful for disease management exist. Sometimes a simple yes/no answer in reference to the presence or absence of pathogens or disease is adequate for the purpose of deciding to apply disease control measures. However, quantitative information may also be obtained; one can detect levels, changes in levels, determine when thresholds of disease have been reached or study levels of pathogens and disease with time.

Searching for and finding disease in an existing crop are aspects of detection that are more familiar for most plant pathologists. This is also detection after the fact, and it involves human activities such as surveying, sampling, scouting, and monitoring. Sampling and scouting with a purpose of disease detection or observation to assist in the management of plant disease will usually be an early warning type of activity. If immediate use is to be made of the information, the methods of detection or observation will reflect this goal. Other sampling and scouting methods for detecting or observing disease—such as those used to gather data for studies of abundance and distribution, survey work, or to provide information feedback on the success of control or management activities—are likely to use different methods. The principles of sampling and detection are the same for all of these applications, but the actual procedures will probably differ to suit the application.

It should be obvious that a tremendous variety of methods and special considerations are possible; these are mandated by a corresponding variety of pathogens, disease situations, and applications. Fortunately, there are often practical guidelines to follow, and many examples to look to for both ideas and methods. The best approach to disease or pathogen detection will be determined by the goals of the project, but the approach may be severely limited by available knowledge of the pathogen and the disease it incites. For this reason, goals or subgoals of the project may often be the development of new knowledge, but if the right questions are identified and asked, the best, or at least a good approach to detection will usually be obvious.

DETECTION OF DISEASE AND PATHOGENS BEFORE THE FACT OF DISEASE

Detection before the fact helps the disease manager to do the right thing by thinking about it in the first place. It allows systematic planning and selection of best options and alternatives that may be available.

Seeds and Propagation Stock

Detection of disease or pathogens on seed or propagation stock before the crop is established offers exceptional opportunities in disease management. Testing of many crop seeds and propagation stock is routine. Testing methods are diverse and effective. Often, they can be adapted easily to new applications or serve as models to develop new and similar detection methods where none now exist. Numerous detection methods are available for bacteria, viruses, and fungi which may be easily adapted for disease management purposes; some are already employed to exclude pathogens or reduce initial inoculum.

Detection methods for pathogens or disease in propagation materials that produce a yes/no result are often helpful, but quantitative estimates of the level or proportion of affected units provide more information. The proportion of units affected as well as the confidence associated with that estimate can be important in helping to select the best available option. For example, if an estimate of the level of infestation is available, a better decision can be made concerning the feasibility of reducing pathogen numbers to acceptable levels or possibly eradicating the pathogen. In some cases, the incidence of disease, number of pathogens, or proportion of units affected may be related to expected crop damage through experimentation, and minimum acceptable levels may be established.

This concept has most successfully been applied to seeds. Significant advancements in the management of lettuce mosaic virus provide an example of one highly successful program. Field experimentation has determined that the small probability of virus-infested lettuce seeds remaining after 30,000 seeds from each seedlot are tested and found to be apparently free of infested seeds, does not materially affect the disease impact in the field. Virus-infected plants can and do occur, but the initial inoculum level is so low in seedlots passing this test that drastically reduced disease damage is sustained where indexed (virus tested) seed is used (Grogan, 1980). Extensive field experimentation was necessary to determine the impact of different levels of virus-infected seed on lettuce production, and an intensive bioassay program has been established to test lettuce seeds for use in California and Florida. Although crop environments are different in these different regions, the basic concept of employing indexed seed is apparently effective wherever large concentrations of serially planted lettuce acreage are encountered. Less than 1 virus-infected seed per 30,000 was decided on as a realistic threshold. New methods of virus detection promise increased efficiency of detection of viruses in lettuce seed and similar or related programs.

If realistic and useful thresholds for pathogen-infested seeds can be established (for example, the lettuce mosaic virus threshold), statistically acceptable detection procedures become desirable. Genge et al. (1983) discussed possible approaches to this problem and provided statistical guidelines for estimating the testing effort that must be expended to detect low-level infestations of seedborne pathogens. Two general approaches were covered. In the first, a large number of seeds or plants grown from seeds can be tested individually and directly. In the second, samples are drawn from a larger population of seeds, and these samples are tested. Useful procedures for calculating the probability of detection for different intensities of effort and sample number are provided.

Clearly, approaches such as the above are greatly influenced by the application and by the host-pathogen system involved. However, the principles of detection before the fact can be extended to other plant disease management activities as well. All of them cannot be discussed here, but some additional examples can provide a better idea of the usefulness of this approach.

Many methods for detection of diseases and pathogens on seed and propagation stock are already operational and have been validated; the information they can provide is greatly underutilized in disease management. Goals as well as problems involved in detection of viruses, bacteria, and fungi are similar. Procedures for detecting seedborne fungal pathogens abound in the literature. Neergaard (1977) conveniently provides numerous examples of detection of fungi on seeds. Much of that literature emphasizes crop protection applications. McGee (1981) discussed important relationships between the detection of fungal pathogens in and on seeds, the goals of general seed-health testing, and the role of the seed pathologist. It is instructive to review these relationships from the point of view of disease management; there are no conflicts here. Goals of seed health are fully compatible with those of disease management.

Schaad (1982) reviewed the detection of seedborne bacterial pathogens and emphasized the need for better detection processes. He pointed out the practical limitations of detection of bacteria in seed and the significance of these limitations in setting tolerance levels of pathogens in or on seeds. The point of this is that a whole array of methods, procedures, and concepts for detection of disease and pathogens on seeds and propagation stock already exist and that they are suitable or adaptable to disease management. Seed pathologists are well aware of needs and goals of disease management and the implementation of present methods as well as development of new ones seem assured.

Potato seed-piece certification programs are good examples of the highly successful use of disease detection in propagation stocks to supply potato seed-piece tubers with acceptably low levels of disease. These programs and their limitations have been described by Shepard and Claflin (1975). A range of detection methods from simple inspection of seed fields to sophisticated sampling and assay are used to gather information that is employed to help reduce levels of pathogens and disease in propagation stock and seed. Because potato seed production is regional and highly concentrated, it is possible to use efficiently the same inspection activities to detect several diseases and pathogens. Potato seed pieces are frequently certified as containing acceptable levels of key diseases and pathogens. Such examples provide insight and serve as useful models for the development or adaptation of new disease and pathogen detection methods for almost any disease management application.

Soil, Water, and Proposed Crop Sites

Detection of pathogens or disease on or near proposed crop sites may imply a probability of future disease damage to crops. This probability must be evaluated at the time the decision to plant the crop is made. If the pathogen or disease is already present, the probability of more disease may be assumed to be fairly high. Of course, possible effects due to disease on a great variety of pathogens and crop systems may be evaluated in this way. However, a few examples will illustrate the potential value of the detection of pathogens or disease in soil, water, and proposed crop sites in advance of the crop itself in plant disease management.

Much progress has been made in the detection of soilborne disease and pathogens. Disease incidence that might be expected in the proposed crop can sometimes be related to the detection of pathogen populations on the site before the crop is planted. Bouhot (1979) has compiled some general considerations for estimating disease risk as related to population densities of soilborne pathogens in proposed crop sites. Bouhot concluded that the risk of disease should not be estimated on the presence of the pathogen, but on inoculum potential and the biological resistance of the soil; the pathogen must have a favorable environment to produce disease. To estimate disease risk, these factors must be examined or known. It is likely that these concepts can be applied with modification to non-soilborne diseases as well; a foliar pathogen may also face a resistive or favorable environment.

Verticillium dahliae has been investigated extensively from the viewpoint of relating population density to disease incidence. A relationship of this type between the population density of *V. dahliae* (race 2 but not race 1) microsclerotia detected in tomato field soils and subsequent disease incidence has been demonstrated by Grogan et al. (1979). They found that numbers of microsclerotia in the soil were well correlated with disease incidence ($r = 0.82$ to 0.88). Their study addressed many other topics, especially the significance of races of the pathogen. In terms of the impact of the soil assay, however, they concluded that if sufficient inoculum was present in the soil and normal weather for crop growth prevailed, serious disease damage could be expected. A complex procedure of soil sampling, wet sieving of soil samples, and assay for fungal propagules on selective medium was employed. Subsequent disease incidence in sampled fields was determined, and a positive correlation was obtained. Sampling and detection were not primary goals of this study, but methods of great potential use in disease management programs were well demonstrated.

As expected, pathogens such as *V. dahliae* which produce propagules that are relatively convenient to sample and count have received more attention than others. Pullman and DeVay (1982) intensively studied *V. dahliae* in two adjacent cotton fields over a 7-year period by sampling soil and recovering microsclerotia. They found a consistent relationship between inoculum density determined near planting time and disease symptoms and damage within the subsequent growing season. They concluded that disease incidence and damage to the cotton crop could be predicted from soil assay at planting. The intensity of this study also demonstrated that less evident but important relationships can be present. For example, air temperature was observed to complicate the relationship between inoculum density detected near planting time and crop damage later in the season.

There are complicated theoretical considerations surrounding inoculum density and disease incidence relationships. Baker (1978) has reviewed and discussed these in detail. In spite of these complications, research-oriented detection methods such as those developed for *V. dahliae* can be adapted for disease management applications.

Adams (1981) used soil sampling and assay methods similar to those developed for *V. dahliae* to find a useful relationship between the density of sclerotia of *Sclerotium cepivorum* in onion field soils and the subsequent incidence of white rot of bunching onions. His detection plan considered both the spatial distribution of sclerotia in the soil and the components of the cultural system (such as field size) used for onion production in New Jersey to forecast disease incidence. Fields were sampled in small (0.2 ha) units, because even when the fields themselves were large, growers usually planted only small, successive plantings of bunching onions in these fields. Therefore, the pathogen was not randomly distributed in onion fields, and small areas of fields with populations of *S. cepivorum* could be identified and avoided. Adams considered both the biology of the pathogen and the onion agroecosystem in developing this application of pathogen detection. This is an important principle which will be pointed out by other examples that follow.

If detection of the pathogen before the fact appears or proves useful in disease management, it is likely that a suitable approach to implementing it can be developed. A great diversity of methods are employed to recover pathogens from soils, such as plating of soil samples on selective media and baiting techniques. Often physical techniques of detection are not only convenient but necessary. Pratt and Janke (1978) employed washing and sedimentation to separate and recover oospores of *Sclerospora sorghi* from soil and were able to identify and count them directly. They found a useful relationship between the number of oospores recovered from soil samples and the incidence of downy mildew damage of the sorghum crops planted in sampled fields. This demonstrates that seemingly difficult problems of detection can sometimes be overcome if the situation requires it. There are many techniques for assaying soil that could be used in disease prognosis; most could be adapted to meet the practical needs of the disease manager.

Soils are not the only source of on-site disease or pathogens that can be detected before the fact. Irrigation water has occasionally been identified as a potential source of initial inoculum; the implications for crop production are similar to those for pathogen-infested soil. Shokes and McCarter (1979) developed an effective method for detecting propagules of pathogenic fungi (*Pythium,*

Fusarium, and *Rhizoctonia* spp.) in irrigation ponds. They employed water sampling methods and equipment that they adapted from those used in general ecological studies of lakes and ponds. Here, water sampling methods from ecology were combined with traditional methods of identifying plant pathogenic fungi to provide a good illustration of the synthesis of effective detection methods from diverse sources.

Often problems of detection that seem incredibly difficult can have simple and effective solutions. Datnoff et al. (1984) were able to detect and estimate numbers of resting spores of *Plasmodiophora brassicae* in bottom sediment samples recovered from ponds that served as local sources of irrigation water. They systematically collected sediment samples, planted cabbage plants in pots containing the sediment samples, grew the plants, and examined them for symptoms of the clubroot disease.

The above examples presented illustrate that disease detection before the fact can often provide valuable aids for disease management. Detection strategies similar to those mentioned are applicable to site selection and evaluation for nurseries, propagation stock facilities, and seedbeds. Thus, disease detection before the fact is feasible for many situations. It is especially useful in disease management because it can help tell the manager of disease what to expect; it allows the manager to assume an offensive approach. Valuable lead time to expected disease situations allows for the planning and careful consideration of options including that of canceling the project. Approaches to detection before the fact are not different in principle from other types of detection. These general considerations will be discussed later.

DETECTION AFTER THE FACT

Detection of disease or pathogens after the fact relies on more easily observable processes of plant disease and pathogenesis and represents a more traditional view of disease detection. The main implication of detection after the fact is that the manager is now in a defensive position; the options are more limited. Disease and pathogen detection after the fact usually involve such activities as repetitive field sampling, scouting, or spore trapping. These are costly in terms of resources, but may detect entities that vary in rather small increments of time such as disease incidence, disease damage, or spores per cubic meter of air.

Detection of Inoculum

Detection of initial inoculum that arrives in an area of interest offers great potential for disease management. Responses to inoculum detection may vary from continued observation or monitoring to initiation of control measures such as fungicide sprays. However, few applications of initial inoculum detection have been utilized for disease management; the primary goal has been to elucidate patterns of pathogen dispersal.

Progress in the detection of inoculum of cereal rusts provides well-documented examples of the development of inoculum detection methods and their potential. Spore detection devices such as sticky-coated slides and rods or spore traps have frequently been used to study the long-distance dissemination of cereal rust. Occasionally, additional goals of these studies included the more efficient scheduling of fungicide sprays (Asai, 1960; Rowell and Romig, 1966). It was soon determined that rust spores could be blown long distances and serve as the initial inoculum in areas where the cereal rusts apparently did not overwinter. With spore traps, it was sometimes possible to detect uredospores well before the disease was detected in local fields (Asai, 1960).

However, spore detection on coated slides, rods, and spore traps was often ineffective and did not always precede detection of disease in the field. Realizing the need for more effective methods, Rowell and Romig (1966) were able to detect uredospores of wheat rusts in rainwater samples. These spores were apparently washed from the atmosphere during spring rains. Rowell and Romig concluded that these spores provided the initial inoculum to infect spring wheat in their area of Minnesota. Later, an efficient device was developed that collected and sampled rainwater which could be examined for rust spores, thus solving some, but not all, of the earlier sampling problems (Roelfs et al., 1970). Although methods such as these are not widely used for disease management purposes, the evolution and development of initial inoculum detection in cereal rusts serves as a valuable model for future work that might be applied to other disease situations where such an approach is indicated.

Sensitivity of detection is always a problem when trying to detect entities such as airborne rust spores. Greater sensitivity can sometimes be provided by bioassay. Wallin and Loonin (1971) adapted methods commonly utilized in entomology to detect early incidence of the barley yellow dwarf virus and its vector, the cereal aphid. They related aphid catches with disease incidence (following an appropriate latent period) while studying small plots of barley planted as a trap crop. Systematic inspection of trap crops and greenhouse assay of trapped aphids demonstrated that aphids were transporting the primary inoculum which resulted in barley yellow dwarf epidemics in Iowa. Similar methods are commonly used in several aspects of entomology, especially in vector studies (Southwood, 1978).

The use of host trap plants in plant disease detection seems potentially great. Young et al. (1978) pointed out some uses of trap crops for plant disease research. Observation plots of cereals are planted for diverse purposes in many parts of the world. Apparently, many of these plots are also used to detect the arrival of cereal disease inoculum in the region. With proper timing, trap crops configured as field plots or perhaps as more versatile and highly movable units such as potted host plants could serve as sensitive and specific detectors of initial inoculum and offer great potential in disease management research. I have often used potted plants placed in pathogen-infested fields at various intervals after the potted plants were sprayed with fungicides to estimate useful fungicide lifetimes. In this case, the event detected was the threshold of protection from the fungicide.

Spore trapping can detect periodic changes in spore abundance or peak levels of spore dissemination as well as the arrival of inoculum. It has often been employed in studies to improve our knowledge of the biology of plant pathogenic fungi. Only occasionally have these methods been routinely employed in disease management. This may be due to the intensive requirements of time, labor, and resources that spore trapping requires. It is seldom undertaken by small growers. Often, intensive spore trapping studies have been initially employed to relate spore production and dispersal to environmental factors or to verify control strategies; later, use of fungicide sprays may be based on observations of weather or other factors alone. The assumption is made that if disease or pathogen is present, favorable weather will result in more spores and more disease and that fungicides are needed.

Apple scab and late blight of potatoes are diseases whose control and management are aided by this concept. The apple scab fungus, *Venturia inaequalis,* is assumed to be present and that following favorable weather, inoculum will be plentiful and effective in inciting new disease. Jones et al. (1980) have described an electronic instrument that evaluates and detects these favorable weather periods. Fungicide applications may be initiated or intensified on the basis of this information. The BLITECAST prediction program provides a similar function for potato late blight (MacKenzie, 1981).

Berger (1969) described how spore trapping could be used in celery and sweet corn production to help schedule fungicide sprays to control *Cercospora apii* and *Helminthosporium turcicum,* respectively. The events that Berger detected were peaks of major spore occurrence; these were used to help schedule the initiation and frequency of fungicide applications. These procedures are still used to some extent by larger growers and scouting services in Florida (author) to schedule fungicide sprays and provide valuable feedback data on the efficacy of the control program as well. For those who can afford the time and expense, such programs are helpful and provide great educational value for disease management personnel concerning disease progress and effects of weather on disease. More recently, models that consider weather and other variables to accurately predict sporulation are replacing spore trapping.

Numerous published reports describe research that employs spore trapping and related activities; most reports list better disease control as a primary goal. I did a small and informal survey that indicated that relatively few of these methods are being routinely applied to aid in disease control or disease management. While it is difficult to determine what practical use is being made of the successful spore detection methods that have been published, it is appropriate to assume that they are of potential use at some level of disease control or management or in the development of disease management methods, and that there are opportunities to apply them more extensively. Most spore trapping methods developed for research purposes are applicable to disease management if needed.

Detection of Disease Incidence and Severity

Often, opportunities for detection are limited to observing disease or disease damage after it has occurred. Usually there is no better alternative than simply going out and looking at the crop to detect disease; however, it can be done systematically. The most common detector available is the human eye (remote sensing and other devices are beginning to challenge this assumption). Although the human eye can be a very efficient detector of disease (Horsfall and Cowling, 1978), it too must be used in a systematic way.

Once it is decided what to look for, sampling and scouting methods become major considerations. It might be thought that detection and sampling methods for disease management or survey purposes are less quantitative or less intensive than those for research purposes. This is not necessarily true. Good scouting and sampling methods for detection or for any purpose are based on intensive research and experimentation and much thought. The methods employed for disease management should obtain the minimum desired information that is commensurate with its equivalent value in time and effort expended. Thus, it is appropriate to consider detection and sampling methods in view of the goals of a project. The methods must have biological significance in terms of the pathogen, the host, and the interactions between them that produce diseases.

Knowing what to look for is very important. The detection of disease and pathogens by visible evidence is somewhat basic but sometimes may be improved upon. For example, it is always desirable to be able to detect disease from the earliest possible external signs or symptoms. Occasionally, a detailed study of the symptomatology of the disease can aid or improve the detection process. In one example, Lundquist and Luttrel (1982) investigated the use of the early and abnormal discoloration of slash pine seedlings as an early indicator of fusiform rust infection. These early signs of disease were found useful in the early detection of diseased pine seedlings. If the goals of the project warrant it, one need not wait for identification of disease or pathogen. Often a strong indication is enough.

A good knowledge of symptomatology can be essential to the success of any project. For example, if detection of the pathogen is a primary goal and this requires isolation or tedious assay, as in virus detection, it is often helpful to expend these limited efforts on samples most likely to contain the pathogen. This would not be a valid approach to estimating the true population mean of infested or infected plants. Timmer and Garnsey (1979) found that the reliability of detection of the citrus ringspot and psorosis viruses was highly dependent on the expression of virus symptoms in portions of budwood source trees which were to be indexed for these viruses. The distribution of virus in citrus trees was irregular; the viruses were usually confined to parts of trees already suspected as virus-infected because of symptoms. Clearly, symptoms are important; knowledge and consideration of symptom expression can increase the sensitivity, precision, and productivity of the detection process.

Detection at Low Levels   The process of detecting disease in the field often involves systematic searching for and observation of symptoms of disease or disease damage at very low levels. Detection at low levels requires proportionally more work and sampling effort, but the information gained is usually justified. Often something may be done about disease when levels are still low.

Detection of low levels of disease in the field by inspection is both feasible and effective. Cereal rust research provides some good examples of the effectiveness of on-site disease detection at low levels by field scouting. In one wheat rust epidemic studied by Rowell and Roelfs (1976), the initial disease density was first detected by field inspection and estimated to be approximately one uredium per 30 m of row or five uredia per 10,000 culms. Low levels of discase such as this are hard to detect. Rowell and Roelfs (1976) stated that their best observers could reliably detect a lower disease density of about one uredium per 10 m of row. Spore traps were also employed but were judged to be less effective than field inspection for detecting initial disease. In the experience of Rowell and Roelfs, uredospores detected in spore traps could generally be attributed to nearby disease which, in turn, could usually be found after a careful search of the area. They felt spore traps were no better than direct search for detecting stem rust. They did not consider the relative cost of labor to inspect fields or to check spore traps.

Spatial Distribution of Disease

Coincident with what to look for is where to look for it. Once again, the approach to detection or observation must have biological significance. Thus, the unit sampled or examined might be a particular plant part where disease is likely to occur first as an economically important but vulnerable plant part, but it could just as well be a region of a forest or portion of a field where disease is thought most likely to be found. Sometimes it is not known where disease is likely to be found; the goal is to see if it is there or not. Moreover, approaches to detect only the presence or absence of disease are likely to be somewhat different than those designed to closely estimate levels of disease. Detection for research or survey purposes may emphasize the distribution and abundance of disease. The point is that all of these considerations reflect the specialized applications or goals of the project. The following examples illustrate some of these points.

Jackson (1981) surveyed peanut fields in Florida for the incidence and severity of peanut leafspot (*Cercospora arachidicola* and *Cercosporidium personatum*). He attempted to relate the incidence and spatial distribution of these diseases which he observed to broad levels of disease control which had been or were being practiced in these fields, including intensive fungicide programs and crop rotation. He evaluated rather small numbers of leaf samples collected systematically from 35 peanut fields. Few effects due to disease control were found. Jackson did find three nonrandom and one completely random distribution pattern of disease that did not seem related to disease incidence. Areas with apparently different distribution patterns existed within the same fields. In this survey, basic information on the distribution and abundance of these diseases over a wide geographic area as well as within individual fields was obtained. The methods Jackson employed reflect these goals. They were not intensive but were designed to survey a large number of fields for the effects he sought.

In contrast, consider an intensive study centered on a few experimental plot areas. Rouse et al. (1981) sampled chosen plot areas at weekly intervals to detect changes in disease incidence and severity of powdery mildew of wheat (*Erysiphe graminis* f. sp. *tritici*). Although a primary goal of this study was to relate disease incidence to severity, their sampling method and data analysis provided useful information on the spatial distribution of diseased plants within fields and on the density and distribution of powdery mildew on individual plant parts over time. Plants were selected at random from replicated plot areas and examined for disease damage on tillers and leaves each week. Numbers of lesions per leaf and per tiller as well as infected leaves and tillers per quadrat were determined and analyzed. The distribution of disease damage on leaves was highly aggregated for the first 4 weeks, but the degree of aggregation decreased and was best described as randomly distributed after 6 weeks. Information on the distribution of disease in time and space was, in turn, used to suggest more effective and efficient sampling procedures. Changes in spatial distribution of disease with time as observed by Rouse et al. (1981) present serious and challenging problems as well as opportunities for epidemiological sampling and for other applications of sampling as well. These opportunities and problems are not restricted to foliar diseases.

The epidemiology of hypocotyl rot of snap bean caused by *Rhizoctonia solani* was extensively investigated by Campbell and Pennypacker (1980). They determined that the spatial distribution of infected bean plants within test fields was random. This could be explained, in part, by the biology of this common soilborne pathogen and by the cultural methods used for bean production which tend to incorporate and disperse infected bean plant residue within field soils. However, the distribution of lesions on individual bean plants was clustered. Thus, at one level of consideration (the field) the hypocotyl rot pathogen may be considered to be randomly distributed; at another level (the plant) it is aggregated. It depends on your viewpoint, but it greatly affects sampling and data interpretation as was pointed out by Campbell and Pennypacker. In a recent evaluation of techniques for assaying potato field soils for *Verticillium dahliae,* Smith and Rowe (1984) found that propagule distribution was nonrandom in only one heavily infested field out of several tested. However, this field was apparently more intensively sampled than other fields. The observed spatial distribution could be compensated for by a revised sampling method.

Studies such as these demonstrate again that pathogens and disease must be dealt with as they occur in the field. Such studies help us to understand how pathogens and disease do occur in the field. Although facts concerning spatial distribution such as those illustrated above may have great biological significance and can be partially explained in that context, they also may have great significance in determining the best approach to detection of the disease or pathogen. It is likely that spatial distributions, once known, can be used to advantage.

The spatial distribution of cabbage black rot, a common bacterial disease of cole crops, incited by *Xanthomonas campestris* pv. *campestris* was studied by Strandberg (1973). The distribution of infected cabbage plants within fields and of lesions per plant was highly aggregated and reflected the way in which the disease spreads. Negative binomial distribution best described the way in which both lesions and infected plants were distributed in the field. A sampling method was developed using methods commonly employed in insect and ecological sampling. The method, known as *sequential sampling,* considers the observed degree of aggregation of the sampled units that is most often encountered to develop sampling guidelines or aids that allow the investigator to stop sampling when a desired precision has been reached. This was done by plotting cumulative counts as they were obtained and numbers of samples taken on a chart or reading from a table. When a certain proportion of counts were obtained for a given number of samples, the chart or aid indicated that the population mean had been estimated with the desired precision and enough samples had been taken. Rouse et al. (1981) developed similar sampling programs for powdery mildew on barley.

The above examples illustrate only a few of the interrelationships of abundance and distribution of diseases and pathogens to the epidemiology and biology of the pathogen. However, they demonstrate some important reasons why detection methods must consider the biology of the pathogen and host and provide some examples of how this knowledge can be used.

Seasonal and Temporal Distribution of Disease

The seasonal or temporal distribution of pathogens and disease is no less important in detection than is spatial distribution. Much of this information may be obtained from experience, previous observation, and biological knowledge of the pathogen or disease, but sometimes it must be obtained through additional research. Weather coupled with host phenology may often be the primary factor that affects when and how detection may be accomplished. For example, Ramos and Kamidi (1981) were able to relate the seasonal periodicity and distribution of bacterial blight of coffee to weather and rainfall patterns that affected both host and pathogen. Rainfall stimulated coffee trees to produce new foliage and flowers (which are very susceptible to infection when they are newly produced), and this response occurred during weather that provided a very favorable environment for the pathogen. In this situation, rainfall patterns greatly affected the distribution and incidence of bacterial blight of coffee and greatly determined the time when surveys of disease incidence could best be accomplished. Thus, events that favor dispersal and disease progress may often influence the scheduling or approach to detection efforts.

As can be seen, the goals and approach to disease detection after the fact also vary according to the biology of pathogen and host and to the purpose of the project. It is important to note that an approach to detection or observation for disease management purposes would not differ in principle from examples presented thus far. Detection methods and procedures for management must be based on results of research efforts in detection and sampling and should be adapted from them.

Estimating Disease Severity

Although it has been emphasized that detection methods must have biological significance for disease management, detection methods must also have significance in terms of what is needed by the manager; what is detected must be useful. The concept of disease incidence is one that has received much attention. Disease incidence is convenient to detect. The question remains, however, what is the significance of the measurement? Several studies have demonstrated that disease incidence can sometimes be related to yield loss or expected damage at harvest in a way that is useful for disease management. Disease incidence may also be related to disease severity. For some applications, such as estimating disease damage to a vegetable or fruit crop, information on the severity of disease may be far more important than information on disease incidence. The specific goal of detection may be to estimate one or the other, or both, and the procedures used will reflect this goal. While this type of information seems potentially useful to growers and managers, the goal of estimating disease severity may also place new constraints and requirements on the detection activity.

Seem and Gilpatrick (1980) studied powdery mildew (*Podosphaera leucotricha*) on apple trees and developed a sampling method to detect the incidence of powdery mildew on apple tree leaves. They examined the relationship between the proportion of leaves infected on each tree (incidence) to the number of lesions and amount of leaf tissue damaged per tree (severity). To do this, they collected 15 to 25 vegetative terminal portions of branches from each tree at three locations and estimated both disease incidence and severity. A simple and useful relationship between disease incidence and severity was found, which was fairly constant for different orchard locations, but varied from season to season; the square root of severity values was closely proportional to incidence. There were also adjustments that had to be made for quantitative severity assessment in different years and for different cultivars. This example demonstrates that very similar sampling and scouting methods can be used to estimate either disease incidence and severity. Often, one may be closely enough related to the other so that one can be calculated from the other.

If estimating severity is a major goal, the sampling and scouting plan may be designed or altered to meet it. James and Shih (1973) conducted an extensive survey of winter wheat fields over a wide geographic area for the incidence of powdery mildew and leaf rust; their goals were also to relate incidence to severity. They were interested in crop losses. To sample fields, they systematically collected wheat culms and later evaluated individual leaves taken from these culms for disease damage. The uppermost leaves were the ones they evaluated because these leaves contributed most to producing dry weight increases of the grain. It is interesting to note that in May, the top four leaves were evaluated; in June, only the top two leaves were evaluated. James and Shih designed and then altered their sampling method because it made sense with what they were trying to achieve. They found a useful linear relationship between disease severity and incidence values below 65%. Sampling procedures must often be changed or modified to meet project goals, especially during development or validation stages. There is no reason why sampling procedures must remain inflexible; they can be changed by need or by design. Flexibility is often a valuable attribute of sampling and scouting procedures.

Teng (1983) has discussed the general subject of estimating and interpreting disease intensity (incidence) and crop loss from the viewpoint of crop loss information and related these goals to the process of gathering and analysis of data. Many of the considerations he mentioned could apply equally well to the development of methods for detection of disease levels for management purposes. These considerations are strikingly similar. Studies of crop loss estimation can provide many methods useful for detection. An extensive discussion of crop loss determination by Main (1977) provides many examples of methods that rely on disease detection to provide information that can be used to determine the impact of disease. The economic threshold concept is one important example. If a level of disease incidence or severity can be established beyond which economic loss is likely, it is useful to be able to detect this level. The manager of disease may or may not be interested in levels of disease greater or less than this value. Sampling and scouting procedures may be designed only to tell if this level is exceeded.

Preceding examples have emphasized the detection of disease, the determination of disease incidence and severity, and often a quantitative estimate of disease or pathogen density; the approaches have reflected these goals. In other situations, the primary goal may be the establishment of the fact of disease or pathogen in a crop where only a yes/no answer is sought. A common situation for which a simple yes/no answer may be adequate is to initiate disease control measures. This concept is probably employed to some degree in many programs both formally and informally; to begin to apply fungicide sprays when disease is detected is a common response to disease detection and probably needs little further explanation here. However, it is often necessary to know if disease is present or not with some assurance and some quantitative estimate of how much disease is always helpful.

Prusky et al. (1983) employed the systematic and preharvest assessment of mango fruit to detect levels of latent infections of black spot caused by *Alternaria alternata*. They concluded that levels of latent lesions detected on inspected fruit could be used to predict the incidence of fruit damage likely to occur during postharvest storage. Prusky et al. (1983) suggested that these levels could also be used as a basis for the application of postharvest fungicide treatments to harvested fruit when critical disease incidence levels detected on the sampled fruit were exceeded. Such an approach would require a very high confidence that latent disease was properly assessed since little could be done once the fruit was in storage. A sampling and detection plan such as this one would have little room for error, but the idea is a good one; the problem is to come up with a detection method that is adequate. Note that now the problem is at least reduced to a manageable level. The research problem is to verify the utility and accuracy of the detection method. This seems feasible and illustrates a point that I have tried to make. Once the intended application is known, the problem becomes the development of a reliable and acceptable sampling or detection plan. Although the basic methods of detection are likely to be much the same as those already discussed, the efficiency in terms of effort expended and the sensitivity of detection methods to produce confident answers are likely to be primary considerations.

Seem et al. (1985) set out to develop a sampling method designed to detect the grape downy mildew disease (caused by *Peronospora viticola*) at a very low incidence (often as low as 0.01% incidence) with a confidence of 99%. Their approach was based on the assumption that once detected, disease could be controlled by fungicide applications that could otherwise be delayed until needed. When lack of disease threatened the success of their project, they simply resorted to a model system that in the final outcome may have provided more information than the original project intended. To accomplish this they simulated closely known levels of disease damage on grape vines by affixing either opaque yellow, lesion-sized plastic disks, which were easily visible to scouts, or light yellow, transparent disks, which simulated the early stages of downy mildew lesions known as the "oil spot stage" and which were not very visible. The disks were placed at different but known densities on replicated units of several grape cultivars. The disks were also placed at different parts on the grape vines. In this way, the effects of different cultivar foliage types and growth habits were also examined.

The ability of scouts to detect this simulated disease was extensively investigated as were the effects of scout experience, cultivar habit, and other factors. Results were highly encouraging and indicated that the approach would probably be adequate for detecting the disease itself. An innovative modeling approach like the one used by Seem et al. is certain to be applicable to a wide variety of similar problems. Although unlikely to become so, such an exercise would be a good first step for many projects that seek to develop detection methods because of the early and valuable insight provided.

It should now be apparent from the examples presented that detection methods for both before and after the fact of disease are highly diverse in their approach and that this is mainly due to the diversity in pathogens, disease, and to the goals and resources of the project. While it has not been possible to point to many of these examples of detection and state that they are being routinely used for disease management, research precedes application, and there is no reason why these methods could not be applied to serve the needs of future disease management programs or to assist in the development of methods that can be applied. Time will tell.

SOME GENERAL CONSIDERATIONS FOR DETECTING DISEASES AND PATHOGENS

A thorough knowledge of the biology of host and pathogen and a consideration of epidemiological principles are needed to manage disease (Fry, 1982). These components are also necessary to detect diseases and pathogens with confidence and efficiency and to decide which life stages or processes of the pathogen are feasible for detection. A pathogen or disease must be identified or recognized to be detected. Any practical method that meets the constraints and criteria of the project can probably be used. As was mentioned, symptoms are also important because they are often the basis for detection. Knowledge of symptom expression will increase the sensitivity and efficiency of the procedure. The sooner and more dependably a disease can be recognized, the earlier it can be detected with confidence. Some of the examples presented have demonstrated the importance of symptoms in disease and pathogen detection. Coincident with what symptoms to look for is where to look for them on the plant or sampled unit. Like the spatial distributions of pathogens and disease in the ecosystem, diseased tissue may or may not be randomly distributed on affected plants. Age of plant parts, susceptibility of tissues, and other factors determine where on the plant disease or pathogens are most likely to be found. Where and when to look are major considerations in detection. Fortunately, there is usually abundant literature that can help in choosing the correct approach to a detection or sampling problem.

Abundance and Distribution of Pathogens and Disease

A knowledge of the temporal and spatial distribution of disease and pathogens can be useful in developing sampling and detection procedures as well as in descriptive studies of pathogen biology. Distribution and abundance of pathogens in time and space are a result of life processes such as reproduction or dispersal, but can sometimes be explained by other factors such as responses to mortality or unfavorable environment (Andrewartha and Birch, 1954; Bliss and Fisher, 1953; Krebs, 1972).

One of the greatest problems in sampling and detection is that pathogens and diseases must be dealt with as they occur in the field. Why pathogens and disease are abundant in one place and rare in another or numerous one year and scarce in another is not always known, but is a fact that must be considered.

A common misconception is the belief that because agroecosystems and especially the crop-host environment are relatively uniform, disease should also be uniform. Environments are seldom as uniform as they appear to be. Plant epidemiologists always consider increases in populations of pathogens, but seldom consider what happens to them. Pathogens may increase under favorable conditions, but must remain the same or decrease under unfavorable conditions. Chance may greatly determine the distribution or persistence of initial inoculum but the environment determines the outcome. Where one colony or individual is successful, or perhaps survives because of a favorable environment, the probability is high that another will be successful also. Disease or pathogens in favorable environments will create population centers where pathogen density will be high, and may result in higher disease densities. These centers are seldom randomly distributed.

However, caution must be observed in interpreting what is observed. While spatial distribution may be used to describe observations or to reinforce hypotheses which account for these observations, it is probably not valid to directly infer specific processes from observed distributions alone (Skellam, 1952). Nevertheless, spatial distribution is important and should be considered in developing sampling and detection methods. Many common statistical methods assume a normal distribution of sampled units, but this may not always be true as we have seen. For in-depth discussions of this topic see Waters (1959), Pielou (1969), and Southwood (1978). Transformation of acquired data for statistical analysis may be one part of the solution, but it does not solve the practical problem of how and when to sample. Determining how sampling will be done is a very difficult part of the project.

Sampling Procedures

The study of spatial distribution and abundance of plants and animals can be, in itself, an exciting and rewarding area of study. Good field plant pathologists consider spatial distribution more than they may be consciously aware of. The pattern of disease is important and becomes a component in identifying the disease and its source and even if it is a plant disease or not.

Spatial distribution can also be employed to advantage in the development of special sampling methods such as sequential sampling, which are more accurate or efficient in terms of time and effort expended (Southwood, 1978). Sequential sampling was originally developed to aid quality control in the manufacture of strategic materials. Basically, the number of samples taken is variable. While examining each sampling unit in sequence, the cumulative results for the sampling effort are continually examined until a previously prepared sampling aid such as a table of values or a nomograph indicates that a confident decision about the population being sampled mean can be made. Such techniques depend on a prior and extensive knowledge of how the items being sampled are likely to be distributed within the sampling units chosen. It is the latter that takes much time and effort to develop, but the resulting sampling plan can be very time saving and cost effective.

To develop a sampling plan, several sets of counts are taken, preferably over a period of time and in environmental conditions typical of those to be encountered during the sampling program, and fitted to one of several possible distributions. For a family of counts fitting the negative binomial distribution, which is commonly the case (other cases are possible), the values of an aggregation parameter k are pooled to estimate a "common k" which, in turn, is used to develop the sampling aid referred to above (see Southwood, 1978, for a detailed explanation). Strandberg (1973) gives an example applied to plant disease.

Generating a sequential or similar sampling procedure requires much effort and data. It is probably justified only when the sampling procedure is routine and will be frequently carried out. Sampling plans for insects have received much attention and are widely used; those for plant disease are not common. Sequential sampling plans do not have to be based on aggregated distributions, and there are other approaches to determining sample size.

Lin et al. (1979) investigated some aspects of general sampling in aggregated disease distributions. Their conclusions are interesting and helpful in designing new projects that will rely on field sampling. The literature of general ecology, plant ecology, and entomology are particularly rich in sampling and detection methods, many of which are highly transportable to plant disease epidemiology (Southwood, 1978; Krebs, 1972; Pilou, 1969). The principles on which these ecological sampling and detection studies are based can also apply to plant pathogens and diseases as well. Many processes, methods, and concepts found in these sources are easily adaptable to disease or pathogen detection; occasionally, analogous processes are suggested.

There are two problems that seem particularly troublesome in adapting ecological methods of sampling and detection to disease and pathogen detection. The first is mean population density of pathogens, which is often very difficult to estimate with confidence. Although the population density of soilborne pathogens can sometimes be estimated with some confidence by direct sampling of soil, population estimates of pathogens found aboveground may be more difficult because pathogen density is ever-changing in time and space, and the time frame may be very short. A generally accepted approach in plant disease epidemiology has been to estimate units or proportions of host tissue damaged or colonized by pathogens. It is probably appropriate to assume that plant disease in its many forms can represent population centers of pathogens that exist currently or did exist in the recent past. These centers of high population density are indicative of once favorable habitat and of the capabilities and limitations of the pathogen. Their distribution and abundance are also characteristic in some way and provide useful knowledge.

The second problem is the differences between the way plant epidemiologists have viewed population dynamics and the traditional concepts of demography such as reproduction, mortality, and fecundity. These concepts may be difficult to apply directly to most plant pathogens, but it may be very useful to do so because of the extensive general theory and knowledge surrounding these concepts as they apply to other organisms. Perhaps traditional methods of plant growth analysis or the concepts of pathogen biomass will suffice here. There are great opportunities for significant and challenging advances in this area.

Sampling frequency is also important. Often a convenient interval is chosen which is a trade-off between time, effort, and knowledge gained. Usually, this is adequate for the purpose; however, samples taken at more closely spaced intervals will more accurately detect time-varying processes or events. Sampling theorems from electronics and information theory suggest that it is possible to periodically sample time-varying processes without losing very much information. This should apply to disease and pathogen life processes as well, but the approach cannot be hit or miss. Sampling intervals should be rigorously chosen.

Additional considerations that may apply to specific projects designed for detection include statistical confidence in the methods and results. It is helpful and perhaps essential that statistical expertise be available in planning the project. Statistical methods not now commonly applied to disease and pathogen detection will require evaluation and testing, but there are great opportunities for their use. Pfender et al. (1981) used a most-probable-number method, commonly used in bacteriology, to improve the quantitative detection of the effective inoculum density of *Aphanomyces euteiches* in pea field soils. The method involves testing aliquots of soil for presence or absence of the pathogen. Using the proportion of the two possible outcomes, it is possible to estimate the numbers present. Pfender et al. (1981) combined this method with a host bioassay to estimate the most probable number of effective units of the pathogen, a quantity they called infective inoculum density. They used the most-probable-number method because it seemed highly appropriate for their problem. New statistical methods like this one will probably find increasing application in problems of this type.

Applications of detection for disease management will be heavy users of time and labor and the methods must be efficient and dependable yet statistically adequate for the task. Provisions for the adaption of scaled-down versions for routine field use should be built into the original project. This can usually be done with little increase in time and effort. Testing and validation of the accuracy and sensitivity of the method may take more time and effort.

It has been difficult to enumerate general considerations for sampling and scouting for plant disease or pathogens. Examples were presented to demonstrate that the goals of the project and the biology of the pathogen greatly influence the methods employed. This is logical and helps explain the great diversity of sampling, scouting, and survey methods that are likely to be encountered in biology. To begin a new project, a thorough evaluation of the problem and establishment of specific, realistic objectives will help to identify what must be done. Possible approaches to the problem as well as ideas for new approaches can then be sought. Good approaches are frequently found by examining approaches and solutions to similar problems that have already been solved. These need not be in plant pathology. Numerous methods for sampling, scouting, and observation are commonly employed in entomology, ecology, and other fields and are adaptable to the study of plant disease (see Southwood, 1978). The main constraint to the application of feasible methods is that they must have biological significance in terms of the host-pathogen system being studied. They can usually be modified to meet these constraints.

Sampling requirements such as frequency, intensity, sampling unit, and quadrant size must be scaled in relation to the life system of host and pathogen. Recall the studies of Campbell and Pennypacker (1980). Hypocotyl rot of snap bean was uniformly distributed in the ecosystem (presumably by cultivation of the field soil), yet the distribution of propagules of *Rhizoctonia solani* was aggregated, as evidenced by the number of hypocotyl lesions per unit area sampled. Thus, assessment of spatial distribution of disease and pathogens, like many things, depends on one's point of view. *Rhizoctonia solani* propagules (as indicated by lesions per infected plant) were clustered when a space closer to the size of the immediate life system of a fungal colony was considered.

Frequently, much preliminary work is required to resolve problems of sampling and scouting. A reasonable approach should consider available knowledge and project goals; then the best possible program should be designed and then tested and verified. Such a program is unlikely to be adequate in all respects, but it may be revised and refined as needed to meet project goals. Precision and accuracy should be sought, but unrealistic precision and goals must be avoided; some measurement, however imperfect, is better than no measurement.

Practical applications should be considered from the outset. Research standards necessitate the careful testing and validation of any proposed sampling or scouting procedure, but the researcher should not lose sight of the practical application of the procedure. It will not be in a research oriented environment. Provisions must be made to determine the minimum effort that will meet the requirements of disease management, and this level of intensity should be tested and validated as well. Complicated and labor-intensive procedures are seldom adopted at an applied level and the procedures that are adopted are usually greatly modified in the process. Thus, it may be pointless to strive for rigid protocols; flexible ones that allow options for two or more levels of scouting intensity are more likely to be adopted or adapted to disease management programs.

An informal survey that I made indicated that few formal sampling and scouting procedures, including most of those mentioned in this discussion have been extensively or routinely employed in disease management. In my personal experience, I have found that results of comprehensive sampling and scouting research projects are of great educational value to disease managers and their staff and often establish basic solutions to sampling and scouting problems. However, these are adapted and modified at various levels of intensity by scouting services or those who manage disease often without verification of the accuracy or precision of the modified procedures. This is seldom a calamity. The disease manager may not expect a high confidence level because so much of what must be done is inherently uncertain. A sampling and scouting procedure used by the disease manager must also be compatible with similar procedures for other pests; commonly, insect pests take precedence.

In Florida celery production, some scouting programs routinely sample leaves for damage from *Cercospora apii*. Although it has been demonstrated that the incidence of leaf blight caused by *C. apii* could be more accurately estimated by sampling celery leaves along a random path through celery fields, the practical application of the method usually requires the evaluation of the same leaves that are collected to monitor damage caused by the vegetable leaf miner *Liriomyza trifolii* which has consistently caused a much larger economic impact than *Cercospora* leaf blight. These leaves are typically collected from sampling stations which are visited a few times each week and where plant growth and petiole damage by *Rhizoctonia* is also estimated. Undeniably a secondary goal of the scouting program, the information gained about *C. apii* is still useful as feedback information for the fungicide control program; sudden increases in *C. apii* damage may indicate that a modification is called for. Even if these methods seem to lack the formal validity of the original sampling method, which was worked out in detail for monitoring *C. apii,* the information is valuable. It is a significant improvement over no information at all. The point is that practical application is likely to be very different than originally anticipated. Clearly, a comprehensive study is still desirable because it establishes what can be done if the need arises, but it may also be useful to identify what minimum effort can be useful.

CONCLUSIONS

Methods and applications of disease and pathogen detection are both diverse and abundant, but so are the opportunities. Future applications in disease management might focus on a more extensive utilization and application of this technology. Although the processes of detection and the information generated may be applicable at the field level to assist in managing disease, they must be in a form usable by extension workers and disease managers. This will not be easy for there is great contrast between the resources and capabilities of the researcher and those of the disease manager. Each application will have to be fabricated in view of what is needed and what is known. This will identify new opportunities in research and development and in bridging the gap between the theoretical and applied aspects of disease and pathogen detection.

This chapter has been directed toward students of plant disease epidemiology and disease management. It may help them to do what they should have done by thinking about it in the first place. Thus, facts have been few and the number of concepts presented has been large. Examples were presented to show that the information and methodology exists. It can be done. Those contemplating work or study in this area are advised to carefully identify the problem of interest. Having done so, a set of values will be generated. With this set of values in mind, the diverse sources of potential ideas can be profitably read and a suitable method can more easily be found.

Innovative research results that are potentially useful to disease detection and management are abundant in the literature, and they are appearing with increasing frequency. The challenge seems to be to adapt what is known and to find out what needs to be known in order to provide new and useful applications for use in the field. More work is needed, but disease and pathogen detection are supported by a broad experience in plant pathology; and there are principles and ideas to guide future research, even if they are sometimes from disciplines other than our own. Disease and pathogen detection, like other concepts of plant pathology, is one of the basic strategies of disease management. Its successful implementation into present disease management programs is likely to increase future opportunities and expectations.

REFERENCES

Adams, P. B. 1981. Forecasting onion white rot disease. *Phytopathology* 71:1178–1181.

Andrewartha, H. G., and Birch, L. C. 1954. *The Distribution and Abundance of Animals,* reprinted 1974. University of Chicago Press, Chicago.

Asai, G. N. 1960. Intra- and inter-regional movement of uredospores of black stem rust in the Upper Mississippi River Valley. *Phytopathology* 50:535–541.

Baker, R. 1978. Inoculum potential, in: *Plant Disease: An Advanced Treatise,* vol. 2 (J. G. Horsfall and E. B. Cowling, eds.), pp. 137–157. Academic Press, New York.

Berger, R. D. 1969. A celery early blight spray program based on disease forecasting. *Proc. Florida State Horticultural Soc.* 82:107–111.

Bliss, C. I., and Fisher, R. A. 1953. Fitting the negatively binomial distribution to biological data. *Biometrics* 9:176–200.

Bouhot, D. 1979. Estimation of inoculum density and inoculum potential: Techniques and their value for disease prediction, in: *Soil-borne Plant Pathogens* (B. Schippers and W. Gams, eds.), pp. 21–34. Academic Press, London.

Campbell, C. L., and Pennypacker, S. P. 1980. Distribution of hypocotyl rot caused in snapbean by *Rhizoctonia solani*. *Phytopathology* 70:521–525.

Datnoff, L. E., Lacey, G. H., and Fox, J. A. 1984. Occurrence and populations of *Plasmodiophora brassicae* in sediments of irrigation water sources, *Plant Dis.* 68:200–203.

Fry, W. E. 1982. *Principles of Plant Disease Management*. Academic Press, New York, 378 pp.

Geng, S., Campbell, R. N., Carter, M., and Hills, J. F. 1983. Quality control programs for seedborne pathogens. *Plant Dis.* 67:236–242.

Grogan, R. G. 1980. Control of lettuce mosaic with virus-free seeds. *Plant Dis.* 64:446–449.

Grogan, R. G., Ioannou, N., Schneider, R. W., Sall, M. A., and Kimble, K. A. 1979. Verticillium wilt on resistant tomato cultivars in California: Virulence of isolates from plants and soil and relationship of inoculum density to disease incidence. *Phytopathology* 69:1176–1180.

Haskell, P. T. 1977. Integrated pest control and small farmer crop production in developing countries. (Preface to article.) *Outlook Agr.* 9:121–126.

Horsfall, J. G., and Cowling, E. B. 1978. Pathometry: The measurement of plant disease, in: *Plant Disease: An Advanced Treatise*, vol. 2 (J. G. Horsfall and E. B. Cowling, eds.), pp. 120–136. Academic Press, New York.

Jackson, L. F. 1981. Distribution and severity of peanut leafspot in Florida. *Phytopathology* 71:324–328.

James, W. C., and Shih, C. S. 1973. Relationship between incidence and severity of powdery mildew and leaf rust on winter wheat. *Phytopathology* 63:183–187.

Jones, A. L. 1976. Systems for predicting development of plant pathogens in the apple orchard ecosystem, in: *Modeling for Pest Management Concepts, Techniques, and Application* (R. L. Tummala, D. L. Haynes, and B. A. Croft, eds.), pp. 120–122. Michigan State University Press, East Lansing.

Jones, A. L., Lillevik, S. L., Fisher, P. D., and Stebbins, T. C. 1980. A microcomputer-based instrument to predict primary apple scab infection periods. *Plant Dis.* 64:69–72.

Krebs, C. J. 1972. *Ecology, the Experimental Analysis of Distribution and Abundance*. Harper and Row, New York.

Lin, C. S., Poushinsky, G., and Mauer, M. 1979. An examination of five sampling methods under random and clustered disease distribution using simulation. *Can. J. Plant Sci.* 59:121–130.

Lundquist, J. E., and Luttrell, E. S. 1982. Early symptomatology of fusiform rust on pine seedlings. *Phytopathology* 72:54–57.

Main, C. E. 1977. Crop destruction: The raison d'etre of plant pathology, in: *Plant Disease: An Advanced Treatise*, vol. 1 (J. G. Horsfall and E. B. Cowling, eds.), pp. 55–78. Academic Press, New York.

MacKenzie, D. R. 1981. Scheduling fungicide applications for potato late blight with BLITECAST. *Plant Dis.* 65:394–399.

McGee, D. C. 1981. Seed pathology: Its place in modern seed production. *Plant Dis.* 65:638–642.

Neergaard, P. 1977. *Seed Pathology*. The MacMillan Press, Ltd., London.

Pfender, W. F., Rouse, D. I., and Hagedorn, D. J. 1981. A "most probable number" method for estimating inoculum density of *Aphanomyces euteiches* in naturally infested soil. *Phytopathology* 71:1169–1172.

Pielou, E. C. 1969. *An Introduction to Mathematical Ecology*. John Wiley & Sons (Interscience), New York.

Pratt, R. G., and Janke, G. D. 1978. Oospores of *Sclerospora sorghi* in soils of south Texas and their relationships to the incidence of downy mildew in grain sorghum. *Phytopathology* 68:1600–1605.

Prusky, D., Fuchs, Y., and Yanko, U. 1983. Assessment of latent infections as a basis for control of postharvest disease of mango. *Plant Dis.* 67:816–818.

Pullman, G. S., and Devay, J. E. 1982. Epidemiology of verticillium wilt of cotton: A relationship between inoculum density and disease progression. *Phytopathology* 72:549–554.

Ramos, A. H., and Kamidi, R. E. 1981. Seasonal periodicity and distribution of bacterial blight of coffee in Kenya. *Plant Dis.* 65:581–584.

Roelfs, A. P., Rowell, J. B., and Romig, R. W. 1970. Sampler for monitoring cereal rust uredospores in rain. *Phytopathology* 60:187–188.

Rouse, D. I., MacKenzie, D. R., Nelson, R. R., and Elliott, V. J. 1981. Distribution of wheat powdery mildew incidence in field plots and relationship to disease severity. *Phytopathology* 71:1015–1020.

Rowell, J. B., and Roelfs, A. P. 1976. Wheat stem rust, in: *Modelling for Pest Management Concepts, Techniques, and Applications* (R. M. Tummala, D. C. Haynes, and B. A. Croft, eds.), pp. 69–79. Michigan State University, East Lansing.

Rowell, J. B., and Romig, A. R. 1966. Detection of uredospores of wheat rusts in spring rains. *Phytopathology* 56:807–811.

Schaad, N. W. 1982. Detection of seedborne bacterial plant pathogens. *Plant Dis.* 66:885–890.

Seem, R. C., and Gilpatrick, J. D. 1980. Incidence and severity relationships of secondary infections of powdery mildew on apple. *Phytopathology* 70:851–854.

Seem, R. C., Magarey, P. A., McCloud, P. I., and Wachtel, M. F. 1985. Sampling for detection of downy mildew of grapes. *Phytopathology* 75:1252–1257.

Shepard, J. F., and Claflin, L. E. 1975. Critical analyses of the principles of seed potato certification. *Ann. Rev. Phytopathol.* 13:271–293.

Sherwood, R. T., and Hagedorn, D. J. 1958. Determining the common root rot potential of pea fields. *Wisc. Agr. Exp. Sta. Bull.* No. 531.

Shokes, F. M., and McCarter, S. M. 1979. Occurrence, dissemination, and survival of plant pathogens in surface irrigation ponds in southern Georgia. *Phytopathology* 69:510–516.

Skellam, J. G. 1952. Studies in statistical ecology: I. Spatial pattern. *Biometrika* 39:346–362.

Smith, V. L., and Rowe, R. C. 1984. Characteristics and distribution of *Verticillium dahliae* in Ohio potato fields and assessment of two assay methods. *Phytopathology* 74:553–556.

Southwood, T. R. E. 1978. *Ecological Methods*, 2d ed. Chapman and Hall, London.

Strandberg, J. O. 1973. Spatial distribution of cabbage black rot and the estimation of disease plant populations, *Phytopathology* 63:998–1003.

Teng, P. S. 1983. Estimating and interpreting disease intensity and loss in commercial fields. *Phytopathology* 73:1587–1590.

Timmer, L. W., and Garnsey, S. M. 1979. Variation in the distribution of citrus ringspot and psorosis viruses within citrus hosts. *Phytopathology* 69:200–203.

Young, H. C., Jr., Prescott, J. M., and Saari, E. E. 1978. Role of disease monitoring in preventing epidemics. *Ann. Rev. Phytopathol.* 16:263–285.

Wallin, J. R., and Loonin, D. V. 1971. Low level jet winds, aphid vectors, local weather, and barley yellow dwarf virus outbreaks. *Phytopathology* 61:1068–1070.

Waters, W. E. 1959. A quantitative measure of aggregation in insects. *J. Econ. Entomol.* 6:1180–1184.

Williams, P. H. 1980. Black rot: A continuing threat to world crucifers. *Plant Dis.* 64:736–742.

Chapter 8

Nematode Population Dynamics and Management

H. Ferris

Nematodes occupy a wide range of habitats in marine, freshwater, and terrestrial environments (Chitwood and Chitwood, 1975; Cobb, 1914). Most research efforts have been applied to those nematodes perceived as having a negative impact on human beings, primarily those parasitic in human beings, animals, and higher plants. However, among nematode genera, parasitic forms are in the minority. The majority of nematodes, whatever environment they occupy, function in decomposer trophic networks. Hence, they are involved in the cycling of nutrients and other resources through the decomposition of animal and plant remains, which is critical to the continuation of life. In this sense, the majority of nematodes can be considered to have a positive impact on the human objectives. However, life-sustaining nematodes have attracted relatively little attention, an unfortunate circumstance since nematodes have proved invaluable models in the study of embryology, development, and the genetic control of cell lineage (Herman and Horvitz, 1980; Sulston et al., 1983; Von Ehrenstein and Schierenberg, 1980). This chapter, as a component of *Plant Disease Epidemiology,* concentrates on nematodes that are parasites of higher plants, in other words, nematodes that function at the primary consumer trophic level. Obviously, there are other parasitic nematode species functioning at secondary and higher consumer levels as parasites of humans and animals.

Plant parasitic nematodes, in the current context, are nematodes that feed on the roots, stems, leaves, or seeds of higher plants. They may function as ectoparasites, remaining physically outside the plant while feeding or endoparasites, moving into the plant tissues during feeding, or they may occupy some intermediate physical location and associated appellation. They may be sedentary in their feeding habit, remaining in one location throughout the feeding cycle, or migratory, retaining the capability of moving to new feeding sites. Both categories convey unique adaptive advantages and are taxed by unique penalties in the sense of community energetics, which are explored elsewhere (Ferris, 1982). The term parasitism implies that at least one of the life stages of the nematode feeds on living plant tissues. The number of life stages that feed varies among genera. Frequently, several life stages exist in the soil or plant tissues (Bird, 1959; Tyler, 1933; Van Gundy, 1958) without feeding. Plant-parasitic nematodes are generally considered obligate parasites. They are functionally and anatomically adapted for parasitism by possession of a protrusible stylet for penetrating and withdrawing the contents of plant cells.

This chapter focuses on the population dynamics and management of nematodes of the genus *Meloidogyne,* the root-knot nematodes, for which we have much information. However, references and comparisons will be made throughout to nematode genera with alternative life history strategies. In review, the life history of the root-knot nematode consists of six definable life stages. The first juvenile stage is passed within the egg and molts to an infective second juvenile stage that emerges from the egg and then locates and penetrates a plant root slightly behind the growing root tip. Within the root system, the second juvenile stage migrates through and between cells to the vascular region and stimulates the production of the large multinucleate physiologically active syncytia as transfer cell-like feeding sites. The nematode feeds and grows before passing through nonfeeding, reorganizational third and fourth juvenile stages, and then emerges as an adult female or male. Adult males are vermiform, nonfeeding, and move out of the root and into the soil. Adult females are sedentary, globose, and adapted to production of large numbers of eggs (Bird, 1959; Tyler, 1933). The majority of root-knot nematode species are mitotically parthenogenic and males are rare except under conditions of population or environmental stress (Triantaphyllou, 1973, 1979). Some genera, usually considered less successful, are miotically amphimictic and produce a larger proportion of males (Triantaphyllou, 1979). This "*k*-strategy" attribute is nonadaptive in managed agroecosystems but would confer advantages in natural systems.

Root-knot nematodes are well adapted for success in agriculture of annual crops. Eggs are deposited in a gelatinous matrix which confers a survival benefit (Wallace, 1966), the majority of eggs hatch concurrent with soil conditions becoming favorable for seed germination. However, a smaller percentage remain in a survival state and hatch at a slower rate (Ferris et al., 1978). In *Meloidogyne naasi* an obligate diapause occurs (Ogunfuwora and Evans, 1976). Root-knot nematodes generally have a wide host range and can parasitize a variety of crops or weeds. They establish a sophisticated physiological relationship with the host, and because of their sedentary nature, transform a high proportion of the assimilated energy into production (Ferris, 1982). Since they are parthenogenic, most adults are females and, therefore, productive. Under population stress conditions the sex ratio may vary, thereby reducing the number of females and decreasing the stress on the plant host (Triantaphyllou, 1973). Nematodes further have survival capabilities relative to adverse environmental conditions in the form of hypobiosis, reduced metabolism as a result of dehydration, anaerobic, and other unfavorable conditions (Evans and Perry, 1976; Freckman, 1978; Freckman and Womersley, 1983).

Other life history strategies of plant parasitic nematodes are somewhat less productive. Ectoparasitic nematodes remain active in the soil, perhaps move after growing root tips and so burn a higher proportion of assimilated energy in the process of feeding. Multiplication rates are consequently somewhat slower. Intermediate cases occur where ectoparasitic nematodes prevent root tip elongation so that the food source is immobile. In general, however, migratory nematodes can be expected to have lower multiplication rates than sedentary nematodes (Ferris, 1982).

REVIEW OF APPLICABLE MODELS

Since population processes of plant parasitic nematodes are a complex function of host status (food source suitability), resource limitations, environmental conditions (temperature, moisture, aeration), and parasitism and predation, it is convenient and productive to explore the biology through the use of models. Models constitute a simplification of reality. They are essentially hypotheses of the functioning of the system that allow study and testing of this understanding of the system. As with other organisms, models descriptive of the population biology of plant parasitic nematodes can be considered to range from wholly empirical to wholly explanatory, with a spectrum of intermediate conditions. They range from critical point models through multiple point to biologically descriptive simulation models. Critical point models attempt to describe population phenomena or plant damage based on a single assessment of the nematode population status and age structure at a critical time. Multiple point models make similar predictions based upon multiple assessments through time of the population structure. Simulation models may be seeded with a single assessment of the population size and structure. They simulate subsequent conditions of the population on a continuous basis based on the environmental conditions experienced by that population. Continued monitoring and assessment of the population during the course of the simulation allows verification of the descriptive or predictive potential of the model. As confidence in the model improves, the effort invested into monitoring can be reduced (Tummala and Haynes, 1977).

Critical point models assume that seasonal multiplication rates and crop damage levels can be predicted from a nematode population measured at one point in time. They are therefore most likely to be successful in nonvolatile biological systems. Since in practical agriculture, particularly in irrigated agriculture, environmental conditions are somewhat manageable for both the crop and the pest, the rates of processes are largely a function of temperature. Temperatures are somewhat dampened by the buffering capacity of the soil environment, and a relatively constant number of heat units is required to grow a crop from one season to the next. The population dynamics of plant-parasitic nematodes are not complicated by either immigration or emigration. Rather the population is relatively immobile in the soil and not subject to self-propelled or mechanically propelled movement over any distance. For more dynamic systems, multiple point approaches become preferable or even necessary. Simulation models generally have greater complexity than the somewhat empirical critical and multiple point models. They are based on the assumption that parameters measured through reductionist research and observations can be integrated in a synthetic model to predict the functioning of a whole system.

Advantages of empirical models are that they are relatively easily constructed from field gathered data using regression techniques, and that they are relatively easily applied for predictive purposes. Simulation models, on the other hand, have the advantage of being more biologically satisfying. They are more transportable in the sense that, when moved from one environment to another, it is possible to make changes in individual parameters descriptive of the environmental shift.

Disadvantages of empirical models are the needs to redefine parameters for different environmental conditions and their lack of adjustment capacity for unexpected variation in weather or environment. Disadvantages of simulation models are their often cumbersome structure, the need for electronic computation, and frequently, the need for a current or historical weather data base.

The Damage Function

The damage function may take several forms. With empirical models it is frequently embodied in the basic structure of the model. That is, the model is descriptive of expected yield or yield loss relative to a nematode population measured at a specific point in time. In simulation models, the damage function represents the hypothesis and functional coupling of a plant and a pest model.

In a management sense, when decisions are made based on the state of the biology of the system at a single point in time (e.g., prior to planting) an empirical model may be valuable for predicting damage. Such a model may also formulate the basis for determining an appropriate sampling intensity for measuring a population, since the appropriate sampling intensity must be a function of the anticipated cash investment in the management procedure (Ferris, 1984). Simulation models, on the other hand, allow greater exploration and understanding of the interactive biology of the plant-pest system. They allow the manipulation of individual parameters in an educational sense to further understanding of system behavior. Studying alternative strategies through simulation models generates an understanding of success probability and risk associated with specific management decisions.

PROGRESSION OF CONCEPTUAL COMPLEXITY

Simple Analytic Models

Classic analytic models of organismal growth are based on the exponential function $N_t = N_0 e^{rt}$. This model indicates that, for a given set of environmental conditions, a population will increase according to some intrinsic rate r, the number of organisms N involved in the increase, and the time period t. The model assumes an unlimited environment, free from competition, pollution, and resource limitation. Limitation can be invoked by the inclusion of a carrying capacity concept. That is, the recognition that the environment has a limited capacity for the organism, either limited availability of resources or limited capacity for disposal of environmental pollution. Such constraints are embodied in the logistic function, which has been adapted for description of nematode population increase (Jones and Kempton, 1978; Seinhorst, 1966; 1967b).

The logistic function implies a fixed amount of food availability for support of the nematode population, but does take into account that the food availability for a parasite population is in a state of dynamic flux. The dynamic nature of the system can be conceptualized by invoking two logistic models. The first describes the rate of nematode growth, that is,

$$\frac{dN}{dt} = \frac{r_1 N (K_1 - N)}{K_1}$$

where r_1 is the growth rate of the nematode population, K_1 is the current carrying capacity of the system for the nematode population, and N is the current size of the nematode population. The second model describes the rate of plant growth:

$$\frac{dP}{dt} = \frac{r_2 P (K_2 - P)}{K_2}$$

where P is the current plant size, r_2 is the growth rate of the plant, and K_2 is the maximum attainable size of the plant. The increase in the nematode population then becomes a function of the current size of the plant, that is,

$$K_1 = P \times C$$

where C is the number of nematodes that can be supported per unit of plant. This model, however, still does not take into account that plant growth may

be impaired by the presence and magnitude of the nematode population. Therefore, the plant growth-limiting term can be modified by the nematode population
growth-limiting term, under the somewhat simplistic assumption that the plant
ceases to grow when the maximum number of nematodes per gram of root is
approached. Hence,

$$\frac{dP}{dt} = r_2P \, ((K_2 - P)/K_2)((K_1 - N)/K_1)$$

Similarly, the carrying capacity for the nematode (K_1) population is adjusted as
the plant grows. The process would involve updating the carrying capacity
parameters of both logistic equations during the growing season at frequent
intervals. Such iterative approaches formulate an appropriate structure for real-
time simulation models in which rates are environmentally determined and calculated for the conditions during each iteration.

Jones and Kempton (1978) embodied notions of host status in the logistic
equation descriptive of nematode population growth. The equilibrium density
(i.e., that population density at which there is just sufficient food available during
the growth of the crop to maintain the population at its current level) is used as
a measure of the seasonal carrying capacity of that environment. Over a period
of several years, all initial population levels stabilize to this density, particularly
in a perennial crop or in monocultured annuals (Jones and Kempton, 1978).
Such phenomena have been exhibited by population studies in microplots with
alfalfa (Noling and Ferris, 1984).

The analytic models discussed consider population growth to be a function
of the number of organisms and the resource limitation of the environment, but
do not account for environment effects on the reproductive potential of the
population. Since the systems under consideration are poikilothermic, metabolic
rates are temperature dependent and have a linear relationship over a wide range
of temperature (Ferris et al., 1978; Jones, 1975; Milne and DuPlessis, 1965;
Tyler, 1933). Other environmental factors may have limiting effects on nematode
population increase. For example, as soil moisture becomes suboptimal, either
too high or too low, the reproduction rate may decrease due to oxygen stress or
moisture deprivation and the consequent inability of motile stages of the nematode
to reinvade the host. Such environmental effects on nematodes are reviewed in
Chap. 5.

Consideration of Age Structure and Time Varying Rates

Time-varying life tables are a formalization through which the effect of current
environmental conditions and the age structure of the population can be reflected
in terms of age-specific mortality/survivorship and age-specific fecundity (Getz
and Gutierrez, 1982; Gilbert et al., 1976; Wang et al., 1977). The life table
approach to describing population biology dictates a finer level of resolution than
is demanded for analytic models. Progression of the population through time

can be determined by using the matrix manipulations introduced into population biology by Leslie (1945, 1948). Rates are respecified at each time interval relative to the current environmental conditions or population development is measured on a physiological time scale, reflective of the heat units experienced by the system through time. The population biology is simulated in small time interval steps involving considerable computation and storage of rates and numbers of organisms of various age groups. This approach is essentially the basis of computer simulation of biological populations and its evolution is well reviewed by Getz and Gutierrez (1982). Computational complexity and precision can vary from discrete to distributed delay consideration of the population development.

In the discrete delay modeling methodology, every age cohort is required to spend a finite amount of physiological time in a life stage before graduating to the next stage. In distributed delays, variability in population processes is recognized and a probability of graduating from one life stage to the next is calculated (Fig. 1) (Ferris and Hunt, 1979; Ferris et al., 1982, 1984). A continuous form of the distributed delay formalism derived from electrical engineering (Manetsch, 1976; Vansickle, 1977) has been used for several pest phenology and population models in recent years (Welch et al., 1978). However,

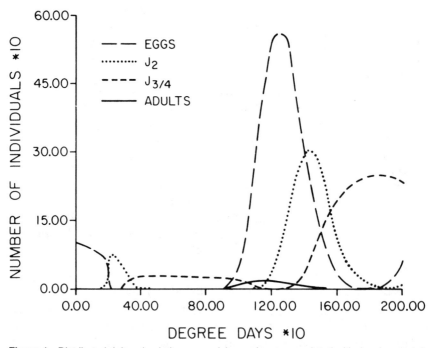

Figure 1 Distributed delay simulation recognizing variance associated with developmental processes of the age class redistribution of a *Meloidogyne* population developing from a single age cohort of eggs.

the resolution afforded by such structures may surpass requirements for population predictions (Blythe et al., 1984).

POPULATION DYNAMICS OF *MELOIDOGYNE* SPECIES

Analytic Models

As reviewed, analytic single-season models of nematode population increase under a host crop have been developed from the familiar logistic equation (Jones and Kempton, 1978; Seinhorst, 1966, 1967a, 1967b). The models redefined the rate of increase and carrying capacity of the logistic expression in terms of the maximum multiplication rate a (P_f/P_i at low population densities, where P_f and P_i are the final and initial population densities, respectively) and equilibrium density E (that initial population at which there is just sufficient food to maintain the population at the same level at the end of the growing season). In general, at higher initial population densities, resources become limited, or the host is damaged, so that the multiplication rate (P_f/P_i ratio) decreases (Fig. 2). The expected multiplication rate of a nematode population decreases with increase in density, consistent with the notion of the associated limitation of resources. Data for several nematode species (Ferris, 1985; Inserra et al., 1983) indicate a negative exponential relationship between the seasonal multiplication rate and the log of the initial population density

$$\frac{P_f}{P_i} = cP_i^{-b}$$

where c is a regression constant and b the rate factor for decrease in multiplication rate with increased P_i. To standardize the models for different crops it is convenient to divide through by a so that P_f/P_i is expressed as a relative multiplication factor on a 0 to 1 scale. The best estimate of a occurs at low population densities (Seinhorst, 1967b). In practice this would be at the lowest population density class observed in the data set, or at the P_i value identified as the tolerance level in the damage function. The slope parameter b is not affected by the adjustments for maximum multiplication rate. For all population densities below the lowest observed P_i, the relative multiplication rate is assumed to be 1.0 (Fig. 2) (Ferris, 1985). This rationale, of course, assumes that underpopulation phenomena (Seinhorst, 1968) do not exist in the data set. Such phenomena (depressed reproduction due to scarcity of mates) are not a problem with parthenogenic species. E is, by definition, that P_i for which the value of $P_f/P_i = 1.0$. Solving the equation for $P_f/P_i = 1.0$ and $P_i = E$, gives $E = c^{1/b}$. The E value provides another indicator of host status and of nematode pathogenicity on the crop (Seinhorst, 1967b).

Parameters of the negative exponential model for seasonal multiplication rates were influenced by crop species, environmental suitability, and geographic

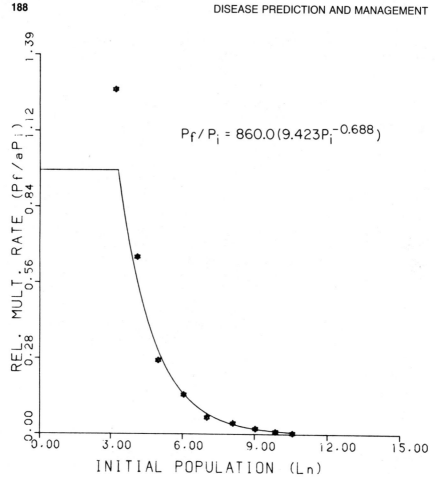

Figure 2 Relative multiplication rate of *Meloidogyne incognita* in relation to P_i, according to the model $P_f/P_i = acP_i^{-b}$. Values for *M. incognita* on tomatoes: $a = 860; c = 9.4; b = 0.688$. (*Adapted from Ferris, 1985.*)

location in California. For processing tomatoes, maximum multiplication rates and equilibrium densities were high, indicating a favorable host status. Maximum multiplication rates were sensitive to soil texture, being highest in coarse-textured soils and in warm climatic regions favorable to the nematode species of *Meloidogyne* (Ferris, 1985). In the model, larger b values indicate increased density dependence of nematode multiplication, that is, greater impact per additional nematode on the decline of the nematode multiplication rate. In tomatoes, the values differed somewhat with environment, being lowest in fine-textured soil and higher in coarse-textured soil, again reflective of environmental suitability. Equilibrium densities varied according to the tolerance of the plant to the nematode. Bell pepper (*Capsicum frutescens*) is tolerant to *M. incognita* and had a high equilibrium density, whereas cantaloupe (*Cucumis melo*) is intolerant,

Table 1   Parameters of a negative exponential model for crop season population increase of *Meloidogyne incognita*

$f = a c P_i^{-b}$ for $P_i > T$; $f = a$ for $P_i \leq T_2$

Crop	California county location	Soil	Parameters[a]					
			1	2	3	r^2	E	t
Processing tomatoes	Orange	Loamy sand	860	9.4	0.699	0.99	207	25
	Kern	Sandy loam	990	23.4	0.828	0.95	245	35
	Riverside	Silt loam	544	108.8	1.056	0.85	210	84
Green beans	Orange	Loamy sand	709	5.7	0.630	0.92	164	15
Bell pepper	Orange	Loamy sand	599	28.2	0.797	0.99	226	65
Cantaloupe	Orange	Loamy sand	586	1.0	0.665	0.96	46	0
Blackeye beans	Kern	Sandy loam	21	6.8	0.590	0.94	23	25

[a]1 = Maximum seasonal multiplication rate; 2 = scaling factor of negative exponential model; 3 = rate determining variable for effect of increase in P_i; r^2 = coefficient of determination for model fit to density-class data sets; E = equilibrium density (P_i when P_f/P_i = 1.0); and t = tolerance limit in damage function $y = m + (1-m)z^{P_i - T}$.

Source: H. Ferris, *J. Nematol.* 17:93–100 1985.

suffering damage at low population densities, and had a very low equilibrium density (Table 1).

Similar critical point analytic models can be used to describe overwinter survivorship of nematode populations. Empirical observation of overwinter survivorship data from experimental plots also indicated a negative exponential relationship to the natural log of the population density at the beginning of the winter (Duncan, 1983; Ferris, 1985). This is somewhat surprising in that pressures of resource limitation in the most obvious sense are not applied to nematodes that are not feeding. Survivorship might be expected to be density independent. However, several possibilities should be considered:

1. Since the nematodes are surviving with an energy component derived from the parent, the population at high density was produced under high competitive stress and may have reduced energy reserves.
2. There is greater prevalence of parasites and predators at high nematode population densities due to the availability of their food, resulting in greater mortality at high densities.

In areas where the nematode population had recently been introduced for experimental purposes, the overwinter survivorship rates were higher than in those areas where the population had been long established (Ferris, 1985).

Meloidogyne Life Table

Creation of a life table for any organism implies knowledge of the development rate or of the expected state of organismal phenology with time. It also implies knowledge of the understanding of age-specific mortality and the survivorship of the original population through time. The third piece of information necessary for construction of a life table is knowledge of the age-specific fecundity of the organism. Considerable information can be pieced together in this framework for plant parasitic nematodes from the literature. However, there is some difficulty in experimental determination of life table parameters because of the obligate parasitism of the organisms, their microscopic size, and their generally subterranean habitat. Life tables can be conveniently determined for many organisms by isolating or selecting a single age cohort of individuals, maintaining them in culture or captivity, and observing the same age cohort on a daily basis to determine developmental state, mortality, and fecundity. For obvious reasons this is not easily accomplished with plant-parasitic nematodes and the life table usually has to be constructed by repeated sampling from a population. Life tables are more readily constructed for organisms that can be maintained on artificial culture in a nonopaque medium. Thus, life tables have been constructed for the bacterivorous nematodes *Caenorhabditis briggsae* and *Plectus palustris* (Scheimer, 1983). For obligate parasites some aspects of the life table are host specific, depending on food quality and host suitability.

Life tables for poikilothermic organisms are appropriately constructed on a physiological time scale (degree days), utilizing the stability of the relationship between temperature and metabolic rates. Otherwise, the time frame of the life table is temperature dependent and the table has to be reconstructed for different temperatures. Another approach, essentially the same as the physiological time scale, is to construct the age classes on the basis of nematode phenology, in other words, to recognize individual life stages as age classes.

Life table data for *Meloidogyne arenaria* have been constructed by hand-picking single age cohorts of eggs based on developmental state followed by microscopic observation at daily intervals under different temperature regimes to determine the developmental time and mortality at various temperatures (Ferris et al., 1978). Similarly, single age cohorts of second-stage juveniles were obtained from freshly hatched eggs and their rate of penetration into grape rootings of different varieties was determined (Ferris et al., 1982). Developmental delays and survivorship of parasitic juveniles were determined for different grape varieties by inoculation with a single age cohort and daily microscopic observation to determine the rate of achievement of adulthood (egg production) (Ferris and Hunt, 1979). Female longevity, survivorship and fecundity were determined by suspending inoculated root systems in a misting apparatus and collecting hatched juveniles on a daily basis under constant temperature conditions. The rate of production of juveniles mirrored the rate of production of eggs at this constant temperature and allowed formulation of the egg production rate per degree day, the length of the productive period and the variances associated with these parameters (Ferris et al., 1984).

A complexity in life tables of obligate parasites is the effect of density dependence. At higher population densities host damage increases so that fecundity and survivorship decrease. Fecundity and survivorship then become complex functions of environmental conditions, nematode population density, and host damage and vary through time.

Simulation Models

Leslie (1945, 1948) applied the formalism of matrix algebra to a life table to allow projection of the population through time (Fig. 3). Thus, the matrix consists of age-specific fecundity in the first row, recognizing that the values will be zero for certain life stages and will probably vary among substages of the adult life stage. Values on the subdiagonal of the matrix are age-specific survivorship. Both survivorship and fecundity values can be considered variables, changing from one time to the next as a function of environmental conditions, host status, and population density. Values in the vector column are the number of individuals in each life stage or age class of the organism. The age classes are selected for uniform length to allow synchronous graduation through the life cycle. The matrix

Age classes of equal length

E_1	E_2	J_2	P_1	P_2	P_3	A_1	A_2	A_3		Age distribution at time 1		Age distribution at time 2
0.0	0.0	0.0	0.0	0.0	0.0	80.0	60.0	40.0		200.0		400.0
0.4	0.0	0.0	0.0	0.0	0.0	0.0	0.0	0.0		50.0		80.0
0.0	0.6	0.0	0.0	0.0	0.0	0.0	0.0	0.0		30.0		30.0
0.0	0.0	0.4	0.0	0.0	0.0	0.0	0.0	0.0		10.0		12.0
0.0	0.0	0.0	0.5	0.0	0.0	0.0	0.0	0.0	×	5.0	=	6.0
0.0	0.0	0.0	0.0	0.9	0.0	0.0	0.0	0.0		4.0		4.5
0.0	0.0	0.0	0.0	0.0	0.9	0.0	0.0	0.0		3.0		3.6
0.0	0.0	0.0	0.0	0.0	0.0	0.9	0.0	0.0		2.0		2.7
0.0	0.0	0.0	0.0	0.0	0.0	0.0	0.9	0.0		1.0		1.8

Figure 3 Matrix structure and manipulation as the basis for a time-varying life table approach to population simulation. Matrix columns represent approximately equal divisions of the life span of *Meloidogyne* spp. (E = eggs; J_2 = infective juveniles; P = parasitic juveniles; A = adults). The first row of the matrix is expected fecundity (egg production) of each age class (m_x values). The subdiagonal represents expected survivorship (l_x) of individuals in that age class. Actual m_x and l_x values are time-varying in that they are environment and density dependent.

multiplication is achieved by summing across each row of the matrix the product of the value in each column and the equivalently subscripted value in each row of the vector. Each resultant sum becomes the next row in a new vector. The first row of the new vector is created by summing the products of the vector and first (fecundity) row of the matrix. The new vector represents the new number of individuals in each life stage/age class of the population after a short time period in which fecundity and survivorship factors have remained essentially constant. Matrix manipulation of life tables formulate the basis of many simulation models and have been used extensively in population biology (Getz and Gutierrez, 1982).

The life table approach outlined with age classes structured on the basis of time is applicable to homoiothermic organisms, whose internal temperature regulation ascertains that physiological processes occur at rates independent of external conditions. In poikilothermic organisms, however, the influence of extrinsic temperatures on physiological rates results in variable fecundity rates, developmental times, and survivorship. That is, the rates vary through time. This can be recognized by updating the life table (multiplying the matrix) after short intervals of time during which environmental conditions have been approximately constant, and restating the appropriate rates as a function of temperature for the conditions during that time interval. Another approach is to restate the time frame of the life table in terms of physiological time, that is, to recognize that through a wide range of temperature there is a linear relationship between metabolic rates and temperature. This results in the familiar notion of degree days as a measure of the age of an organism and effectively results in a time-varying life table since the matrix multiplication process is performed after a constant degree day interval, which will be a nonconstant time interval.

Many organisms exhibit adaptive and behavioral phenomena during periods of adverse temperature conditions, or during periods of adverse conditions of other extrinsic factors. Such phenomena include diapause, cryptobiosis, migration, sex reversal, sex ratio change, and aestivation. By monitoring conditions other than temperature during a temperature-determined update interval of the Leslie matrix manipulation, fecundity and survivorship rates can be modified as appropriate for the interval. Alternatively, the interval can be prolonged by causing the accumulation of "effective degree days" as a reflection of the physiological activity of the organism relative to other environmental conditions.

Essentially, the evolution of a dynamic simulation model has been outlined. The structure of such a model with recognition of age-specific characteristics of biological events and phenomena allows superimposition of management effects. These would include the effects of cultural events and application of pesticides on age-specific mortality of the pests and similarly, the effects of biological antagonism (parasitism, predation, pathogenesis, competition) on the pest. At the same time the effect of the host status and its influence on survivorship and fecundity can be applied to the pest after each physiological time update interval. The influence of resource limitation resulting from host damage or competition among pest individuals can be applied in terms of its age-specific effects to different life stages of the pest.

THE DAMAGE FUNCTION

Analytic Models

It is common in annual crops to develop critical point analytic models of the influence of preplant population densities of plant parasitic nematodes on crop yield and damage (Barker and Olthof, 1976; Oostenbrink, 1966; Seinhorst, 1965). The importance of such models is that most nematode management decisions, given current technology in annual crops, must be made prior to planting. Such decisions include the use of a phytotoxic nematicides, the selection of crop species or variety, the decision to enter a crop rotation sequence, or the decision that production of the crop under the influence of the existing nematode community would be nonprofitable. Hence, it is very convenient to make such projections on the basis of a preplant assessment of the nematode population density.

The basis for a rational approach in management decisions for nematodes lies in the notion that, as with most pests, there is some relationship between the numbers of pest organisms and the yield or value of the crop. For nematodes, it is convenient to use a critical point model of yield or crop values relative to preplant population densities (Fig. 4). It would be theoretically possible to make a more precise prediction of expected crop loss based on multiple point model with more intensive assessment of the biological situation during the crop production period, or even a simulation model using input of either real time or historical weather data. However, these may be of less value in an applied sense because of the preplant nature of management decisions. Also supporting the use of critical point models is that the nematode community which is affecting the crop is in the soil prior to planting. In other words, the timing of infestation is predictable. Further, the life cycle of the nematode is not explosive. For root-knot nematodes under most conditions there are only about three generations during a single crop season.

The management decision basis for critical point models applied to annual crops is the economic threshold. This is simplistically defined as the population density of nematodes that will result in a loss of crop value equal to the cost of the management alternative under consideration (Ferris, 1978). Consequently, if the population density is higher than this threshold, the potential crop loss will be greater than the cost of controlling the nematode population and management is economically justified. Conversely, if the population density is less than the economic threshold, the expected crop loss is less than the cost of managing the population and the management is not justified. Decision models of this type are applicable to single season consideration but may not be the optimal long-term solution. A decision not to manage may result in the population increasing to high levels and exacerbating the problem for subsequent years.

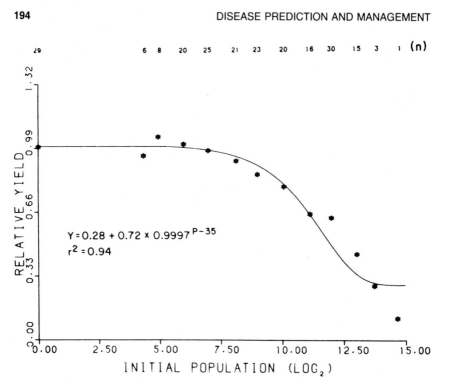

Figure 4   The relationship between tomato yield and preplant population density (eggs and juveniles per 1000 g soil) of *Meloidogyne incognita* on a loamy sand in Orange County, California. Graph represents averages over a 4-year period. Data are described by the model $y = m + (1 - m)z^{P-T}$ (Seinhorst, 1965) where y = relative yield; m = minimum yield; T = tolerance limit; z is a constant reflecting nematode damage; and P is the initial population density.

The confidence with which the management decision is made depends, at least in part, on the accuracy with which the population has been measured. Part of the cost of the management decision is the cost of the population assessment process. The greater the precision required in population assessment, the greater the associated effort and associated cost. It is rational, therefore, that the intensity of the population assessment process be a function of the magnitude of the management cost under consideration. In an applied sense, we are not interested in measurement of the absolute population density, rather assessment of it as an indicator of expected yield or monetary loss. The relationship between the precision of population estimation and sampling intensity is a function of the mean and variance of the population:

$$n = \frac{t^2}{d^2} \cdot \frac{s^2}{x^2}$$

where n is the number of sample units, s^2 is the variance, x the population mean, t is Student's t value, and d the acceptable range of the estimate as a proportion of x (Karandinos, 1976; Wilson and Room, 1982). By definition, in an aggregated population the variance is greater than the mean, and further, the variance increases at a greater rate than the mean (Elliott, 1977; Taylor, 1961, 1971, 1984). This results in an exponential relationship between variance and mean. If this relationship can be reliably described by the parameters of Taylor's power law, $s^2 = ax^b$ (Taylor, 1961, 1971), and the economic threshold level is used as the target population mean, an associated variance can be determined. Hence, the number of samples required to obtain a certain level of precision can be calculated (Ferris, 1984a, 1984b; Wilson and Room, 1982). If sufficient sampling intensity is applied to assess the population if it is at the economic threshold level, and an assessment is obtained below this threshold, the assessment will have greater precision than required. This is appropriate since the choice is not to manage. If the population is above this threshold, it has been measured with less than required precision, but this is of little consequence since it invokes the decision to manage (Ferris, 1984a).

The relationship between crop loss and nematode population density is such that the influence per nematode decreases at higher population densities (Seinhorst, 1965). Consider then, the relationship between unit change in the crop value relative to change in nematode population. A management option costing 20% of the expected crop value would only be applied if the population density of the nematode was greater than the economic threshold density. To make the decision, it is appropriate to ensure that the crop loss estimate is within an acceptable range (damage interval) of the economic decision value (the break-even point between crop loss and the cost of the management option). Projecting such a damage interval onto the nematode damage function, the projected d value (proportional range of population assessment) in the intensity/precision relationship is considerably greater than the proportion represented by the damage interval (Fig. 5). In the example portrayed in Fig. 5, a damage interval of 10% may result in a projected d value about the economic threshold population of around 60%, recognizing that the population axis is on a geometric scale. That is, $d = 0.6$, then $(1/d)^2 = 2.8$, a large difference from the 100-fold effect of applying the 10% decision range to the population estimate. Population assessment for management decisions in this light is a more palatable process (Ferris, 1984a).

A further extrapolation is consideration of risk. Confidence intervals can be drawn around the damage function with a known probability level. Similarly, the t_α in the intensity/precision relationship dictates a probability level associated with the sampling precision. Since the two probabilities are independent, the probability level associated with the damage interval estimate is their product. If each has a probability level of 0.8, the probability associated with the damage interval estimate is 0.64. If the user is risk-averse, a higher probability on the damage interval may be required. This could be achieved by selecting a t_α value with greater probability which would increase the required sampling intensity.

Consider the problem of optimization of a nematode management decision.

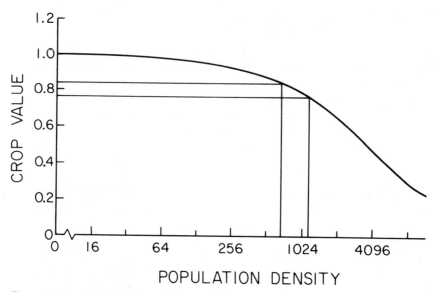

Figure 5 Determination of sampling intensity as a function of management cost and crop value. Relationship between expected damage interval and predictive population interval. (*Source: Ferris, 1984a.*)

Assume that the management strategy under consideration is rotation to an alternative crop of lesser value which is not damaged by the nematode population (Fig. 6). If the population density in the field is at level P_1 (Fig. 6), the short-term management decision is to grow the alternate crop. If during the period that the alternate crop is grown, the population in the absence of the host will be reduced to the level P_2, the optimal decision the second year is again to grow the alternate crop. After the third year, the optimal decision is to grow the primary crop. This optimization, however, as with the previous example, may exacerbate the problem by providing a food source to the nematode population and allowing it to increase to high levels. A solution yielding higher profits over the whole cropping sequence might have been to grow the alternate crop for 2 more years, reducing the nematode population to very low levels, before reverting to the primary crop (Duncan, 1983; Duncan and Ferris, 1983).

Multiple crop sequence decisions require not only critical point models of the damage function, but also critical point models of seasonal multiplication rates for nematodes (Duncan, 1983; Ferris, 1985) and overwinter survivorship previously discussed (Fig. 2, Table 1). They allow projection of the expected consequences of a management decision on the nematode population (Duncan and Ferris, 1983), determination of preplant population density for a subsequent crop, and the projection of profits through time. These three critical point models set the stage for projecting expected crop yields and values associated with nematode populations through several growing seasons. Nematodes may be one

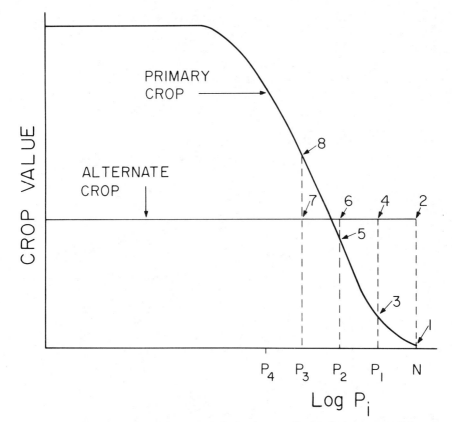

Figure 6   The relationship between crop value and nematode density for a susceptible and resistant crop, as a basis for crop rotation decisions. Intervals on the population axis indicate expected population decline with time. (*Source: Ferris, 1978.*)

of the pest groups where multiseason cropping sequence optimization is most readily achieved. Linear programming models are of direct application in this process. More simple multiple runs of the serial critical point models can be performed for selecting the approach yielding maximum profits (Duncan, 1983; Duncan and Ferris, 1983).

Coupling Structures in Simulation Models

Coupling structures between plant and nematode simulation models may be based on a metabolite or photosynthate pool concept of plant growth (Wang et al., 1977; Ferris et al., 1984). Energy flowing through the plant is a function of supply and demand. Supply is a function of PAR (photosynthetically active radiation), temperature, leaf area, photosynthetic efficiency of the current leaf-age structure and physiological efficiency, and the metabolic costs of the photosynthetic process. Demand for available energy is given priority to satisfy the respiratory demands of the existing plant biomass (vegetative and propagative) followed by the growth and the growth costs of the appropriate tissues of the plant for the current state of plant phenology.

The ratio f = supply/demand (modified by stored reserves), provides an indication of plant physiological stress such that growth processes will proceed at a maximum if $f \geq 1.0$, and be constrained if $f < 1.0$. A nematode population affects this supply/demand accounting in at least three ways:

1. It reduces the physiological efficiency of the supply process (Fig. 7). This is a function of the number and age structure of nematodes per

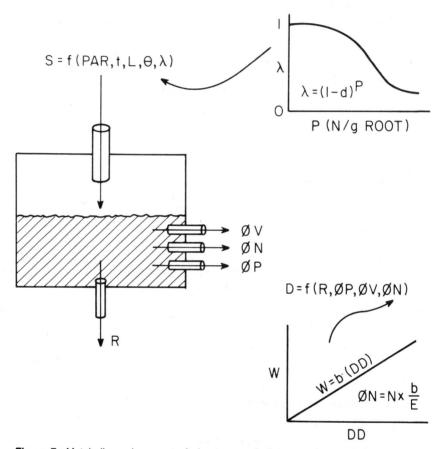

Figure 7 Metabolic pool concept of plant/nematode linkages. Supply S is a function of radiation PAR, temperature t, leaf area L, leaf age structure θ, and physiological efficiency λ. Physiological efficiency is influenced by nematode population density P and virulence d (inset top right). Demand D is a function of maintenance respiration R and the cost of new tissue (vegetative ϕV or propagative ϕP), including growth respiration and nematode demand ϕN. Nematode demand is a function of nematode productive efficiency E and growth W, a product of the growth rate b and elapsed physiological time DD (inset, lower right). (*Adapted from Wang et al., 1977; Ferris et al., 1984.*)

unit of root and the consequent effect on the uptake of water and nutrients by the plant.

2. It withdraws photosynthate, reflecting that the nematode is a primary consumer and, in effect, is a sink for metabolites produced through photosynthesis. This sink effect is exacerbated by the development of sophisticated nematode feeding sites that are effectively transfer cells (Jones and Northcote, 1972) and rapidly deliver food material to the nematode (McClure, 1977). The quantity of metabolites withdrawn from the cell is a function of the growth rate of the nematode. Since assimilation is partitioned into production and respiration in biological organisms, food demand by the nematode probably approximates a 4- to 5-fold product of its growth rate, reflective of the approximate efficiency of these sedentary systems (Fig. 7).

3. Since growth processes generally have higher respiratory costs than maintenance respiration, hypertrophy, hyperplasia, regrowth and repair resulting from root-knot nematode penetration and infection, produce an increase in plant respiratory demand in the infected tissues. The resulting changes in the plant supply/demand ratio will, in the sense of a simulation model, effectively reduce the metabolite pool available for growth and propagative processes, so reducing size and yield of the plant.

Another aspect of the linkage of nematode and plant simulation models is the impact of the host on the nematode. Such effects are also mediated by the available metabolite pool of the producer. Even though the parasite may adopt a higher demand priority for the available resources than the various plant growth and life processes, there is no doubt that as the resources become increasingly limited due to plant damage and increased nematode population density, the nematode system will be affected. The influence on the nematode system may be manifested through decreased growth rate, fecundity, change in sex ratio, and increased mortality. Reduction in the growth rates of the nematode reflect that assimilation can be partitioned into respiration and production. The first priority of the feeding nematode will be for respiratory maintenance of the existing tissues. As the available supply is reduced, the amount of material available for production may decrease. Effectively, this should result in the development of smaller nematodes and when these nematodes begin to produce offspring, it should result in the production of fewer eggs. Another impact of resource limitation on root-knot nematodes is alteration of the sex ratio. Generalizing, in healthy, vigorous, nonlimited populations, males are very rare, reproduction is by parthenogenesis and characteristics of *r* strategists are exhibited. As resources become limited, the sex ratio may change resulting in more males, fewer females, lower propagative output, and consequently, reduction of stress population (Triantaphyllou, 1973). Increase in mortality may be exhibited as reduced survivorship of developing parasitic stages within the root system, reduced longevity of productive females, or reduced infectivity of second-stage juveniles unable to locate suitable or attractive feeding sites. Fecundity may be lowered through reduction in the number of eggs produced per female based on reduced food intake, and reduction in total egg production in the population because of the lower egg output per female and the altered sex ratio.

CONCLUSIONS

As with many other groups of organisms, examples can be found in the Nematoda that represent the full range from r to k life history strategies. Those plant parasitic nematodes possessing k-type strategies exhibit long life cycles (6 months to 2 years). They tend to have large body size and relatively low reproduction rates. Such nematodes are successful in undisturbed ecosystems, natural woodland, ditch banks, and perennial crops. Plant parasitic nematodes successful in annual agriculture tend to exhibit the characteristics of r strategists: high reproduction rates, relatively rapid life cycles, some tendency to the shortening of the life cycle (for example, first stage passed in egg and short third stage), and a wide host range. The somewhat wasteful reproductive strategies compensate for the high mortalities incurred in the disturbed annual agricultural system (perhaps 90% overwintering mortality in root-knot nematodes). In some cases, the strategy is modified in that the host range is narrower but there are survival structures and mechanisms, and a host recognition signal required for emergence from the survival stage.

A variety of approaches to the modeling of nematode populations and their impact on plants has been discussed in this chapter, ranging from empirical through descriptive or explanatory models. Empirical models are relatively easy to construct, often by least squares fit of individual data sets to an analytic model selected for its form. However, such models are usually descriptive only of the data sets for which they are developed, and may be totally inapplicable with another data set or another environment. Explanatory models, by contrast, require greater knowledge of the biology and ecology of the system. They generally require the modeler to state a formal hypothesis of the interactions and functional relationships within the system. Such hypotheses, if biologically rational, should be descriptive of the system in other environments, or of other data sets. The validation process testing such models reveals shortcomings in understanding of the biology and dictates the direction of appropriate avenues of research.

In an applied sense, there are some advantages and disadvantages to modeling the biology, ecology, and life history strategies of nematodes as a basis for rational management decisions. Plant-parasitic nematodes in the soil are relatively nonmotile and may undergo from one to four life cycles during the course of an annual crop growing season. Since the population of concern is in the soil prior to planting and the life cycle of the nematode is not explosive, it is convenient to use critical point preplant models as predictors of crop yield or damage, and of nematode population increase. Such models are significant because they can be applied to preplant management options. Similarly, overwintering survivorship of plant-parasitic nematodes in annual agriculture appears predictable for various locations based on a critical point model. The stability and predictability of these systems, especially in intensely managed irrigated agriculture, promotes the selection of optimal cropping sequences across several seasons. Nematodes may be one of the pest groups where multiseasonal cropping sequence optimization is most readily achieved. Some of the same principles apply to the nematode management decision process in perennial crops; however, the problem is conceptually more complex in that a critical point preplant model is inappropriate. Instead the management decision may be based on the rate of increase in damage relative to the rate of increase of the nematode population, as determined by a multiple point population assessment program (Noling and Ferris, 1984).

A disadvantage in the development of management systems for nematode populations is the difficulty involved in assessing the population density. The nature of the soil medium dictates that samples be transported to a laboratory for extraction and analysis of the nematode population. The process is physically and technically laborious and is expensive. As understanding of the factors affecting nematode population dynamics and distribution increases, the relationship between sampling intensity or effort and the associated precision of the population assessment is better understood. Knowledge of this relationship formulates a basis for understanding the risks associated with a nematode management decision based on an assessment of the population using a critical point model. Advances in information transfer systems, including computer networks and distribution of floppy disks for microcomputers, provide ideal media for the dissemination of models and decision rules for rational management of nematode pests in crop production systems.

REFERENCES

Barker, K. R., and Olthof, T. H. A. 1976. Relationships between nematode population densities and crop responses. *Annu. Rev. Phytopathol.* 14:327–353.

Bird, A. F. 1959. Development of the root-knot nematode, *Meloidogyne javanica* (Treub) and *Meloidogyne hapla* Chitwood in the tomato. *Nematologica* 4:31–42.

Blythe, S. P., Nisbet, R. M., and Gurney, W. S. C. 1984. The dynamics of population models with distributed maturation periods. *Theor. Pop. Biol.* 25:389–411.

Chitwood, B. G., and Chitwood, M. B. 1975. *Introduction to Nematology.* University Park Press, Baltimore, 334 pp.

Cobb, N. A. 1914. North American free-living fresh-water nematodes, in: *Contributions to a Science of Nematology* (N. A. Cobb, ed.), pp. 35–99. Williams and Wilkins, Baltimore, 490 pp.

Duncan. L. W. 1983. Predicting effects of plant-parasitic nematode communities on crop growth. Ph.D. thesis, Univ. of California, Riverside, 160 pp.

Duncan. L. W., and Ferris, H. 1983. Effects of *Meloidogyne incognita* on cotton and cowpeas in rotation. *Proc. Beltwide Cotton Prod. Res. Conf.,* pp. 22–26. Nat. Cotton Council, Memphis.

Elliott, J. M. 1977. Some methods for the statistical analysis of samples of benthic invertebrates. *Freshwater Biol. Assoc.* Publ. No. 25. Ambleside, England, 160 pp.

Evans, A. A. F., and Perry, R. N. 1976. Survival strategies in nematodes, in: *The Organization of Nematodes* (N. A. Croll, ed.), pp. 383–424. Academic Press, New York, 439 pp.

Ferris, H. 1978. Nematode economic thresholds: Derivation, requirements, and theoretical consideration. *J. Nematol.* 10:341–350.

———. 1982. The role of nematodes as primary consumers, in: *Nematodes in Soil Ecosystems* (D. W. Freckman, ed.) pp. 3–13. University of Texas Press, Austin, 206 pp.

———. 1984a. Probability range in damage predictions as related to sampling decisions. *J. Nematol.* 16:246–251.

———. 1984b. The stability and characteristics of variance/mean relationships for nematode populations. Proc. 1st Int. Cong. Nematol. 27–28.

———. 1985. Density dependent nematode seasonal multiplication rates and overwinter survivorship: a critical point model. *J. Nematol.* 17:93–100.

Ferris, H., DuVernay, H. S., and Small, R. H. 1978. Development of a soil-temperature data base on *Meloidogyne arenaria* for a simulation model. *J. Nematol.* 10:198–201.

Ferris, H., and Hunt, W. A. 1979. Quantitative aspects of the development of *Meloidogyne arenaria* larvae in grapevine varieties and rootstocks. *J. Nematol.* 11:168–174.

Ferris, H., Schneider, S. M., and Semenoff, M. C. 1984. Distributed egg production functions for *Meloidogyne arenaria* in grape varieties, and consideration of the mechanistic relationship between plant and parasite. *J. Nematol.* 16:172–183.

Ferris, H., Schneider, S. M., and Stuth, M. C. 1982. Probability of penetration and infection by root knot nematode, *Meloidogyne arenaria*, in grape cultivars. *Am. J. Enol. Vitic.* 33:31–35.

Freckman, D. W. 1978. Ecology of anhydrobiotic soil nematodes, in: *Dry Biological Systems* (J. H. Crowe and J. S. Clegg, eds.), pp. 345–357. Academic Press, New York.

Freckman, D. W., and Womersley, C. 1983. Physiological adaptations of nematodes in Chihuahuan desert soils, in: *New Trends in Soil Biology* (P. Lebrun, H. M. Andre, A. de Medts, C. Gregoire-Wibo, and G. Wathy, eds.), pp. 395–403. Dieu-Brichart, Louvain-la-Neuve., 709 pp.

Getz, W. M., and Gutierrez, A. P. 1982. A perspective on systems analysis in crop production and insect pest management. *Annu. Rev. Entomol.* 27:447–466.

Gilbert, N., Gutierrez, A. P., Frazer, B. D., and Jones, R. E. 1976. *Ecological Relationships.* W. H. Freeman, San Francisco, 157 pp.

Herman, R. K., and Horvitz, H. R. 1980. Genetic analysis of *Caenorhabditis elegans*, in: *Nematodes as Biological Models*, vol. 1 (B. M. Zuckerman, ed.), pp. 227–261. Academic Press, New York, 312 pp.

Inserra, R. N., O'Bannon, J. H., DiVito, M., and Ferris, H. 1983. Response of two alfalfa cultivars to *Meloidogyne hapla*. *J. Nematol.* 15:644–646.

Jones, F. G. W. 1975. Accumulated temperature and rainfall as measures of nematode development and activity. *Nematologica* 21:62–70.

Jones, F. G. W., and Kempton, R. A. 1978. Population dynamics, population models and integrated control, in: *Plant Nematology* (J. F. Southey, ed.), pp. 333–361. H.M.S.O., London, 440 pp.

Jones, M. G. K., and Northcote, D. H. 1972. Nematode induced syncytium: A multinucleate transfer cell. *J. Cell Sci.* 19:789–809.

Karandinos, M. G. 1976. Optimum sample size and comments on some published formulae. *Bull. Entomol. Soc. Am.* 22:417–421.

Leslie, P. H. 1945. On the use of matrices in certain population mathematics. *Biometrika* 33:183–212.

———. Some further notes on the use of matrices in population mathematics. *Biometrika* 35:213–245.

Manetsch, T. J. 1976. Time-varying distributed delays and their use in aggregative models of large systems, *IEEE Trans. Systems. Man and Cybernet.* 6:547–553.

McClure, M. A. 1977. *Meloidogyne incognita:* A metabolic sink. *J. Nematol.* 9:88–90.

Milne, D. L., and DuPlessis, D. P. 1964. Development of *Meloidogyne javanica* (Treub) Chit on tobacco under fluctuating soil temperatures. *S. Afr. J. Agr. Sci.* 7:673–680.

Noling, J., and Ferris, H. 1984. Epidemiological analysis of the cumulative effects of *Meloidogyne hapla* on crop losses in alfalfa. Proc. 1st Int. Cong. Nematol. 62.

Ogunfuwora, A. O., and Evans, A. A. F. 1976. Factors, affecting the hatch of eggs of *Meloidogyne naasi,* an example of diapause in a second stage larva. *Nematologica* 23:137–146.

Oostenbrink, M. 1966. Major characteristics of the relation between nematodes and plants. *Meded. Landbouwhogesch. Wageningen* 66(4):1–46.

Scheimer, F. 1983. Comparative aspects of food dependence and energetics of free living nematodes. *Oikos* 41:32–42.

Seinhorst, J. W. 1965. The relation between nematode density and damage to plants. *Nematologica* 11:137–154.

———. 1966. The relationships between population increase and population density in plant parasitic nematodes. I. Introduction and migratory nematodes. *Nematologica* 12:157–169.

———. 1967a. The relationship between population increase and population density in plant parasitic nematodes. II. Sedentary nematodes. *Nematologica* 13:157–171.

———. 1967b. The relationships between population increase and population density in plant parasitic nematodes. III. Definition of terms host, host status and resistance. IV. The influence of external conditions on the regulation of population density. *Nematologica* 13:429–442.

———. 1968. Underpopulation in plant parasitic nematodes. *Nematologica* 14:549–553.

Sulston, J. E., Schierenberg, E., White, J. G., and Thomson, J. N. 1983. The embryonic cell lineage of the nematode *Caenorhabditis elegans. Dev. Biol.* 100:64–119.

Taylor, L. R. 1961. Aggregation, variance and the mean. *Nature London* 189:732–735.

———. 1971. Aggregation as a species characteristic, in: *Statistical Ecology,* vol. 1 (G. P. Patil, E. C. Pielou, and W. E. Waters, eds.), pp. 357–372. Pennsylvania State University Press, University Park.

———. 1984. Assessing and interpreting the spatial distributions of insect populations. *Annu. Rev. Entomol.* 29:321–357.

Triantaphyllou, A. C. 1973. Environmental sex differentiation of nematodes in relation to pest management. *Annu. Rev. Phytopathol.* 11:441–462.

———. 1979. Cytogenetics of root-knot nematodes, in: *Root-Knot Nematodes (Meloidogyne Species): Systematics, Biology and Control* (F. Lamberti and C. E. Taylor, eds.) pp. 85–109. Academic Press, London, 477 pp.

Tummala, R. L., and Haynes, D. L. 1977. On-line pest management systems. *Environ. Entomol.* 6:339–349.

Tyler, J. 1933. Development of the root-knot nematode as affected by temperature. *Hilgardia* 7:391–415.

Van Gundy, S. D. 1958. The life history of the citrus nematode *Tylenchulus semipenetrans* Cobb. *Nematologica* 3:283–294.

Vansickle, J. 1977. Attrition in distribution delay models. *IEEE Trans. Systems, Man Cybernet.* 7:635–683.

Von Ehrenstein, G., and Schierenberg, E. 1980. Cell lineage and development of *Caenorhabditis elegans* and other nematodes. in: *Nematodes as Biological Models,* vol. 1 (B. M. Zuckerman, ed.), pp. 1–71. Academic Press, New York, 312 pp.

Wallace, H. R. 1966. Factors influencing the infectivity of plant parasitic nematodes. *Proc. Roy. Soc. B* 164:592–614.

Wang, Y. H., Gutierrez, A. P., Oster, G., and Daxl, R. 1977. A population model for plant growth and development coupling cotton-herbivore interaction. *Can. Entomol.* 109:1359–1374.

Welch, S. M., Croft, B. A., Brunner, J. F., and Micheles, M. F. 1978. PETE: An extension phenology modeling system for management of a multi-species pest complex. *Environ. Entomol.* 7:487–494.

Wilson, L. T., and Room, P. M. 1982. The relative efficiency and reliability of three methods for sampling arthropods in Australian cotton fields. *J. Austral. Entomol. Soc.* 21:175–181.

Chapter 9

Studies of Plant-Pathogen-Weather Interactions: Cotton and Verticillium Wilt

A. P. Gutierrez
J. E. DeVay

Plant pathology is the study of host-parasite relationships, and the impact of parasite populations on the growth and development of populations of plants or plant parts (Gutierrez et al., 1983). In recent years, modeling has played an increasing role in the analysis of plant-pathogen interactions. Many early models of disease effects on plants have been Malthusian growth models (i.e., Van der Plank models) for pathogen growth when weather conditions are favorable (Waggoner, 1981).

Zadoks (1979) reviewed the state of plant disease modeling and categorized pathogen models as:

- Statistical in nature (Royle and Butt, 1979)
- Dynamic computer simulations (Kranz, 1979; Rabbinge and Rijsdijk, 1983; Rapilly, 1979; Rijsdijk, 1975; Waggoner and Horsfall, 1969; Waggoner et al., 1972; Zadoks, 1971; Zadoks and Rijsdijk, 1972)

- Spatial models (Kiyasawa and Shigomi, 1972; Shrum, 1975; Zadoks and Kampmeijer, 1977)

Statistical models often are used to predict disease potential, while the major thrust of dynamic simulation models is to understand disease processes and, in a few cases, predict disease. Spatial models are designed to elucidate not only the rate of disease increase but also its potential dynamics over a large area. Zadoks (1979) did not review analytic models such as those developed by Kiyosawa (1972a, 1972b) and Kiyosawa and Shigomi (1972). Of the dynamic simulation models, the most widely implemented one in Europe is EPIPRE (Rabbinge and Rijsdijk, 1983) which models the interaction of winter wheat, certain wheat pathogens, and cereal aphids. This model has had a very positive impact on pest management practices in Europe.

This chapter reviews the biology of both verticillium wilt (*Verticillium dahliae* Kelb.) and cotton (*Gossypium hirsutum* L.), and describes a mathematical model of their interaction (Gutierrez et al., 1983, 1984). The verticillium wilt problem in cotton has both biological and economic aspects, which are complicated by the interactions of inoculum density, pathotype virulence, cotton cultivar, environmental factors affecting the progression of the disease, soil type and moisture holding capacity, plant density, and many other factors that affect yields and profits.

Examining the effects of these interactions on yields using more traditional methods of experimentation and analysis has had limited success. For this reason simulation methods of systems analysis were used here to guide the research on verticillium wilt and examine the dynamics of the problem. The analysis has provided unexpected insight into physiological processes associated with verticillium wilt development including dry matter allocation and fruit retention patterns, plant water relations, osmotic adjustment, and ethylene metabolism of diseased tissues (Gutierrez et al., 1983; Tzeng and DeVay, 1985; Tzeng et al., 1985).

BIOLOGY OF VERTICILLIUM WILT

Verticillium wilt in cotton and other crops is a worldwide problem. It was first reported in California cotton by Shapovalov and Rudolph (1930). The disease becomes a major problem when the density of soilborne propagules reaches high levels (DeVay et al., 1974). In California cotton, average yearly yield reductions have been as high as 7%, with losses up to 100% in individual fields (Pullman and DeVay, 1982b). The problem is most severe on small farms where the resources to diversify to other crops are limited.

Characteristically, *V. dahliae* is a nonaggressive soil resident which seldom ventures more than a few millimeters from its propagule base. Its persistence in soil is associated with the formation of microsclerotia which may remain viable for several years. Dormancy is caused by a bacterium, and can be overcome by surface sterilization of the microsclerotia or by air drying the soil (30 to 40% RH) for 6 weeks (Butterfield and DeVay, 1975; DeVay et al., 1974). Under natural conditions the inhibitory effect of the bacteria is overcome by an apparent nutritional effect of young host roots on microsclerotia. In the absence of hosts, low soil water matric potential (-1 mb to air dry) and high soil temperature (28° C) enhances the decline of microslerotia, with no viable microsclerotia remaining after 3 to 4.5 years, depending on soil type (Green, 1980).

Infection Process and Symptom Development

Many factors can influence the infection of cotton roots by *V. dahliae*, the systemic spread of the pathogen, and the development of foliar symptoms. These factors include potassium and nitrogen nutrition and wilt susceptibility of cotton cultivars, plant spacing (volume or density of roots), water management or availability, air temperature, and most important, the density of soilborne propagules and their virulence (Grimes and Huisman, 1984; Gutierrez et al., 1983; Hafez et al., 1975; Pullman and DeVay, 1982a, 1982b; Ranney, 1973).

Microsclerotia of *V. dahliae* consist of a few to 30 or more cells, and germination is triggered by interaction with roots of susceptible and nonsusceptible (e.g., grass) plant (DeVay et al., 1974). Infection hyphae emerge from microsclerotia about 16 h after germination begins, and tend to penetrate directly into uninjured areas of young roots. The hyphae grow both intercellularly and intracellularly through the root cortex. Eventually some hyphae penetrate the endodermis and become established in the xylem tissues. The reproductive conidial stage of the fungus moves rapidly into the shoot via the transpiration stream (Garber, 1973).

The root cortex of both wilt resistant and wilt susceptible cotton cultivars is colonized by *V. dahliae*. Resistance mechanisms of host plants are usually not expressed until the pathogen has breached the endodermis and entered the xylem (Charudattan and DeVay, 1972). Further development and establishment of the pathogen depends on the structural and chemical changes that are induced within vessels and the associated vascular ray cells (VanderMolen et al., 1983). The degree to which these changes or stress factors develop are related to virulence factors of the pathogen and the dynamics of photosynthate partitioning to vegetative and reproductive growth of the host (Tzeng and DeVay, 1985; Tzeng et al., 1985).

Pathotypes of *V. dahliae*

Ecotype development is evident within the species *V. dahliae*. Puhalla (1979) found that isolates from various sources can be grouped into at least 16 population subgroups based on heterokaryon incompatibility. Isolates of *V. dahliae* from cotton of different compatibility subgroups may have quite different virulence in tomato and strawberry.

Isolates that are pathogenic on cotton can be divided into two major pathotypes: those that cause defoliation and those that are nondefoliating but that can cause mild to severe pathogenesis. Within each pathotype there is a continuum of strains ranging in virulence from high to low. The defoliating pathotypes are more aggressive as measured by their relative sporulation and growth performance over a wider range of temperatures (Schnathorst and Mathre, 1966a; DeVay et al., 1974; Friebertshauser and DeVay, 1982). Also isolates of *V. dahliae* from defoliated plants, without regard to the cultivar from which they came, are more aggressive than isolates taken from randomly selected, infected plants (Ashworth, 1983). There is evidence, however, that a nondefoliating pathotype will cross-protect against a defoliating pathotype (Schnathorst and Mathre, 1966b). In the field plots used to develop the model, the mixture of strains of *V. dahliae* favored the nondefoliating pathotypes. The relative percentages of pathotypes were unchanged through four growing seasons of continuous cotton (Pullman and DeVay, 1982a).

Assay and Sampling for *V. dahliae*

Microsclerotia are distributed mainly in the upper 45 cm of field soils with the greatest concentrations in soil zones mixed with crop debris; their density is usually estimated before cotton is planted. Two main methods of soil assay for propagules of *V. dahliae* have been used by various workers: dry soil plating and wet sieving. Butterfield and DeVay (1977) compared the two methods, and although both are reliable, the plating of soil, air-dried for 6 weeks, gave consistently higher propagule counts per gram soil. Air-drying the soil also overcame the dormancy often associated with propagules isolated from fresh soil, and eliminated from consideration conidia and hyphal fragments that are short-lived in drying soil (Schreiber and Green, 1962).

In cotton, no significant correlation was found between the numbers of microsclerotia of *V. dahliae* in air-dry soil samples collected at planting time in 13 fields and the incidence of cotton plants with foliar leaf symptoms of verticillium wilt just prior to defoliation (DeVay et al., 1974; see also Smith and Row, 1984). However, the incidence of foliar wilt symptoms was highly correlated to inoculum densities in individual fields over a period of 7 years.

Sampling procedures used to estimate propagule density in soil samples often affect the interpretation of results. For example, Marois and Adams (1984) found that quadrat size is a major consideration in developing soil sampling procedures because it affects the frequency distribution of the samples and hence the interpretation of the data. They emphasized that biological rather than logistic factors should determine quadrat size. Smith and Rowe (1984) found with their methods that *V. dahliae* propagules had a clustered distribution.

BIOLOGY OF COTTON

Domestic cultivars of cotton were selected from wild cottons collected in Central America and Mexico (Fryxell, 1981). The general growth characteristics of all varieties are similar (Gutierrez et al., 1975, 1977, 1984). Data on dry matter

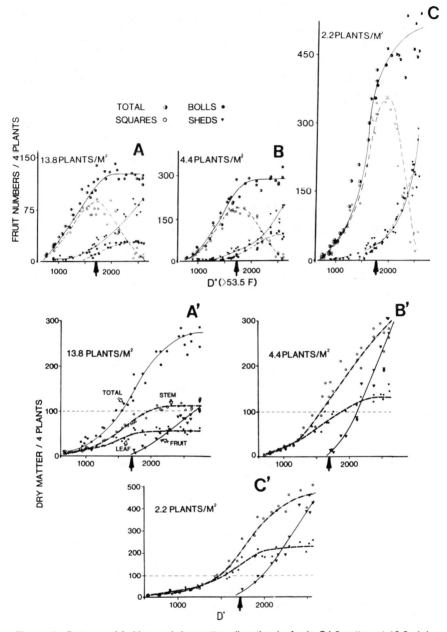

Figure 1 Patterns of fruiting and dry matter allocation in Acala SJ-2 cotton at 13.8, 4.4, and 2.2 plants per square meter. (A), (B), (C) fruiting patterns, and (A′), (B′), (C′) corresponding dry matter allocation patterns. Data are plotted on degree days above 53.5° F. Upturned arrows indicate beginning of plant stress (i.e., t_s in the text). (*Source: Gutierrez et al., 1983.*)

accumulation in plant parts and age structure of fruits in Acala SJ-2 cotton at three planting densities grown in a pest-free minimal stress environment are presented in Fig. 1 (cf. Gutierrez et al., 1975). The number of bolls set per plant, plant size, and fruit numbers increased as plant density decreased. Dry matter allocation to vegetative growth (i.e., leaves and stem, plus tap root) slowed, while that to bolls (fruits) increased rapidly after the time of plant stress (the upturned arrows in Fig. 1), while square (small fruit) production decreased and abscission of squares and small bolls increased. These are reasonably optimal, disease-free phenology and growth patterns for Acala cotton as influenced by planting density, and form a basis for consideration of the impact of verticillium wilt.

The Cotton-Verticillium Wilt Model

Gutierrez et al. (1975, 1977, 1984) and Wang et al. (1977) modeled the growth and development of cotton using both discrete and continuous models (Leslie, 1945; von Forester, 1959) derived from animal demography. A brief description of the mathematical details of these models are presented in the appendix of this chapter, and more completely by Wang et al. (1977). For model development, five parameters plus initial plant density were found to characterize the growth development of different varieties of cotton sufficient to mimic all of the data presented in Fig. 1:

1. The position of the first fruiting branch
2. Square and boll age-dependent growth rates
3. Square and boll maturation times
4. Leaf, stem, and root growth rates
5. Fruit bud production rates

Despite its demographic origin and concise mathematical form, the model easily incorporates detailed physiology (Wang et al., 1977; Gutierrez et al., 1984; appendix). For example, the original functions for estimating photosynthetic rates and metabolic costs was derived from SIMCOT II (McKinion et al., 1974) but have now been replaced by more general functions. In addition, the model accommodates the observations of Harper (1977) that populations of plants consist of individual plants with age structure (Fig. 2A), and that within each plant there exist populations of plant parts (e.g., leaves, stems, root, and fruit) also with age structure (Fig. 2B). In domestic cotton, the entire crop is planted at one time, hence age structure of whole plants can be ignored in our applications.

Plant population dynamics can be viewed as shown in Fig. 2B, wherein cohorts of leaves (L_i) of age $i = 1, \ldots, k$ produce photosynthate at age-specific, environmentally modified rates, and this, plus a fraction of reserves (Q), is distributed during the time interval t to meet demands for respiration, growth of leaves (L) themselves, stem and root tissues (S,R), and fruit (M)

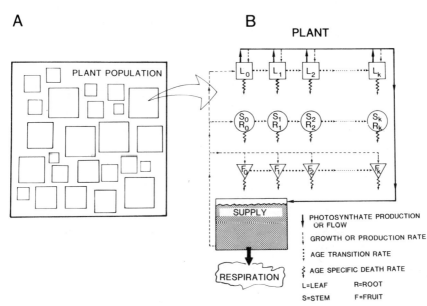

Figure 2 (A) Conceptual model of a plant population with age structure, and (B) conceptual model of the within-plant age structure of plant parts illustrating photosynthate production and allocation, respiration, aging, and age-specific birth rates.

growth. The plant-part growth dynamics in relation to the flow of photosynthate and the priorities for allocation are shown in Fig. 3 and are mathematically formalized in the appendix. The effects of virulence V of *V. dahliae* on severity of verticillium wilt and on photosynthate P production and allocation to growing plant parts are also shown, but are discussed more fully in the following sections. Each plant-part array in Fig. 2B can be viewed as a population with age structure (i.e., a Leslie-von Foerster model) births flowing into the first age category ($\rightarrow$), transition rates ($\alpha = 1 \dashrightarrow$) between age categories, and age-specific death rates flowing out of each ($\leadsto$) category. If $\alpha < 1$, then the resulting distribution of maturation times has an Erlang distribution with characteristic mean and variance specified by the number of age categories (Manetsch, 1976). Gutierrez et al. (1984) applied this model to cotton growth and development.

The metabolic pool model in Fig. 3 integrates all aspects of cotton growth and development, with actual rates determined by the interplay between photosynthate supply (daily photosynthesis plus some fraction of reserves) and the sum of the *maximum* current demands for plant part growth and for respiration. The levels of the outlets from the metabolic pool imply that the photosynthate is allocated first to respiration, which must be met or the plant dies, second to fruit if the plant is mature enough, and, if photosynthate remains, to vegetative growth and reserves (see appendix).

The model predicts that the photosynthate supply becomes less than photosynthate demand when many fruits enter the very rapid boll growth stage (Fig.

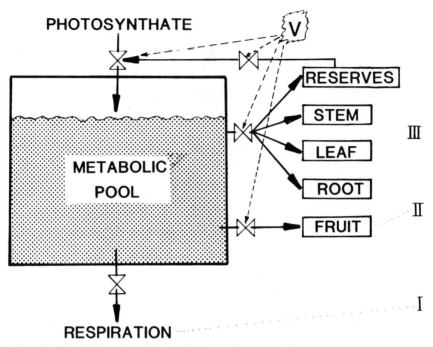

Figure 3 Metabolic pool model showing priorities and allocation rates of photosynthate to respiration, fruit growth, and vegetative growth and reserves. (*Adapted from Gutierrez et al., 1983.*)

4A). For example, the per fruit demand rate in the Acala cotton cultivars may increase by 16-fold after flowering, as can be seen in the stress-free and disease-free field data (Fig. 1). Most of the dry matter available for assimilation at this time (t_s) is shunted to boll growth at the expense of leaf, stem, and root growth which decline or cease altogether. Fruit point production and vertical growth as measured by mainstem node production (not shown) are also reduced. Peak squaring and increased shedding of small to medium sized squares and small bolls at rates proportional to the shortfall of the photosynthate supply coincide with the beginning of the boll set period (cf. Gutierrez et al., 1975). The fruit are shed with a small time delay of approximately 30 degree days (D°) above the base of 12° C (i.e., 55 D° above 53.5° F) (Gutierrez et al., 1975) in shed windows (age windows) which increase with increasing stress (see Fig. 4A). In the model, all of the growth, death, or production rates are scaled by the supply-demand ratio.

Effects of Pests on Cotton Growth and Development

In general, weather conditions and water and nutrient availability affect the supply side of the supply/demand ratio, and determine the underlying patterns of cotton plant growth and development (Gutierrez et al., 1975, 1984). Pests and diseases

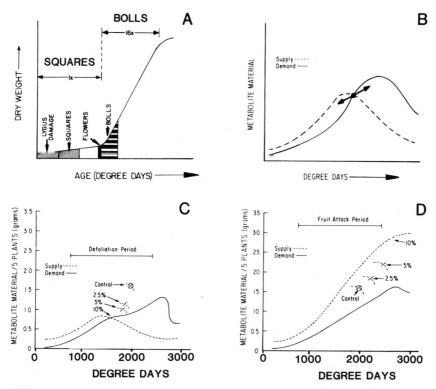

Figure 4 Interplay of photosynthate supply and demand. (A) Pattern of maximum growth by individual Acala cotton fruits and shedding window due to damage by plant bug *Lygus hesperus* and to carbohydrate stress (shaded and striped areas); (B) hypothetical patterns of supply and demand; and simulated pattern (C) in the presence of increasing leaf damage and (D) for increasing rates of square depletion. (*Adapted from Wang et al., 1977.*)

affect either or both sides of the ratio. Organisms that cause premature abscission of fruit affect the demand side, while those that increase leaf mortality (e.g., defoliating insects) or decrease photosynthesis, affect the supply side (e.g., nematodes and verticillium wilt). The reduction in demand due to leaf loss is negligible in comparison to the loss in potential supply. The vectors in Fig. 4B illustrate the direction of displacement of the stress point from the pest-free condition caused by the two kinds of pest damage. Pests that effect the supply side cause the plants to be stunted, while those that affect the demand side delay the time of stress and cause the plants to be large. The interplay of supply and demand as influenced by organisms that cause defoliation and premature abscission of squares is shown in Fig. 4C and 4D.

If demand is reduced (e.g., fruit depletion by the plant bug *Lygus hesperus* Knight), the plants allocate the photosynthate to vegetative growth, producing large, rank-growing plants. The observed low productivity of such plants may be due not only to fruit depletion, but also to competition for light, increased respiration demands of larger plants, and crowding effects, which restrict photosynthesis to a constant proportion of the canopy. Diseases such as verticillium will affect the supply side, and hence stunt that plants and increase shading competition from more vigorous neighbors.

Effects of Verticillium Wilt on Cotton Growth and Development

In our model, verticillium wilt slows the rate of photosynthesis and allocation rates to growing plant parts as a linear scalar of virulence ($V \varepsilon [0,1]$) (see Fig. 3), while environmental influences are included in the photosynthesis function. In our approach, *no demographic model of verticillium wilt is required,* as most of the interactions are with the physiology of cotton. This explains the absence of a specific model for verticillium wilt in the appendix. In the model, the season is divided into *n* periods during which cohorts of plants may show symptoms. The growth of each cohort of plants is simulated separately, and their contribution to the overall yield is summed.

Pullman and DeVay (1982b) followed the growth and development of cohorts of tagged field cotton plants during the 1978 and 1979 growing seasons to determine the effects of time of symptom expression on growth and development. Their data were collected in a manner similar to that reported by Gutierrez et al. (1975) and are illustrated in Figs. 5 and 6. Data were collected on leaf, stem, root, fruit, and total dry weight as well as plant height, mainstem node numbers and square and boll number patterns.

Several comparisons of plant growth and development parameters can be made between cohorts of plants showing no symptoms and those showing symptoms on different dates. Plants developing foliar symptoms earlier in the season were shorter (Fig. 5A and 5B), produced fewer mainstem nodes (Fig. 5C and 5D), and had fewer squares (Fig. 5E and 5F) and fewer bolls (Fig. 5G and 5H). The dry matter data presented in Fig. 6 show that compared to symptomless plants, those developing foliar symptoms earlier in the season accumulated lower amounts of leaf (Fig. 6A and 6B), stem (Fig. 6C and 6D) and root (Fig. 6E and 6F) dry matter. The simplest comparison to make is in total dry matter (Fig. 6G and 6H) where the same trends hold. The growth patterns of plants in both years were very similar until 1500 degree days, when growth rates slowed in 1978, but not 1979. The observed between-year differences in growth patterns are due to differences in the weather patterns (i.e., 1979 was a longer season), small differences in planting density and in agronomic practices such as the timing of irrigation (see Gutierrez et al., 1984, for an analysis). Such factors often are not easily separated in more formal agronomic experiments, but can be investigated using simulation (see Gutierrez et al., 1984).

Verticillium dahliae Pathotypes

In field studies, approximately 9% of the isolates tested were defoliating pathotypes, 67 to 77% were intermediate types, and the remainder were nondefoliating types (Pullman and DeVay, 1982a). Stem inoculation of disease-free plants

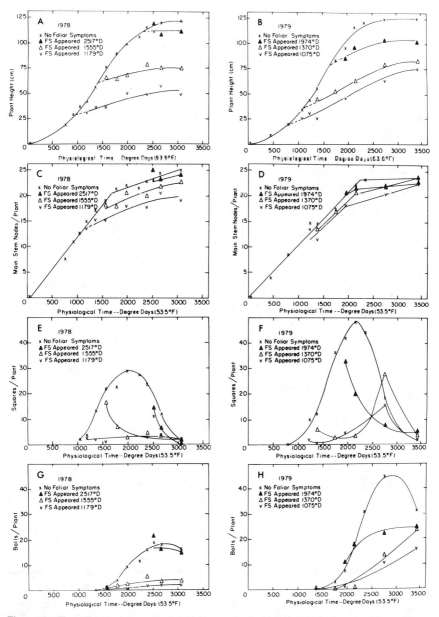

Figure 5 The effects of time of verticillium wilt symptom expression in field cotton plants during the 1978 and 1979 seasons on: (A,B) plant height; (C,D) mainstem node production; (E,F) square retention patterns; and (G,H) boll set patterns at the University of California West Side Field Station. (*Adapted from Pullman and DeVay, 1982b.*)

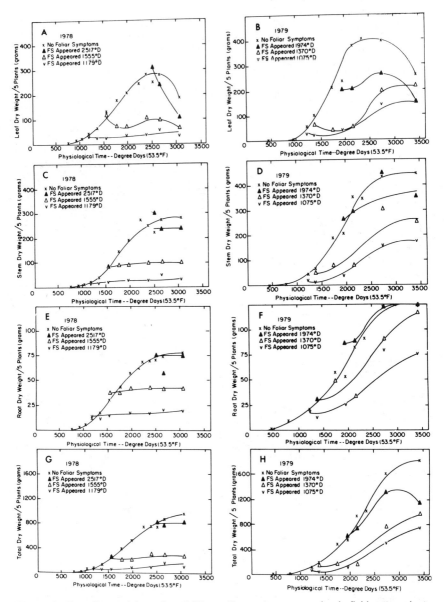

Figure 6   The effects of time of verticillium wilt symptom expression in field cotton plants during the 1978 and 1979 seasons on: (A,B) leaf dry weight; (C,D) stem dry weight; (E,F) tap root dry weight; and (G,H) total dry weight at the University of California West Side Field Station. (*Adapted from Pullman and DeVay 1982b.*)

with defoliating and nondefoliating pathotypes during the 1977 season at 50, 70, and 90 days after planting confirmed the relationships shown in Figs. 5 and 6 between time of symptom appearance and growth and fruitfulness (Fig. 7A to 7C). In addition, the studies showed the effect of pathotype disease severity on growth reduction (Friebertshauser and DeVay, 1982). Plants inoculated early, including those injected only with water, exhibited marked reductions in growth. The physiological explanation for the growth inhibition in control plants was related to wound stress, and not to decreased leaf water potential (Tzeng and DeVay, 1985). The reductions in biomass production in the control plants relative to the 90-day inoculation were 0.5 and 0.15 for the 50- and 70-day treatments (Fig. 7A to 7C).

In general, plants inoculated with the defoliating pathotype had the greatest inhibition of growth. The effects of time of infection, growth compensation by healthy neighbors as well as shading on end-of-season plant size are illustrated in Fig. 8. Competition between neighbors also occurs for water, nutrients, and other resources further slowing the growth of smaller plants.

Estimating the Effects of Shading and Plant Compensation

The degree of shading experienced by a cohort of plants may be approximated over time t as the ratio $\gamma(t) = L_I/L_H$ of the leaf mass L of diseased plants (subscript I) to that of healthy H neighbors. A more accurate measure of this competition would be to compare L_I to the average leaf mass of all plants showing symptoms later. As shading of plants affects their photosynthetic rates, the amount of photosynthate $P_I(t)$ produced by each diseased cohort at time t is scaled by $\gamma(t)$ (Fig. 8).

$$\gamma(t) = \frac{L_I(t)}{L_H(t)} \ \varepsilon \ [0,1] \tag{1}$$

Disease Progression in the Field

Infection of plants in the field, as measured by vascular discoloration, occurs approximately 200 degree days before foliar symptom expression, the exact time being a function of the number of root tips penetrating the soil, plant size, inoculum density (ρ_v), and pathotype virulence V (Pullman and DeVay, 1982b). Pullman and DeVay (1982a) showed that the percentage of plants showing foliar symptoms was a linear function of degree days, and propagule density $\rho_v = $ ppg (propagules per gram) of soil and probably pathotype virulence (Fig. 9). Foliar disease expression, however, may be suppressed when maximum temperatures are above 100° F (Fig. 10).

The prevalence of leaf symptoms is reduced as the number of plants per unit area are increased (Garber et al., 1981; Grimes and Huisman, 1984), but this factor is not included in the model. For the same pathotype, the slope $d\phi/dt$

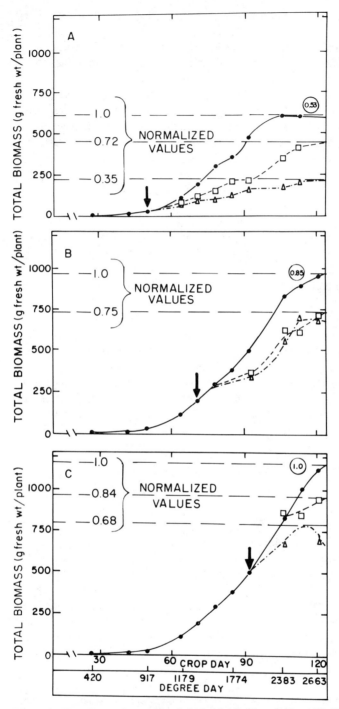

Figure 7 The effects of stem inoculation of field cotton plants at 50, 70, and 90 days after germination with defoliating (△) and nondefoliating (○) pathotypes of *Verticillium dahliae* and sterile water (●). (*Adapted from Friebertshauser and DeVay, 1982.*)

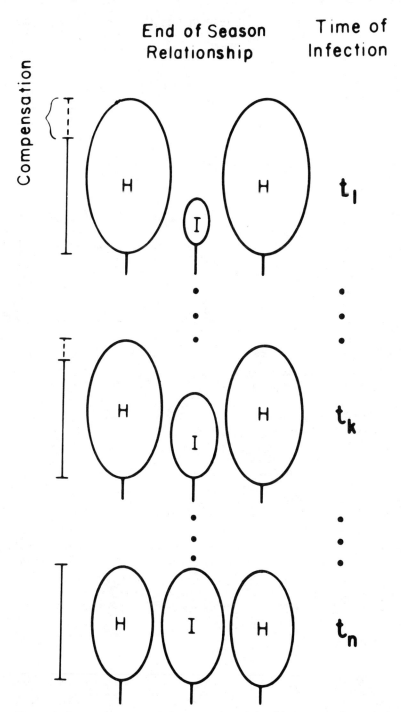

Figure 8 The conceptual end of season size relationship between plants showing verticillium wilt symptoms at different times in the season ($t_i = t_1, \ldots, t_k, \ldots, t_n$). The dotted line indicates compensation by healthy plants. (*Source: Gutierrez et al., 1983.*)

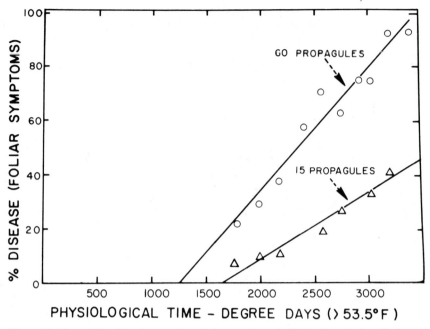

Figure 9 The relationship of percentage foliar symptoms in field cotton plants with degree days and propagule density of *Verticillium dahliae*. (*Source: Gutierrez et al., 1983.*)

of the disease progression function is upward convex with increasing propagule density saturating at 55 to 60 ppg soil (Fig. 11) and pathotype virulence. Thus,

$$\frac{d\phi}{dt} = f(\rho_v, V) = b_1\rho_v + b_2\rho_v^2 + b_3V^* \qquad (2)$$

where $d\phi/dt$ is the rate of change of the proportion of plants showing foliar symptoms at time $t > t^* = (1440 \text{ degree days} - 6 \, \rho_v) \, V^{*-1}$, and $V^* = V/V_s$ (Gutierrez et al., 1983). t^* is the time of first appearance of foliar symptoms, and V^* in Eq. 2 is an index relating the virulence of a pathotype V to a standard nondefoliating one (V_s). The slopes of plant infection and foliar symptom development are expected to be equal, but the latter is displaced by the 200 degree days time lag. The proportion of plants showing foliar symptoms is $0 < \phi(t) = (d\phi/dt)\Delta t^* < 1$, in which $\Delta t^* = t - t^*$.

The density of healthy plants at t is thus

$$\rho_{p,t+\Delta t} = \rho_{p,t} - C_t\rho_{p,0}\Delta t^* \frac{d\phi}{dt} \qquad (3)$$

where $\rho_{p,0}$ is the density of plants at t_0 and C is the plant compensation scalar described below.

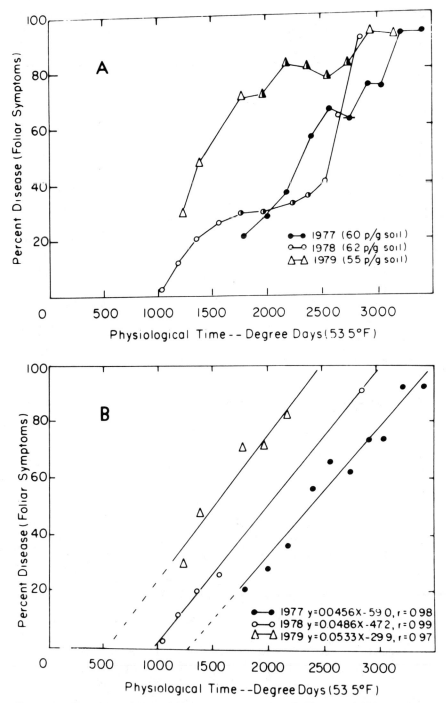

Figure 10 The effect of high ambient temperatures (>100° F) on verticillium wilt foliar symptom expression. (A) Uncorrected data, and (B) the same data corrected for high temperatures. (*Source: Pullman and DeVay, 1982a.*)

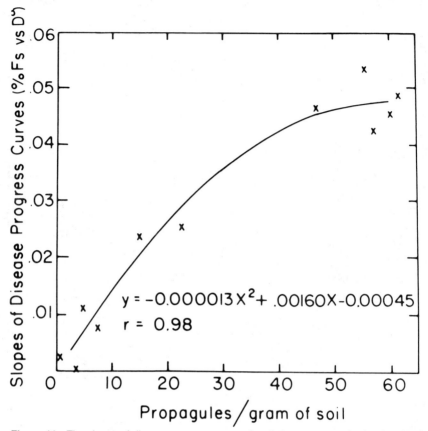

Figure 11   The slopes of disease progress curves in relation to propagule density of *Verticillium dahliae*. (*Source: Pullman and DeVay, 1982a.*)

Larger plants may compensate in terms of yield for diseased neighbors until the crop canopy closes (LAI $\approx$ 2.5), reflected in the model by a relaxation of density-dependent constraints on growth.

$$C_t = (1 - e^{-k,\text{LAI}}) \; \varepsilon \; [0,1] \text{ for LAI}_t \; \varepsilon \; [0,2.5] \qquad (4)$$

The convex scalar C_t slows the rate of compensation as the canopy closes by decreasing the rate of decline of ρ_p. Note that $C_t(d\rho_p/dt)$ equals μ_P in Eq. A1a (see appendix). Hence, plant compensation in the model occurs via the state variable ρ_p, as leaf, stem, root, and fruit growth rates increase as ρ_p decreases. Crop compensation for diseased plants is always incomplete due to time lags in growth and maturation. In areas where the growing season is very short, it may be relatively unimportant.

ECONOMICS OF PEST CONTROL

Economics of Verticillium Wilt Control

After the introduction and establishment of *V. dahliae* in a cotton field, propagule densities increase under continuous cotton culture at a rate of about 14 ppg soil per year (Fig. 12). However, this rate may be suppressed by cultural practices (Grimes and Huisman, 1984), fallowing or soil solarization (Pullman et al., 1981), extended soil flooding or paddy rice rotation (Pullman and DeVay, 1981), chemical methods, or a combination of these techniques. Increases in percentages of plants showing foliar symptoms, decreases in yield, and changes in population composition of the pathotypes (Pullman and DeVay, 1982; Ashworth, 1983) occur with increasing propagule densities. These and other factors strongly influence the economics of cotton production and grower profits.

Profit maximization is one of the major driving forces determining crop production methods. The simplest single-season (T), profit-maximization model we can propose for the cotton verticillium wilt system is

$$\Pi_{max}(T) = B(T) - U(T) - O(T) \tag{5}$$
$$X_1, X_2, \ldots, X_n$$

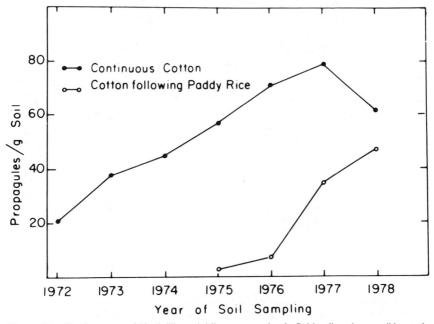

Figure 12   The increase of *Verticillium dahliae* propagules in field soil under conditions of continuous cotton and continuous cotton after paddy rice. (*Source: Pullman and DeVay, 1982a.*)

where Π is profit; $B(T) = P_c Y(T, \cdot)$; P_C is price; $Y(T, \cdot)$ is the yield function; $U(T)$ are the summed production costs; and $O(T)$ is the cost of controlling verticillium wilt subject to (s.t.) one or more of the X_i management or control practices described above. The farmer may decide not to control the pathogen if revenue increases are less than control costs (i.e., economic threshold), or if the profit from continued cotton production becomes less than profits from growing some alternative crop.

The function $Y(T, \cdot)$ above implies that yields are a function of plant density (ρ_v), variety (var), weather (W^*), propagule density (ρ_v), pathotype (V), agronomic practices (e) and other pests (ρ^*). Figure 13 shows the expected yield response surface $Y(\cdot) = Y(\cdot, W^*)$ using 1979 weather (W^*) from the University of California Westside Field Station in the southern San Joaquin valley for the cotton growing period of April 1 to September 15. The response surface is a least-squares fit of several hundred simulated yields for Acala SJ-2 cotton across the relevant ρ_p, ρ_v, V categories and weather W^* (see Gutierrez et al., 1983). Using a different weather pattern, the function would be different, but the results would be qualitatively the same.

While the model provides an intuitive feeling for the relative importance of each factor on yield, it tells us little about the dynamics determining the path of yields, $\rho_v(T)$ and $V(T)$ over a several year time horizon (T^*). For example, propagule densities increase in the soil over time when cotton is grown continuously (Gutierrez et al., 1983), and pathogen virulence is thought to attenuate if too high and to increase if virulence is too low. Concentrations of pathotypes are known to be influenced by their virulence (Schnathorst and Mathre, 1966b) and by the susceptibility of the cotton cultivar (Ashworth, 1983).

Models for yield, pathogen propagule density and virulence, and the effects of the various control practices on verticillium wilt can be linked simply in a simulation model with a 1-year time step, and used to evaluate possible paths of maximum profits over several years. Unfortunately, this is nothing more than a series of linked, myopic, single-season profit-maximization solutions, and the result is not likely to be optimal.

To find the optimal long-term solution we must reformulate the problem and use more sophisticated techniques of analysis (e.g., dynamic programming) from operation research. The long run ($T^* = 25$ years) profit maximization problem may be written as follows:

$$\max_{X_1, X_2, \ldots, X_n} \sum_{T=1}^{T^*} \Pi(T) = \sum_{T=1}^{T^*} [B(T, \rho_p, \rho_v V) - U(T) - O(T, X_i)]e^{-\delta T} \tag{6}$$

where X_1 is the set of $i = 1, \ldots, n$ possible pest control alternatives (i.e., solarization, fumigation, no control, or alternate crop) and $e^{-\delta T}$ is the discount rate of future profits. The solution is highly dependent on the prices of cotton,

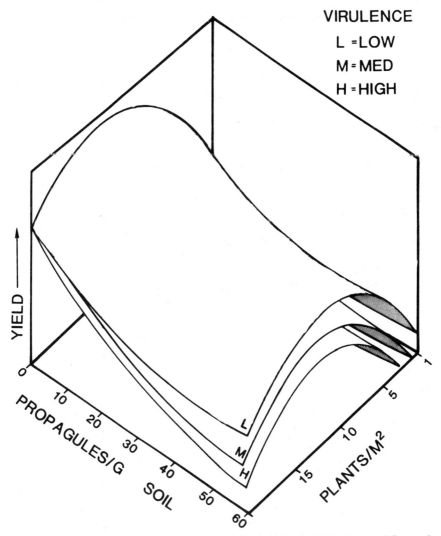

Figure 13 A general model showing expected yields in Acala SJ-II cotton as influenced by propagule densities and virulence of *Verticillium dahliae,* and plant density. (*Adapted from Gutierrez et al., 1983.*)

U and O, the effectiveness of the various X_1, the evolution of V, the discount rate, weather, and other factors. The solution provides not only the optimal path of profits, but also the paths of ρ_v, V, and X_1 (Gutierrez, Regev, Ellis, and DeVay in preparation). The ability to generate the benefit function (i.e., the cotton-verticillium wilt simulation model) enable us to examine not only the biology in detail, but also the economic consequences.

APPENDIX: Cotton Growth and Development Model

A population of cotton plants ρ_p consists of individual plants, and each plant has a population of leaves L, stem S, roots R, and fruits (F = number, M = mass) of varying age. The deterministic, discrete plant-population model proposed by Gutierrez et al. (1975) and Gutierrez and Wang (1976) and the continuous versions by Wang et al. (1977) and Gutierrez et al. (1984) provide the bases for modeling the effects of verticillium wilt on cotton growth, development, and yield (Eq. A1).

The population model for cotton growth and development is composed of separate models for the population of individual plants $\rho_p(t,a)$ per meter of row and for $L(t,a)$, $S(t,a)$, $R(t,a)$, $M(t,a)$, and $F(t,a)$ within plants at time t and age a. Because the crop is planted within a brief span of time $\rho_p(t,a)$ reduced to $\rho_p(t)$. The models and linkages (brackets) between plant parts and to an insect pest [e.g., *Lygus hesperus*, $H(t,a)$] are also shown. The model for the yearly dynamics of verticillium wilt shows this year's (T) dependence of ρ_v on the previous years ρ_p, ρ_v, V, and the effects of between-season management practices (X). The effects of disease on cotton growth and development are incorporated in a physiological way in the various birth-death rate functions (μ_i, $i = P,L,S,R,M,F$) described further on. The incorporation of the insect pest in the model points out the need for interdisciplinary research. In addition, several pests may affect the dynamics of the plant simultaneously.

$$\text{Plants} \qquad \frac{\delta \rho_p}{\delta t} + \frac{\delta \rho_p}{\delta a} = -\mu_p(\cdot)\rho_p(t,a) \tag{A1a}$$

$$\text{Leaves} \qquad \frac{\delta L}{\delta t} + \frac{\delta L}{\delta a} = -\mu_L(\cdot)L(t,a) \tag{A1b}$$

$$\text{Stem} \qquad \frac{\delta S}{\delta t} + \frac{\delta S}{\delta a} = -\mu_s(\cdot)S(t,a) \tag{A1c}$$

$$\text{Root} \qquad \frac{\delta R}{\delta t} + \frac{\delta R}{\delta a} = -\mu_R(\cdot)R(t,a) \tag{A1d}$$

$$\text{Fruit (mass)} \qquad \frac{\delta M}{\delta t} + \frac{\delta M}{\delta a} = -\mu_M(\cdot)M(t,a) \tag{A1e}$$

$$\text{(number)} \qquad \frac{\delta F}{\delta t} + \frac{\delta F}{\delta a} = -\mu_F(\cdot)F(t,a) \tag{A1f}$$

$$\text{Lygus bug} \qquad \frac{\delta H}{\delta t} + \frac{\delta H}{\delta a} = -\mu_H(\cdot)H(t,a) + I_H \tag{A1g}$$

$$\text{Verticillium wilt } \frac{d\rho_v}{\delta T} = f(\rho_P(T-1), \rho_v(T-1), V(T-1), X) \tag{A1h}$$

Brackets: Type — Predation; Supply/Demand — Density Dependence

where $C(t,a)$, $F(t,a)$, and $H(t,a)$ are number density functions; $L(t,a)$, $R(t,a)$, $S(t,a)$, and $M(t,a)$ are mass density functions; and all require initial conditions (e.g., $L(t,0)$ and $L(0a,)$ to guarantee the uniqueness of the solution. Also, t and a are in physiological time units and may differ between species. The various $\mu_i(\cdot)$ are complicated functions containing all factors affecting population birth and death rates (see below), while I_H is the net immigration rate of *Lygus* bugs. The models in Eq. A1 are the continuous form (von Forester, 1959) for the well-known Leslie matrix model (Leslie, 1945), and each can be written in discrete form as

$$N_j, t + \Delta t = A_j N_j, t \qquad (A2)$$

where the N_j are the vectors of plants, plant parts, or *Lygus* bugs of various ages, and the A_i are the appropriate time-varying matrices of birth and survivorship rates for each of the j populations.

For the plant, photosynthesis, $P(\ )$ and new plant-part production are analogous to birth. $P(\cdot) = P(\text{age, LAI}, \rho_v, V, \cdot)$ and is apportioned to the various plant demands according to the priority scheme of the metabolic pool model described later. Verticillium wilt affects P directly via propagule density ρ_v and pathogenicity V, and affects the birth and death rates of new plant parts (e.g., F) and plant part growth (e.g., M) by slowing the flow of photosynthate to them. As mentioned above, these dynamics are included in μ_i.

The effects of *Lygus* bugs or any other fruit pest occurs via linkages of its population $L(t,a)$ to the fruit dynamics model. The *Lygus* feeding rate is part of $\mu_F(\cdot)$ and $\mu_M(\cdot)$ (see Gutierrez et al., 1977).

METABOLIC POOL MODEL AND THE EFFECT OF VERTICILLIUM WILT ON COTTON PLANT GROWTH AND DEVELOPMENT

In the model, growth and development of the plant(s) are controlled by priority allocation of photosynthate P from a metabolic pool that goes first to respiration θ, then fruit growth M, and lastly to vegetative growth ($L, S, R,$) and reserves that has a maximum value $Q_{max} = 0.2 \int (L + S + R) da$ (see Fig. 3). The value 0.2 is the maximum proportion of dry matter that may be reserves in cotton. The first priority of plants (infected I, and healthy H) is to meet basic respiratory cost, hence if

$$Q(t) = Q(t - \Delta t) + \gamma(t) \cdot V \cdot P(t) - \theta \int (L + R + S + M) da < 0$$

then the plant dies. The constant θ is the respiration per unit mass of plant per Δt and $\gamma \epsilon [0,1]$ is the shading scalar (see later). In the last two priority levels, the remaining photosynthate is allocated as follows: if $Q = Q(t) > 0$,

1. Q is allocated first to fruit if $t > t_{FFB}$, *where* t_{FFB}, is the time of first fruit (see Fig. 1).
2. and then proportionally to leaf, stem, and root if $Q - \Delta M^*(t) > 0$, in which ΔM^* is the maximum (*) demand for fruit population growth during Δt.

The effects of $Q - \Delta M^* > 0$, and verticillium wilt (V) on growth and production rates of plant parts (i.e., the birth rate) are modeled as follows:

$$(mass) \ \Delta M_t = \frac{dM^*(a)}{dt} \cdot r_1 \cdot \Delta t \ V \qquad 0 \le r_1 = \frac{Q}{\Delta M^*} \le 1 \qquad \text{(A2a)}$$

$$(numbers) \ \Delta F_t = \frac{dF^*(\rho_p)}{dt} \cdot r_2 \cdot \Delta t \cdot V \qquad 0 \le r_2 = \frac{Q - \Delta M^*}{\Delta W^*} \le 1 \quad \text{(A2b)}$$

$$(Leaf + stem + root) \ W_t = \left[\frac{dL^*(\rho_p)}{dt} + dS^* \ (\rho_p) + \frac{dR^* \ (\rho_p)}{dt} \right] \cdot r_2 \cdot \Delta t \cdot V \quad \text{(A2c)}$$

in which r_1 and r_2 are supply/demand ratios for photosynthate allocation to fruit and vegetative growth, respectively, ρ_P is plants per meter of row and W is the change in vegetative mass. Note that, above, the augmentation of fruit numbers $F(t,0)$ as well as mainstem node production N are regulated by r_2 and modeled as follows:

$$\text{Mainstem nodes} = N_t = \frac{dN(\rho_p)}{dt} \cdot r_2 \cdot \Delta t \cdot V \qquad \text{(A3)}$$

The maximum growth rates for all plant parts except those for mass accumulation by fruits f (age) are functions of plant density.

Excess immature fruit (F,M) are shed when $r_1 < 1$ (i.e., t_s in Fig. 1) at rates proportional to the deficit (i.e., death rates); see Wang et al. (1977) for details. Thus we postulate that verticillum wilt affects not only the flow rate of photosynthates within the plant, but also photosynthate supply which, interacting with demand, controls all aspects of plant growth and development (Fig. 3).

REFERENCES

Ashworth, L. J., Jr., 1983. Aggressiveness of random isolates of *Verticillium dahliae* from cotton and the quantitative relationship of internal inoculum to defoliation. *Phytopathology* 73:1292–1295.

Butterfield, E. J., and DeVay, J. E. 1975. Association of a bacterium with microsclerotia of *Verticillium dahliae* inhibited by soil fungistasis (Abstract.) *Proc. Am. Phytopathol. Soc.* 2:41.

———. 1977. Reassessment of soil assays for *Verticillium dahliae. Phytopathology* 67:1073–1078.

Charudattan, R., and DeVay, J. E. 1972. Common antigens among varieties of *Gossypium hirsutum* and isolates of *Fusarium* and *Verticillium* species. *Phytopathology* 62:230–234.

DeVay, J. E., Forrester, L. L., Garber, R. H., and Butterfield, E. J. 1974. Characteristics and concentration of propagules of *Verticillium dahliae* in air-dried field soils in relation to the prevalence of Verticillium wilt in cotton. *Phytopathology* 64:22–29.

Friebertshauser, G. E., and DeVay, J. E. 1982. Differential effects of the defoliating and nondefoliating pathotypes of *Verticillium dahliae* upon the growth and development of *Gossypium hirsutum. Phytopathology* 72:872–877.

Fryxell, A. A. 1981. The natural history of the cotton tribe (Malvaceae, tribe Gossypieae). Texas A&M Univ. Press, College Station, 245 pp.

Garber, R. H. 1973. Fungus penetration and development, in: *Verticillium Wilt of Cotton, Proc. Work Conf.* C. D. Ranney, ed. pp. 69–77. National Cotton Pathology Research Laboratory, College Station, Texas. USDA-ARS-S-19, 134 pp.

Garber, R. H., DeVay, J. E., and Wakeman, R. J. 1981. Effect of plant spacing and symptom expression in Verticillium wilt of cotton, p. 30. *Proc. Beltwide Cotton Prod. Res. Conf. Nat. Cotton Council, Memphis,* 321 pp.

Green, R. J., Jr. 1980. Soil factors affecting survival of microsclerotia of *Verticillium dahliae. Phytopathology* 70:353–355.

Grimes, D. W., and Huisman, O. C. 1984. Irrigation scheduling Verticillium wilt interactions in cotton production, in: *California Plant and Soil Conf. Proc. 1984.* pp. 88–92. American Society of Agronomy, Sacramento, 179 pp.

Gutierrez, A. P., DeVay, J. E., Pullman, G. S., and Friebertshauser, G. E. 1983. A model of Verticillium wilt in relation to cotton growth and development. *Phytopathology* 73:89–95.

Gutierrez, A. P., Falcon, L. A., Loew, W. and Leipzig, P. A. 1975. An analysis of cotton production in California: A model for Acala cotton and the effects of defoliators on its yield. *Environ. Entomol.* 4:125–136.

Gutierrez, A. P., Leigh, T. F., Wang, Y., and Cave, R. D. 1977. An analysis of cotton production in California: *Lygus hesperus* injury—an evaluation. *Can. Entomol.* 109:1375–1386.

Gutierrez, A. P., Pizzamiglio, M. A., Dos Santos, W. J., Villacorta, A. M., and Tennyson, R. E. 1984. A general distributed delay time varying life table plant population model: Cotton (*Gossypium hirsutum* L.) growth and development as an example. *Ecol. Model.* 26:231–249.

Gutierrez, A. P., and Wang, Y. H. 1976. Applied population ecology: Models for crop production and pest management, in: *Proc. Conf. Pest Management* (G. P. Norton and C. S. Holling, eds.), pp. 255–280. International Institute for Applied Systems Analysis, Laxenburg, Austria.

Hafez, A. A. R., Stout, P. R., and DeVay, J. E. 1975. Potassium uptake by cotton in relation to Verticillium wilt. *Agron. J.* 67:359–361.

Harper, J. L. 1977. *Population Biology of Plants.* Academic Press, London, 892 pp.

Kiyasawa, S. 1972a. Mathematical studies on the curve of disease increase: A technique for forecasting epidemic development. *Ann. Phytophathol. Soc. Japan* 38:30–40.

————. 1972b. Theoretical comparisons between mixture and rotation cultivation of disease resistant varieties. *Ann. Phytophathol. Soc. Japan* 38:52–59.

Kiyasawa, S., and Shigomi, M. 1972. A theoretical evaluation of the effect of mixing resistant variety with susceptible variety for controlling plant diseases. *Ann. Phytophathol. Soc. Japan* 38:41–51.

Kranz, J. 1979. Simulation of epidemics caused by *Venturia inaequalis* (Cooke) Aderh. *European Plant Protection Organization Bull.* 9:235–240.

Leslie, P. H. 1945. On the use of matrices in certain population mathematics. *Biometrika* 35:213–245.

Manetsch, T. J. 1976. Time varying distributed delays and their uses in aggregate models of large systems. *I.E.E.E. Trans. Sys. Man. and Cybernet.* 6:547–553.

Marois, J. J., and Adams, P. B. 1984. Frequency distribution analysis of lettuce drop, caused by *Sclerotinia minor* as a function of quadrat size. *Phytopathology* 74:839.

McKinion, J. M., Jones, J. W., and Hesketh, J. D. 1974. Analyiss of SIMCOT: Photosynthesis and growth, in: *Proc. Beltwide Prod. Res. Conf.,* pp. 117–124, National Cotton Council, Memphis.

Puhalla, J. E. 1979. Classification of isolates of *Verticillium dahliae* based on heterocaryon incompatibility. *Phytopathology* 69:1186–1189.

Pullman, G. S., and DeVay, J. E. 1981. Effect of soil flooding and paddy rice culture on the survival of *Verticillium dahliae* and incidence of Verticillium wilt in cotton. *Phytopathology* 71:1285–1289.

―――. 1982a. Epidemiology of Verticillium wilt of cotton. A relationship between inoculum density and disease progression. *Phytopathology* 72:549–554.

―――. 1982b. Epidemiology of Verticillium wilt of cotton: Effects of disease development on plant phenology and lint yield. *Phytopathology* 72:554–559.

Pullman, G. S., DeVay, J. E., Garber, R. H., and Weinhold, A. R. 1981. Soil solarization: Effects of Verticillium wilt of cotton and soil-borne populations of *Verticillium dahliae, Pythium* spp., *Rhizoctonia solani,* and *Thielaviopsis basicola. Phytopathology* 71:954–959.

Rabbinge, R., and Rijsdijk, F. H. 1983. EPIPRE: A disease and pest management system for winter wheat, taking account of micrometerological factors. European Plant Protection Organization Bull. 13:297–305.

Ranney, C. D. 1973. Verticillium wilt of cotton, Proc. Work Conference, Aug. 30 to Sept. 1, 1971. National Cotton Pathology Research Laboratory, College Station, Texas, 134 pp. Publication ARS-S-19. ARS-USDA, P.O. Box 53326, New Orleans, LA 70153.

Rapilly, F. 1979. Simulation of an epidemic of *Septoria nodorum* Berk. on wheat in relation to the possibility of horizontal resistance. *E.P.P.O. Bull.* 9:243–249.

Rijsdijk, F. H. 1975. A simulator of yellow rust of wheat. *Bull. Rech. Agron. Gembloux.,* pp. 411–418.

Royle, D. J., and Butt, D. J. 1979. The place of multiple regression analysis in modern approaches to disease control. *European Plant Protection Organization Bull.* 9:155–163.

Schnathorst, W. C., and Mathre, D. E. 1966a. Host range and differentiation of a severe form of *Verticillium albo-atrum* in cotton. *Phytopathology* 56:1155–1161.

Schnathorst, W. C., and Mathre, D. E. 1966b. Cross-protection in cotton with strains of *Verticillium albo-atrum. Phytopathology* 56:1204–1209.

Schreiber, L. R., and Green, Jr., R. J. 1962. Comparative survival of mycelium, conidia, and microsclerotia of *Verticillium albo-atrum* in mineral oil. *Phytopathology* 52:288–289.

Shapovalov, M., and Rudolph, B. A. 1930. *Verticillium hadromycosis* (wilt) of cotton in California. *Plant Dis. Rep.* 14:9–10.

Shrum, R. 1975. Simulation of wheat stripe rust (*Puccinia striiformis* West) using EP-IDEMIC, a flexible plant disease simulator. *Penn. State Univ. Coll. of Agr. Agr. Exp. Sta. Prog. Rep.* 347, 41 pp.

Smith, V. L., and Rowe, R. C. 1984. Characteristics and distribution of propagules of *Verticillium dahliae* in Ohio potato field soils and assessment of two assay methods. *Phytopathology* 74:553–556.

Tzeng, D. D., and DeVay, J. E. 1985. Physiological responses of *Gossypium hirsutum* L. to infection by defoliating and nondefoliating pathotypes of *Verticillium dahliae* Kleb. *Physiol. Plant Pathol.* 26:57–72.

Tzeng, D. D., Wakeman, R. J., nd DeVay, J. E. 1985. Relationships among Verticillium wilt development, leaf water potential, phenology, and lint yield in cotton. *Physiol. Plant Pathol.* 26:73–81.

VanderMolen, G. E., Labavitch, J. M., Strand, L. L., and DeVay, J. E. 1983. Pathogen-induced vascular gels: Ethylene as a host intermediate. *Physiol. Plant.* 59:573–580.

Van der Plank, J. E. 1963. *Plant Diseases: Epidemics and Control*. Academic Press, New York, 349 pp.

von Forester, H. 1959. Some remarks on changing populations, in: *The Kinetics of Cellular Proliferation* (F. Stohlman, Jr., ed.), pp. 382–407. Grune and Stratton, New York, 456 pp.

Waggoner, P. E. 1981. Models of plant disease. *BioScience* 31:315–319.

Waggoner, P. E., and Horsfall, J. G. 1979. EPIDEM, a simulator of plant disease written for a computer. *Conn. Agr. Exp. Sta. Bull.* 698, 80 pp.

Waggoner, P. E., Horsfall, J. G., and Lukens, R. J. 1972. EPIMAY, a simulator of southern corn leaf blight. *Conn. Agr. Exp. Sta. Bull.* 729, 84 pp.

Wang, Y., Gutierrez, A. P. Oster, G., and Daxl, R. 1977. A population model for plant growth and development: Coupling cotton-herbivore interaction. *Can. Entomol.* 109:1356–1374.

Zadoks, J. C. 1971. Systems analysis and the dynamics of epidemics. *Phytopathology* 61:600–610.

———. 1979. Simulation of epidemics: Problems and applications review. European Plant Protection Organization Bull. 9:227–233.

Zadoks, J. C., and Kampmeijer, P. 1977. The role of crop populations and their deployment, illustrated by means of a simulator EPIMUL 75. *Ann. N.Y. Acad. Sci.* 287:164–190.

Zadoks, J. C., and Rijsdijk, F. H. 1972. Epidemiology and forecasting of cereal rusts, studied by means of a computer simulator normal EPISIM. *Proc. Eur. Mediterr. Cereal Rust Conf. Prague* 1:293–296.

Chapter 10

Systems Analysis in Epidemiology

Robert C. Seem
Douglas A. Haith

The essence of the systems approach, therefore, is confusion as well as enlightenment. The two are inseparable aspects of human living. . . .There are no experts in the systems approach.

C. West Churchman
The Systems Approach

Systems analysis, or the systems approach, is a procedure to facilitate logical and coherent decision making. It is a methodology for dealing with complex problems that involves *systems*. Following Churchman, "a system is a set of parts coordinated to accomplish a set of goals." The systems approach includes (1) definition of total systems objectives and measures of performance, (2) description of the system's environment, (3) determination of system resources, (4) analysis of system components and interactions, and (5) management of the system (Churchman, 1968). A key element of systems analysis is the notion of *optimality*. Any realistic problem has at least several possible solutions, and optimization is the selection of the solution that appears to be most consistent with system goals. Note that optimization is relative; it depends on the analyst's perception of the system and his or her awareness of potential solutions.

Other descriptions of the systems approach vary somewhat from Churchman's. However an emphasis on decision making, system interactions, management, control, or regulation are common to a range of systems applications in ecology (Jeffers, 1978), environmental quality (Haith, 1982), agricultural production (Agrawal and Heady, 1972), pest management (Ruesink, 1976) and plant disease management (Zedoks and Schein, 1979).

Epidemiology is the science of disease in populations (Van der Plank, 1963). Systems analysis in epidemiology refers to the *management* of plant disease through use of the systems approach. Zadoks and Schein (1979) indicate that plant disease management is related to the concepts used in integrated pest management, and involves an appreciation of system interactions and alternative management options. The goal is a plant disease management program that regulates plant disease in an economically optimal fashion.

AGRICULTURAL SYSTEMS

Management Applications

Following World War II and the digital computer revolution of the 1950's and 1960's, system analysis, otherwise known as operations research or management science, became a popular and increasingly standardized approach to industrial engineering and business management. The computer played an essential role, since description of systems components and evaluation of management options often require massive manipulations of large data sets by mathematical models. Without high-speed computers, the analysis of most real systems would be infeasible.

Management applications of systems analysis in agriculture followed the industrial experience. Linear programming models have become standard tools for farm planning and development of animal nutrition programs (Agrawal and Heady, 1972). Optimization and mathematical modeling have many applications in food processing and distribution (Bender et al, 1976). Machine scheduling, forage management, waste disposal, and environmental pollution are only several of the many agricultural problems to which systems analysis has been applied. In general, these applications followed the broad outline of the systems approach and were comparable to industrial situations. The emphasis was on decision making, and models were used to evaluate alternative management procedures with the goal of maximizing profits or minimizing costs.

Biological Applications

Agricultural production requires the management of biological systems. These systems, involving plants, animals, soil microbes, insects, and pathogens, are often less well-defined and understood than the physical systems managed by the industrial engineer. Their study is clearly the subject matter of ecology, which explores the relationships among organisms and their environment. Ecology is very much a systems science, and ecosystems, or, as related to agriculture, agroecosystems, are ecology's fundamental unit of analysis.

In spite of the systems nature of ecology, the application of the decision-making and optimization features of the systems approach to biological problems has not been simple. Although such applications have been demonstrated (e.g., Dent and Anderson, 1971; Norton et al., 1983; Shoemaker, 1973), the biologist's implementation of the systems approach has often neglected decision making. Thus, although Jeffers (1978) provides a description of systems analysis that is remarkably similar to that of Churchman (1968), the overall thrust of his book of ecological applications is on the mathematical modeling of ecological processes and not on decision making or control. Similarly, Ruesink (1976), in his review of systems analysis and pest management, discusses system objectives and control, but observes that often the only objective of systems analysis is the building of a model. He concludes that

Nearly all the effort put into the systems approach to pest management has concentrated on developing models to describe the existing system and that little time has been spent on evaluating alternative strategies.

Kranz and Hau (1980) in their comparable review of systems analysis and epidemiology again note that the purpose of systems analysis is to "ensure an optimal choice between alternatives," but they go on to describe systems analysis as primarily a modeling activity. It is scarcely surprising that they observe that "plant pathologists thus tend to think that systems analysis is synonymous with simulation of epidemics."

Mathematical Modeling and Decision Making

Systems analysis generally requires a mathematical model to describe system interactions and to estimate the system response to management or control alternatives. However, modeling per se is not equivalent to systems analysis. It appears that with the notable exception of certain integrated pest management programs, most "systems studies" of plant growth, insect behavior, and plant disease epidemiology are fundamentally modeling studies with little emphasis on decision making (Mumford and Norton, 1984; Norton, 1982).

The lack of a decision-making thrust in biological models is hardly surprising. It is not clear that many ecosystems have "objectives" that can be defined in any meaningful way. Furthermore, the concepts of management and optimal control are hardly integral to biological research.

Systems Analysis as a Unique Research Area
in Plant Disease Epidemiology

If the plant disease management philosophy of Zadoks and Schein (1979) is to be considered an important direction in future epidemiological research, then the systems approach, with its emphasis on decision making, alternative control strategies, and optimization should be of great value. However, systems analysis should not be viewed as universally applicable to all plant disease research, nor should it be considered a vague general philosophy that encourages "systems thinking." Rather, it is a highly specialized area of plant pathology that is the province of the plant scientist who understands both biological systems and decision making.

SYSTEMS ANALYSIS IN PLANT DISEASE MANAGEMENT

To the extent that modern plant disease epidemiologist work to manage the dynamic interactions of plant, pathogen, and environment, they may be de facto systems analysts. Furthermore, sound disease management programs may be produced by combinations of experience, intuition, and creativity without the benefit of systems analysis. Nevertheless, there are advantages in a structured approach to both research in, and operational management of, plant disease. In principle, systems analysis should provide a framework for the development of combinations of measures that most effectively control disease.

This section first discusses how the systems approach, as described by Churchman (1968), may be applied to problems in epidemiology. Such an undertaking involves more than a simple translation of Churchman's five steps to the context of disease management. For example, the testing or evaluation of system models is a much more critical component in the systems analysis of biological problems than it is in industrial applications. The second part of the section illustrates aspects of the systems approach by means of several plant disease management examples.

STEPS OF THE SYSTEMS APPROACH

Steps of the systems approach, as applied to plant disease management, are outlined in Fig. 1. Whether viewed as research activity or an operational procedure for the plant pathologist, systems analysis is applied to a specific disease control problem that requires solution. The ultimate output of the approach is a disease management program that meets a set of defined objectives.

Objectives

The first several steps of the systems approach are what Norton (1982) terms *descriptive analysis*. Thus, we must first describe the problem in a logical, structured fashion before we can solve it. The initial step is definition of objectives. This is a far more difficult task than it may appear. For example, control of disease x on crop y is a totally inadequate objective since "control" is ambiguous. Eradicate disease x on crop y is much more specific, but it may be both impractical and unnecessary. In the context of plant disease management, the control or regulation of disease at a subeconomic level would appear to be a

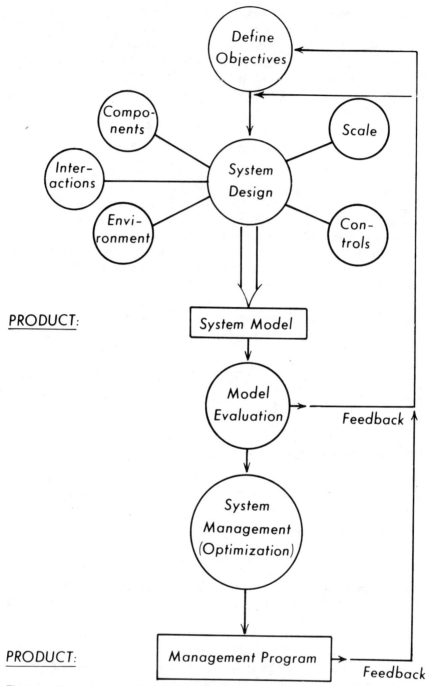

Figure 1 General steps in the systems approach.

more feasible, and probably sounder, long-term objective (Zadoks and Schein, 1979; Robinson, 1976). However, even this objective lacks specificity, since "subeconomic levels" must be defined. Does a subeconomic level of disease imply no economic damages or, perhaps, is it a control level at which further regulation would be more expensive than the savings due to avoided damage? Furthermore, objectives related to environmental quality, safety, risks of crop failure, and management practicality may also be important.

Objectives are important because they provide the focal points for the entire systems analysis. If they do not reflect the real issues of a disease control problem, the systems analysis will not produce an appropriate management program.

System Design

System design involves the detailed desription of the problem to be solved using those factors that the analyst feels are most influential in the behavior relative to the objectives. An important element of systems design is the scale of the system both in time and space. Does the problem express itself in a minute, a month, an infection period, or a growing season? Similarly, does the problem express itself on a leaf of a single plant or in a whole geographic region? The scale of the system is usually dictated by the original objectives. It is often convenient to describe the spatial bounding of a system in the form of a spatial hierarchy where the particular level of the hierarchy represents an area of interest for the problem. Many different hierarchies have been described but they usually include atoms, molecules, cells, tissues, organs, organisms, populations, etc. The advantage of using a hierarchy is that it serves as a logical framework on which to organize information about the system.

System *components* are identified that are appropriate to the problem scale. Components might include a plant part, a whole plant, or plant populations. Related farming activities such as irrigation, fertilization, and planting method may be system components that impinge on disease incidence or severity and control procedures. The system *environment* refers to a set of factors such as weather, soil type, and economic resources and conditions, which influence the plant disease system, but are not directly manageable. These factors are external to the problem and hence the system being designed.

The selection of system boundaries, i.e., the determination of what is internal and external to the system, is subjective, and hence arbitrary. A fundamental limitation of systems analysis is that a system is always embedded in a larger system (Churchman, 1968).

A key to the success of a systems analysis effort is the selection of *controls* or *management options*. These are the possible measures that could be taken to solve the plant disease problem. Controls will depend greatly on the problem and system scale. Solution of a short-term problem in a farmer's field may limit controls to selection of chemicals and application strategies, whereas long-term regional problems permit a wider range of measures to be considered, including development of new varieties and changes in cropping systems.

The final element of systems design is a description of *interactions* among components, controls, and the environment. Such descriptions are basically accounting procedures which allow one to trace the impacts of a management measure through its effects on system components and ultimately on objectives. This accounting process is generally, but not always, a mathematical model that simulates the response of the designed plant disease system to both environmental and management factors.

System Model

A classic view of systems analysis considers the system model that results from system design to be only a means to an end. It is an efficient computational procedure that allows us to evaluate alternative management programs. However, this narrow perspective is not realistic for biological systems. A biological system model may be a very important, and sometimes the only, product of the analysis. The reduction of a biological problem to a system design must generally be considered a remarkable accomplishment. Given the limited understanding, data, and management sophistication that we often bring to biological processes, the integration of knowledge into a system model may be a powerful research and educational exercise that expands our fundamental understanding of a plant disease problem.

It sometimes happens that the systems approach may be truncated well short of the development of a management program. This may be due to many factors, including a lack of attention to objectives, failure to properly bound a system design, development of a system model that is either too elementary to accurately describe the problem or too complex to be computationally feasible, or simply an exhaustion of resources. Such truncated systems analyses should not always be considered failures, particularly when they lead to improved understanding.

Model Evaluation

The final process in the descriptive-analysis phase of systems analysis is to determine whether the simplifications of real-world events have been identified and related in a plausible form in the system model. Actually the evaluation process should be continuous in that each step of the systems analysis development should evaluate what has transpired to that point and should determine if any correction or changes are necessary. This is often called a feedback process. If the design is satisfactory then the system model should behave in the same manner as its real-world counterpart. Failure of this to happen indicates deficiencies in the design (selection of the major components) or in the interrelationships within the system and/or its environment.

Evaluation of systems and their models is one of the more poorly documented activities of systems analysis. Evaluating if the system model behaves as the analyst intended is often called verification (Teng, 1981). Verification is usually an empirical comparison of the system model to its original design and intent. It is an important step in evaluation since acceptable system output may conceivably arise from a biologically untenable design flaw. More rigorous procedures have been developed to test a system's fidelity or trueness to the real world. Teng (1981) presents some useful methods of model validation that can be generally adapted for systems validation. The whole evaluation process should be based on the original objectives of the systems analysis. While some may feel elegance lies in a model with realistic complexity, others seek parsimony (the simplest system to get the job done) for management decision making. The divergence in modeling goals leads to a seemingly contradictory observation that because system models are based on scientific hypotheses, they can never be validated, only invalidated (Weigert, 1975). Yet Welch et al. (1981) warn analysts that because their models usually arise in scientific environments, there is a tendency to expend inordinate efforts to invalidate models based on scientific criteria when in fact, the models may be quite acceptable for management decision making.

System Management (Optimization)

System management is the use of the system model to evaluate alternative disease control strategies. These strategies are combinations of the controls identified in the system description and modeling phase, and it is at this point that information for decision making is produced. The main difference from descriptive analysis is that in system management the system model becomes a tool *for* (not *of*) experimentation. By rationally altering the design and management structure of the system, different expected outcomes can be derived as well as the probabilities of those outcomes. Thus we can compare various management strategies by their impacts on problem objectives.

   This phase of systems analysis may also be called optimization, since ideally we wish to identify that management strategy that best meets our objectives. As we noted earlier, optimization is a relative concept, since it depends on a particular statement of objectives and system design. A number of optimization techniques are available, including analytic optimization, simulation or search, and decision theory. The choice of method depends on such factors as system objectives, complexity of the described system (number of variables), and personal preferences.

   Simulation   Management decisions can be based on the experience "learned" by trying different policies under different conditions of a simulated system. Thus successive "experiments" could be conducted to evaluate a tactic—like fungicide timing—while holding other factors, such as weather patterns or pathogen inoculum, precisely constant. Onstad and Rabbinge (1985) differentiate between two types of simulation analyses: trial-and-error simulation and exhaustive stimulation. Trial-and-error simulation uses the knowledge and experience of the analyst as a guide to find a *good* management policy without expending too much money and time. Exhaustive simulation involves many more simulations of the system and will usually find a *better* management policy but it is impossible to perform on complex systems because of time and money limitations.

Analytic Optimization   This is a numerical method that calculates the optimal management decisions without recalculating the system model for each possible alternative (Shoemaker, 1981). For example, if one of three different fungicides can be applied to a crop 5 different times during a growing season then there are 3^5 or 243 alternative schedules. There are several different types of optimization methods, but we will only consider one of the more commonly used methods: dynamic programming (Shoemaker, 1981). In the above example there were 5 times during the growing season when a decision concerning fungicide application had to be made. Dynamic programming breaks the problem down by considering a single decision at each of the successive stages in time. Not all combinations of the decisions are analyzed; rather, a set of possible decisions is made according to the principle of optimality (Bellman, 1957). The method allows incorporation of new information (observations of random events such as weather) at any time, it incorporates fixed costs of management activities, and it can handle a wide range of initial values (Onstad and Rabbinge, 1985). Dynamic programming algorithms usually consist of a unique set of instructions and rules for the particular model being solved. The model of the real system must also be incorporated in the dynamic programming algorithm in order to "transfer" information about the system from one decision point to the next. The model can also be used to calculate economic cost and benefit functions. The major restriction of dynamic programming is that large models with many variables can be too complex for numerical analysis. Because many crop-pest systems have a large number of variables to describe the interactions in the system (analyzed best by simulation) and they also present a large number of decision options (analyzed best by analytic optimization), Shoemaker (1981) suggests using a combination of the two methods.

Decision Theory   Decision theory is an analytic method that attempts to incorporate uncertainties into the decision process. The method is a form of gaming theory that has wide application in business and industry (Halter and Dean, 1971). It is now finding useful application in crop production and protection. For example, the decision to apply a protective fungicide is plagued with all sorts of uncertainties: Will the pathogen be present after the fungicide is applied? Will a rain storm wash the fungicide off the plant? Will an infection period occur while the fungicide is still active? In the context of system analysis, uncertainty arises because the decision maker does not know the future state of the system. It is possible, however, to assess the probability of some future event occurring either from experience or by analyzing the system. Having completed the description of a system, a researcher can compute expected outcomes, and, if the system is stochastic, the probabilities of the expected outcomes can be determined by repeated analyses. The outcomes can then be weighted by their expected probabilities to form the basis for choosing the optimal solution. The computations for decision theory are relatively straightforward; however, it is often difficult to define or develop the probabilities of expected outcomes.

We should point out an important difference between probabilities associated with classical stochastic processes and Baysian-type probabilities. The former is assumed to be uncertainty due to some random process whereas the Baysian probabilities can be thought of as uncertainty based on *willingness* to take a risk. Although the two probabilities are often interchanged, one should always keep in mind these subtle differences.

Management Program

The ultimate product of the systems approach is a management program to control the disease. However, this may not end the analysis. The program may not work as intended or may be unacceptable to growers. There are many possible causes for this unfortunate outcome. Relevant objectives may have been neglected and/or system design may be incomplete. Sometimes, however, program failure is due to the changing nature of the disease problem. Since the analysis began, objectives and disease characteristics may have changed. Thus, here, as in other steps of the systems approach, feedback is often necessary.

EXAMPLES

Systems analysis is a new approach to plant disease epidemiology, and it is not surprising that there have been few documented case studies. None of the following examples cover all the steps of the systems approach shown in Fig. 1. Nevertheless, the examples demonstrate the feasibility of systems approach to plant disease problems. All four examples strive to manage disease by determining the best time to apply a fungicide. However, the approaches to achieve the best decision are quite different. Readers who desire greater detail concerning these examples are referred to the cited literature.

Cedar Apple Rust System

Seem and Russo (1984) described the problem of cedar apple rust, *Gymnosporangium juniperi-virginianae,* on apple with the objective of providing simple decision aids for growers to determine when a control tactic is necessary. Although not presented as a systems analysis, the process they went through had all the components necessary for a formal systems approach to decision making. They designed the system to be centrally focused on the apple tree, the production unit of a grower. Impinging on the apple tree were the fungus (and only that stage of the fungus life cycle that precedes infection), the portion of the environment that creates favorable conditions for infection, and the grower's previous and current management actions. They bounded the cedar apple rust system by describing the problem in both time and space. They concerned themselves with that portion of the growing season when the apple tree was susceptible to infection and the period of time when the pathogen was producing viable inoculum. The spatial framework was identified as the farm level because from a decision-making perspective it would be impractical to attempt to process information from lower levels (higher resolution) of the hierarchy. However, most biological information on the pathogen (spore liberation and germination) and the relation with the host (infection) has been derived at the organ level. Seem and Russo resolved this conflict by explicitly stating certain assumptions they were making about the system: that all organs on a tree respond similarly, that all trees of the same cultivar within an orchard respond similarly, and that all orchards on a farm respond similarly.

Having bounded the system, they then turned to defining key stages in the system that affected disease development. These stages were the periods of inoculum availability, host susceptibility, and environmental conditions favorable for infection. The start of each period was keyed to an easily observable event, typically some phenological or environmental event that could be recognized by a grower with a minimum input of time or expense. The system design was complete because they had included all of the components they considered important for the particular problem. They clearly defined both the time and space framework relevant to the problem. And when necessary they clearly stated any assumptions that were necessary for the system "pieces" to fit together. The final step was the development of a series of decision aids (loosely defined as graphic models). When certain key stages occurred, the grower was guided through a series of graphic steps that resulted in a recommendation to spray or not to spray.

The cedar apple rust example was developed as a demonstration of how system design can be used in a very simple manner to organize and structure a problem in practical pest control. Most problems cannot be dealt with in such simple terms and, in fact, there is much more to be gained by including more complexity into a problem and expressing that complexity in quantitative form (mathematical models).

Potato Late Blight System

A good example of trial-and-error simulation analysis of a plant disease system has been conducted by Fry et al. on the potato late blight system. Their objective was to determine better management strategies for the use of fungicides to control late blight. Although late blight was being controlled successfully with calendar spray schedules or schedules modified by environmental conditions favorable for the development of the disease (BLITECAST), they rationalized that additional factors such as cultivar resistance and fungicide weathering dynamics could be used to optimize spray schedules by minimizing control costs while maintaining tuber quality. Such an ambitious undertaking could have easily consumed decades of field evaluations in order to achieve their desired optimization. Instead, they turned to simulation and over a period of years (a good systems analysis can often take years) the pathogen, *Phytophthora infestans,* its host,

Solanum tuberosum, and their environment were described in the form of a system model (Bruhn and Fry, 1981, 1982a, 1982b; Bruhn et al., 1980). Having satisfied themselves that the system simulator had sufficient fidelity, they started to evaluate different management policies by incorporating different amounts of cultivar resistance and fungicides with different residue dynamics (Fry et al., 1983; Spadafora et al., 1984). Their choice of simulator inputs were always linked to rational, real-world tactics and the system solutions were constantly evaluated by real-world experimental conditions in field trials.

The potato late blight system illustrates several key points in systems analysis. First, the researchers not only evaluated the system management (altering the system: weather favorability, inoculum availability, etc.) but they altered the description of the system by successively incorporating cultivar resistance and then fungicide residue dynamics as major components of the system. By experimenting with design changes to the system they became innovators rather than observers (Ruesink, 1976). Second, they performed simultaneous experimentation with system development. This had the 2-fold purpose of filling knowledge gaps that were identified by the descriptive analysis while at the same time functioning as a feedback mechanism that allowed for minor corrections and improvements to the system. Third, the system was used to evaluate two types of management policies. At times the system model is evaluated using current, real-world input data in order to derive short-term, but highly accurate output for "on-line" (Haynes et al., 1980) disease forecasts. At other times the model is used to evaluate many different policies in order to aid long-term planning and strategy developement.

Yellow Stripe Rust System

Dynamic programming has been used in the analysis of the *Puccinia striiformis*/wheat system in the Netherlands (Onstad and Rabbinge, 1985). The main objective of this study was to utilize dynamic programming in a plant disease system with the added benefit of gaining insight into needed areas of research and an initial attempt at developing optimal control policies.

The researchers developed a dynamic programming solution to the problem of if and when to spray a fungicide to control a yellow stripe rust epidemic. At each decision point the program considered the development of the crop, the amount of disease, and whether a fungicide had been applied previously to determine if it was economically justifiable to apply another fungicide spray. The proportion of leaf tissue containing visible lesions (disease severity) was a state variable along with a variable identifying the presence or absence of a fungicide. The dynamic program also contained: 6 to 10 decision periods when a decision to spray or not spray a fungicide was made; penalty functions, which computed losses and control costs at each decision stage; and transfer functions, which simulated the state variables. The number of state variables was kept small because all the relevant information was contained within those variables. For example, severity was expressed as a function of development rate dependent on host resistance and host growth stage, which in turn, was a function of physiological time.

Results of the optimizations containing the disease severity state variable produced economic injury levels for yellow stripe rust at various stages of host development. Figure 2 shows the approximate severity level where the optimal decision switched from no spray to spray. Between the heading and flowering stages the economic injury level became very high. This indicated that sprays to control stripe rust were not needed after flowering because the costs of control would exceed the preventable damage.

The dynamic programming study of the yellow stripe rust system demonstrated the utility of using optimization for decision making in complex disease management systems. Onstad and Rabbinge (1985) pointed out that to guarantee the optimal solution for the cropping period under study (May through July) would have required over 31,000 simulations. However, the dynamic program

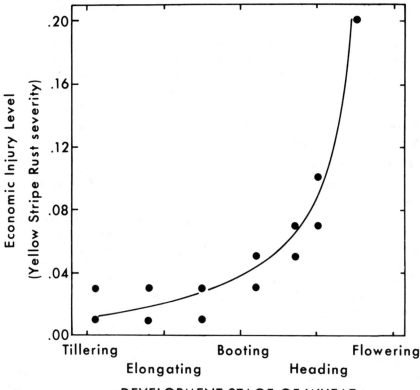

Figure 2 Results of analytic optimization of the yellow stripe rust control on a partially resistant cultivar in the Netherlands. Data points represent results of the optimization. Fitted line represents the economic injury level; below the line no spray is needed, above the line a fungicide spray is required. (*Data from Onstad and Rabbinge, 1985; Onstad, 1984, personal communication.*)

required relatively few executions to determine the solution, with the added benefit that the solutions were the economic injury levels. The authors carried the study further to include a more detailed description of disease severity by differentiating between presymptom, infectious, and postinfectious lesions. This additional complexity, while providing useful insight into the management system, pushed the dynamic programming procedure toward its useful limit by increasing the number of state variables to four. Onstad and Rabbinge (1985) also noted that for the procedure to work, the damage function or yield loss model must be able to predict loss at each decision point. Relatively few loss models are capable of producing this type of information.

Peach Brown Rot System

The first application of decision theory to a pest management problem was developed by Carlson (1969, 1970). He attempted to determine the best decision for fungicide application to peaches that were sporadically attacked by the brown rot fungus, *Monilinia fructicola*. Initially he determined growers' belief that they would incur various amounts of loss due to brown rot. These subjective probabilities represented the growers' feelings, based on their years of farming experience, that events would occur allowing disease and subsequent loss to develop. Standard control options were evaluated for their relative ability to control brown rot in field and laboratory trials, and a payoff matrix was developed (Table 1) that showed all possible outcomes for each control action and potential loss level. The payoff was expressed as net return in dollars per acre. Expected outcome for each control action could then be determined by multiplying each payoff times the growers' subjective probabilities that a particular loss will occur and then summing across all loss levels (Table 1, right-hand column). According to Baysian decision theory the optimal decision is the one that produces the best

Table 1 Payoff matrix for different brown rot control actions at various potential loss levels for extra-early peaches

Dollars per acre

Control action	Potential loss from brown rot					Expected value[a]
	0–4.9%	5–9.9%	10–19.9%	20–29.9%	30–100%	
No spray	76.21	39.48	−15.61	−89.06	−382.85	41.98
One captan	74.89	60.93	40.00	12.08	−99.57	61.88
Two captan	63.48	52.09	35.01	12.24	−78.85	52.86
One sulfur	77.10	53.96	19.25	−27.03	−212.14	55.53
Two sulfur	74.50	57.98	33.18	0.13	−132.09	59.10
Probabilities[b]	.760	.128	.034	.032	.046	1.000

[a]Expected values are based on the summed products of probabilities that a loss level will occur and the income for that loss level.

[b]Probabilities are based on subjective estimates by growers of the likelihood that a particular loss level will occur.

Source: G. A. Carlson, Ph.D. thesis, University of California, Davis, 1969.

Table 2 Optimal strategies for control of brown rot based on maximum expected returns for extra-early peaches

Area[a]	Expected return (dollars per acre) based on percent forecast loss[b]							Opportunity gain (dollars per acre)[c]
	0–4.9%	5–9.9%	10–19.9%	20–29.9%	30–49.9%	50–69.9%	70–100%	
Marysville	72 (1S)	69 (1C)	67 (1C)	62 (1C)	46 (1C)	−9 (2C)	−99 (2C)	67.25
Modesto	74 (1S)	27 (1S)	71 (1C)	67 (1C)	53 (1C)	−16 (2C)	−99 (2C)	69.92

[a]California peach production areas.

[b]Return based on optimal strategy for that level of loss. Loss level determined by a forecast system using fruit maturity and rainfall occurrence.

[c]Overall expected returns using expected returns from optimal control strategy for each level of loss and the likelihood that the loss will actually occur. Optimal control strategy: 1S = one sulfur spray; 1C = one captan spray; 2C = two captan sprays.

Source: G. A. Carlson, Ph.D. thesis University of California, Davis, 1969.

expected value. In this case one application of captan produced the best expected return of approximately 62 dollars per acre.

Of course there are other ways of deriving more objective probabilities that loss will occur. Infection and subsequent loss depends on such factors as inoculum availability, fruit maturity, and rainfall. This is where a systems model can be incorporated into a decision procedure. Carlson added a simple system model to his procedure, thus allowing the amount of loss to be forecast at the time of the decision based on fruit maturity and rainfall occurrence. Additional uncertainties were incorporated, such as the probability that a weather prediction would be correct. The optimal strategies and their associated expected returns are presented in Table 2. Each entry in the expected value row is the maximum expected value among the five control actions (listed in Table 1) for each possible loss forecast. The calculations of these expected values include both objective probabilities generated from the forecast or system model as well as the subjective probabilities generated from grower experience (Carlson, 1970).

The decision theory paradigm need not be limited to the assumption that a grower desires to maximize profits. In fact, a grower may be willing to forego some profits to ensure a more likely income. Such aversion to risk can be presented in quantitative terms as a reduction in the standard deviation of an expected outcome. Carlson (1970) utilized this method and determined that for a forecast of 0 to 5% yield loss a grower may choose to make one application of sulfur (mean return of 74 dollars per acre; standard deviation of 14 dollars) instead of the optimal choice of no spray (expected return of 75 dollars; standard deviation of 20 dollars). The 1 dollar reduction in expected income would be offset by a 6 dollar reduction in the standard deviation in the sulfur application outcome compared to no spray. Carlson found that 38% of the growers he surveyed were averse to risk.

Several other examples of decision theory for plant diseases exist including *Septoria nodorum* on wheat (Webster, 1977) and sugar beet yellow virus (Mumford, 1981).

CHALLENGES FOR THE FUTURE: PROBLEMS AND OPPORTUNITIES

Pathosystems and the Global View

To this point our discussion of systems analysis has emphasized the pathogen as the central focus of the system. Obviously production agriculture focuses on the crop, and the pathogen must be viewed as only a component of a larger system. Geier and Clark (1978) classify pest control as an ancillary enterprise because it is used only as a means of furthering the production system. Geier (1981) further states

that he who undertakes pest control should do so with a 'global' mental picture that connects all aspects of the proposed enterprise—commercial, technological, managerial, sociopolitical, biological and economic. Such pictures comprise the sum total of relevant information about pest events, their causes, consequences and management, that exists at one time. A global picture necessarily underlies one's perception of risks, determines one's subsequent actions, and supplies one's measure of achievement in pest control.

Along with the global view comes the necessity to interact with a wide diversity of disciplines. Plant epidemiologists, because of their quantitative skills and "systems thinking," are often cast in the role of disciplinary representative. On the surface, the idea sounds stimulating and exciting. However, to achieve the high goals of multidisciplinary work one must overcome a series of hurdles that arise from years of emphasis on unidisciplinary thinking. The entire strategy of research management must be altered because common objectives and goals are set for the entire multidisciplinary team (Smith and Adkisson, 1979). Individual creativity, the measure of academic excellence, as well as disciplinary pride must be subjugated to the needs and ambitions of the team. Individuals participating in interdisciplinary research who do not embrace the common goals may often find themselves in direct opposition to the group's objectives (Miller, 1983). Commitment to multidisciplinary activities and the global perspective derived from them is largely the responsibility of the individual participants. However the administrators overseeing such projects must serve as facilitators and must suppress the divisive tendencies of unidisciplinary thinking.

Many of the problems encountered in multidisciplinary interaction can be overcome by utilizing special techniques of group interaction. One method that employs these techniques is adaptive environmental assessment and management (AEAM). It has recently been employed to develop ecosystems analysis, assessment, and policy making through interaction of scientists, managers, policymakers, and public interest groups (Posey, 1982). Leadership responsibilities are given to persons familiar with quantitative sciences and systems analysis as well as the psychosocial sciences of group dynamics. The leadership facilitates successful interaction by minimizing conflict and competition within the group. We are not advocating that the plant pathologist who is interested in systems analysis in epidemiology must also be a psychologist. However, there will be times when a little psychology on the part of the systems analyst will prove highly beneficial.

The Microcomputer Revolution

The pervasive adoption of inexpensive personal or desk-top computers is producing a new generation of farmers with sophisticated data processing skills. Basic modeling tools are available in the form of spreadsheet and database management software packages with impressive graphic capabilities. Furthermore, microprocessor-based monitoring systems are now feasible for even the small grower. The farmers of the near future will have considerable skills in quantitative analysis. They will be accustomed to manipulating data sets to explore the financial implications of alternative management decisions. On-line data collection facilities can be used to constantly update the decision-making information base and management decisions will be adjusted accordingly. Farmers will thus become *practicing systems analysts*.

We noted earlier that systems analysis is a relative novelty in plant disease epidemiology, and this leads to the uncomfortable conclusion that farmers, who are the ultimate users of plant disease research, may become more advanced in their use and understanding of systems analysis than the average plant disease researcher. The farmer with data management skills will be increasingly concerned with the effects of a disease management strategy on both the farm system as a whole as well as financial objectives. Unless the researcher can provide this information, research results may be seen as essentially irrelevant.

Artificial Intelligence

Another impact of computers on systems analysis in plant epidemiology will be in the area of artificial intelligence (AI). In its simplest conceptual form, AI is the effort to mimic the thought, analysis, and decision-making process of the human mind. Unlike the analytic decision-making procedures described earlier in this chapter, AI is an attempt to capture and store knowledge in a computer by special programs. Decisions are then made based on that knowledge and the inference drawn from it. Specialists in AI have found that as knowledge about a topic increases, the knowledge rapidly becomes amorphous and is very difficult to structure within the confines of a computer program. As the number of decision points and options increase, the number of choices or combinations increases exponentially (see the discussion of analytic optimization). Whereas analytic optimization overcomes this problem by simply limiting the number of possible combinations, AI uses heuristics (rules of thumb) to shortcut the decision process. Researchers, in the specialized area of AI called expert systems, are finding that true experts take a global perspective of a problem and rapidly solve it by matching it to similar problems and successful solutions previously encountered (Waldrop, 1984).

Although the development of expert systems is made easier by high-level languages and programs for large and small computers, the real task faced by plant pathologists is the structuring of the knowledge base necessary for expert systems. The methods of systems analysis outlined in this chapter can be used for portions of that knowledge base, but the vast majority of information must be characterized by nonanalytic methods that summarize observational data bases in ways useful for expert systems. While the need for characterization of data bases for agricultural systems has been recognized (Russo and Seem, 1985), the precise methodologies are yet to be developed and will become one of the major challenges of agricultural systems science.

Education in Plant Disease Systems Analysis

How does the plant pathologist learn enough about the systems approach to be able to apply it to a research program? If systems analysis is more than a superficial set of concepts, a reasonable expertise cannot be acquired without significant investment in time and effort. In engineering, for example, training in systems analysis often requires several years of study. Moreover, there is reason to believe that the systems approach is somewhat antithetical to the plant pathologist's basic training in the biological sciences. The construction of a useful system model often requires broad generalizations, acceptance of half-truths, and frequent substitution of judgment for knowledge. It has the appearance of bad science, or at least the perversion of science to satisfy the manager's needs. It is no accident that systems analysis has been developed largely in engineering, economics, and business management, since these disciplines have strong problem-solving emphases and are not generally considered sciences. The apparent lack of scientific rigor is inversely related to the amount of effort put into the analysis of the system. As more relative time is spent developing, validating, and refining a system, the early suppositions give way to more rational descriptions of a complex system. One must pragmatically balance the need for good science with the timely need for a decision-making process.

A suitable approach for grafting systems analysis skills onto a plant pathologist is certainly not obvious. When available, a university course on "systems analysis and plant disease management," which would emphasize the systems aspects of epidemiology and provide a background in mathematical modeling, would be a reasonable way to instill an awareness of systems analysis. However, it would not be adequate preparation for a research program based on systems analysis. For this purpose, the scientist must first have a good working knowledge of mathematics and statistics and then must look to those disciplines such as engineering and agricultural economics that place premiums on decision making and problem solving. Optimization, mathematical programming, simulation modeling, and microeconomic analyses are study areas that could be most useful. Equally important is the management perspective that they represent, for although the plant pathologist cannot and should not discard scientific principles, he or she cannot be an effective systems analyst without understanding the practical necessities of decision making.

CONCLUSIONS

In the brief span of this chapter we have attempted to describe the subdiscipline of epidemiological systems analysis. It is a new field and researchers have various opinions about how it can and should be used. As a tool it is an amplification of the general scientific method. As a concept it embraces holistic and transdisciplinary thinking. We also support the idea that it is a highly specialized area of plant pathology which is the province of the scientist who understands both biological systems and decision making.

While many of the components of this subdiscipline are addressed more directly in other chapters of this volume, we have emphasized the general concept of systems analysis as it relates to biological systems and the methods of linking the system to a decision-making process. We have only scratched the surface, but we hope that in the process we have piqued the interest of those who are and will be supportive of epidemiological systems analysis.

ACKNOWLEDGMENTS

The authors thank D. W. Onstad for his critical review and helpful suggestions regarding this chapter.

REFERENCES

Agrawal, R. C., and Heady, E. O. 1972. *Operations Research Methods for Agricultural Decisions*. Iowa State University Press, Ames.

Bellman, R. 1957. *Dynamic Programming*. Princeton University Press, Princeton.

Bender, F. E., Kramer, A., and Kalian, G. 1976. *Systems Analysis for the Food Industry*, Avi Publ., Westport, Connecticut.

Bruhn, J. A., and Fry, W. E. 1981. Analysis of potato late blight epidemiology by simulation modeling. *Phytopathology* 71:612–616.

———. 1982a. A statistical model of fungicide deposition on potato foliage. *Phytopathology* 72:1301–1305.

———. 1982b. A mathematical model of the spatial and temporal dynamics of chlorothalonil residues on potato foliage. *Phytopathology* 72:1306–1312.

Bruhn, J. A., Fick, G. W., and Fry, W. E. 1980. Simulation of daily weather data using theoretical probability distributions. *J. Appl. Meteorol.* 19:1029–1036.

Carlson, G. A. 1969. A decision theoretic approach to crop disease prediction and control. Ph.D. thesis, Univ. of California, Davis.

———. A decision theoretic approach to crop disease prediction and control. *Am. J. Agr. Econ.* 52:216–223.

Churchman, C. W. 1968. *The Systems Approach*. Dell Publ. Co., New York.

Dent, J. B., and Anderson, J. R. (eds.). 1971. *Systems Analysis in Agricultural Management*. John Wiley & Sons Australia Pty. Ltd., Sydney.

Fry, W. E., Apple, A. E., and Bruhn, J. A. 1983. Evaluation of potato late blight forecasts modified to incorporate host resistance and fungicide weathering. *Phytopathology* 73:1054–1059.

Geier, P. W. 1981. Adequate knowledge: An editorial policy. *Prot. Ecol.* 3:1–6.

Geier, P. W., and Clark, L. R. 1978. The nature and future of pest control: Production process or applied ecology? *Prot. Ecol.* 1:79–101.

Haith, D. A. 1982. *Environmental Systems Optimization*. John Wiley & Sons, New York.

Halter, A. N., and Dean, G. W. 1971. *Decisions under Uncertainty*. Southwestern, Cincinnati.

Haynes, D. L., Tummala, R. L., and Ellis, T. L. 1980. Ecosystem management for pest control. *BioScience* 30:690–696.

Jeffers, J. N. R. 1978. *An Introduction to Systems Analysis with Ecological Applications*. University Park Press, Baltimore.

Kranz, J., and Hau, B. 1980. Systems analysis in epidemiology. *Annu. Rev. Phytopathol.* 18:67–83.

Miller, A. 1983. Integrated pest management: Psychosocial constraints. *Prot. Ecol.* 5:253–267.

Mumford, J. D. 1981. A study of sugar beet growers' pest control decisions. *Ann. Appl. Biol.* 97:243–252.

Mumford, J. D., and Norton, G. A. 1984. Economics of decision making in pest management. *Annu. Rev. Entomol.* 29:157–174.

Norton, G. A. 1982. A decision-analysis approach to integrated pest control. *Crop Prot.* 1:147–164.

Norton, G. A., Sutherst, R. W., and Maywald, G. F. 1983. A framework for integrating control methods against the cattle tick, *Boophilus microplus* in Australia. *J. Appl. Ecol.* 20:489–505.

Onstad, D. W. and Rabbinge, R. 1985. Dynamic programming and the computation of economic injury levels for crop disease control. *Agr. Sys.* 18:207–226.

Posey, C. 1982. AEAM upon AEAM; Adapt! *Options,* Spring, pp. 1–8, Int. Inst. for Applied Systems Analysis, Laxenburg, Austria.

Robinson, R. A. 1976. *Plant Pathosystems.* Springer-Verlag, New York.

Ruesink, W. G. 1976. Status of the systems approach to pest management. *Annu. Rev. Entomol.* 21:27–44.

Russo, J. M., and Seem, R. C. 1985. Development of an apple production model for IPM, in: *Proc. National Symposium* for *Integrated Pest Management* on *Major Managed Agricultural Systems,* (R. E. Frisbie and P. L. Adkisson, eds). Texas A & M Univ. Press, College Station (In press).

Seem, R. C., and Russo, J. M. 1984. Simple decision aids for practical pest control. *Plant Dis.* 68:656–660.

Shoemaker, C. A. 1973. Optimization of agricultural pest management III. Results and extensions of a model. *Math. Biosci.* 18:1–22.

———. 1981. Applications of dynamic programming and other optimization methods in pest management. *IEEE Trans. Automat. Contr.* AC-26:1125–1132.

Smith, R. F. and Adkisson, P. L. 1979. Expanding horizons of integrated pest control in crop protection, in: *Proc. IX Int. Congr. Plant Prot.:* (Opening Session and Plenary Session Symp.), Washington, D. C. pp. 29–30.

Spadafora, V. J., Bruhn, J. A., and Fry, W. E. 1984. Influence of selected protectant fungicides and host resistance on simple and complex potato late blight forecasts. *Phytopathology* 74:519–523.

Teng, P. S. 1981. Validation of computer models of plant disease epidemics: A review of philosophy and methodology. *Z. Pflanzenkr Pflanzenschutz* 88:49–63.

Van der Plank, J. E. 1963. *Plant Diseases: Epidemics and Control.* Academic Press, New York.

Waldrop, M. M. 1984. The necessity of knowledge. *Science* 223:1279–1282.

Webster, J. P. G. 1977. The analysis of risky farm management decisions: Advising farmers about the use of pesticides. *J. Agr. Econ.* 28:243–256.

Weigert, R. G. 1975. Simulation models of ecosystems. *Annu. Rev. Ecol. Sys.* 6:311–338.

Welch, S. M., Croft, B. A., and Michels, M. F. 1981. Validation of pest management models. *Environ. Entomol.* 10:425–432.

Zadoks, J. C., and Schein, R. D. 1979. Epidemiology and Plant Disese Management. Oxford University Press, Oxford.

Part Four

Modeling Approaches

Chapter 11

The Potential of Analytic Compared with Simulation Approaches to Modeling in Plant Disease Epidemiology

M. J. Jeger

The extent to which mathematical models are used in the biological sciences varies greatly over the range of disciplines to be found. In some disciplines, such as population genetics, the need for mathematics is recognized and permeates teaching and research. In other disciplines, such as plant pathology, the need for mathematics appears only in rather esoteric applications, and that which is done is usually characterized by a signal lack of cross reference to modeling in other disciplines.

The comparison of disciplines in which mathematics does or does not play an important role is not necessarily one of numerate as opposed to nonnumerate subject matter. The use of mathematics in modeling biological phenomena has a closer affinity with abstraction, the ability to extract from the empirical world simple statements concerning basic processes and structures, rather than with numbers per se. This does not relegate modeling to a perhaps futile search for perfect architecture in the biological world. Instead it reflects at least one view of science, which is that underlying complexity are general principles of sufficient simplicity and elegance to be stated in mathematical terms. Accordingly, models have a very proper role to play in assisting our steps to understand the empirical world. In my view, this motivation for modeling should not be denied or undervalued by comparison with more utilitarian views. According to these views models are only useful if they have application and lead to material benefits of some kind; especially in disciplines as practical and of such human relevance as agriculture.

This chapter is directed to the plant pathologist who is motivated to understand the nature of disease in plant populations, whether in agricultural or natural systems, and the role that mathematical models can play in this process. The chapter is intended to be instructional rather than a complete review of modeling in epidemiology. The author's bias is made clear but, hopefully, this does not lead to too unbalanced a treatment.

EPIDEMIOLOGY: DISEASE IN POPULATIONS

The biological world is characterized by hierarchical structure. There are different levels of biological integration that have a great influence on our attempts to understand living organisms. A broad classification of these levels of integration corresponds to the molecule, cell, organism, population, and community. The divisions between these levels are usually intuitively clear, although their logical status can be difficult to justify completely. The question to be asked is whether the principles that govern processes at any one level of integration are sufficient to explain processes at a higher level. If so, and if taken to the extreme, the implication would then be that the biosphere is wholly explicable in terms of DNA and other biological molecules. On the other hand, can the principles that govern processes at one level be in conflict with those at a lower level?

The answers to these questions depend on our philosophical inclinations, but there are undoubtedly ways to circumvent these problems. Thornley (1980) proposed that an explanation of a phenomenon at one level of integration depends on a description at the lower level. Hence researchers at a particular level of integration in the plant sciences should be concerned with using descriptions from lower levels as a means of explanation and with providing descriptions for higher levels of integration. According to Thornley, the latter is rarely of concern to researchers: plant physiologists spend their time being cell biologists, and cell biologists being biochemists, and so on, without applying their results to higher levels of integration. Similar views on the relationship between description and explanation have been expressed by Passioura (1979) and Waggoner (1983).

Epidemiology is concerned with disease in populations, that is, with host populations interacting with pathogen and possibly vector populations. Accordingly, an explanation of plant disease epidemics depends on descriptions of the host population and the pathogen population. In homogeneous agricultural crops the former usually involves a description of events at the whole plant level. In natural systems, temporal and spatial heterogeneity in the genetic structure of populations, and in the environment, become increasingly important. Description of the pathogen population does present a problem in that the unit of the population is less obvious (Jeger, 1985). In practice most epidemiologists study the diseased population (e.g., pustules per unit area, proportion of leaf area diseased, proportion of host plants diseased) and largely ignore the host and pathogen populations except when recording host growth and development, or trapping spores. Seldom have epidemiologists considered ways of describing events at the level of the diseased population in order to provide an explanation of events in communities (Hau and Eisensmith, 1983).

The consideration of hierarchical structure is certainly germane to the topic of this chapter and suggests boundaries that those concerned with modeling epidemics should be aware of even if they are not content to stay within them. Only rarely need the epidemiologist be concerned with processes occurring beneath the level of the individual plant part.

USES OF MODELS IN EPIDEMIOLOGY

An important consideration when comparing different approaches to modeling is the use to which models are are put. Models in epidemiology have assisted in the development of theory, in the undertaking of experimental programs, and in disease management. These uses of models will now be discussed in some detail and two approaches toward realizing them compared.

Models in the Development of Theory

Most experimental or observational studies are done within some conceptual framework or overall frame of reference, although this frame of reference may be loosely defined. Few experimental programs aim solely to accumulate a mass of empirical data. The role of theory in experimental science is a distinctive feature of the hypothetical deductive method. Insofar as epidemiological studies are concerned with explanation at the population level, then population dynamic theory should be critical in its development as a scientific discipline. The power of theories is that they provide explanations and, by so doing, enable predictions which are open to empirical tests. Mathematical models are then indispensible in epidemiology, as in other disciplines, in formalizing theories and deducing implications that can be tested. Of course not all mathematical modeling aims at explanation in this sense, but rather explores the range of possible outcomes given a basic set of suppositions (Jeger, 1985). The ultimate objective, however, should be prediction and testing by experimental means. Thus theory and modeling are inextricably linked to experimentation, whether by giving impetus and direction to an experimental program, or by making predictions that can be tested. In passing it should be noted that although explanation, by definition, implies prediction, the converse implication does not hold (i.e., predictions can be made without the antecedent of explanation). The simplistic dichotomy of the explanatory vs. the predictive model (Fleming and Bruhn, 1983) should, however, be avoided.

Models as Aids to Calculation

A further way in which mathematical models can assist in an experimental program is to serve as a means of calculation. For example, experiments on enzyme kinetics are inconceivable without an underlying assumption of Michaelis-Menton or some other model of reaction rate. Calculated values of the parameters of mathematical models may be used to make experimental comparisons between treatments. There are many examples of such use in epidemiological studies, especially in the calculation of infection rates or asymptotic levels of disease. The most common approach is to use the transformation equations of Van der Plank (1963) to estimate the rate parameters. When one does this, one is implicitly using a model.

Growth curves have often been used to compare disease progress curves. The most commonly used have been the logistic, Gompertz, and generalized Richards functions (Jeger, 1982a; Jowett et al., 1974; Madden, 1980; Waggoner, 1977, and Chap. 1). In many cases these curves may be equally good in providing fits to experimental data but the rate parameters differ numerically. There are several limitations to the growth curve approach that have led to some abuse. First, it is only possible to make comparisons using a rate parameter where the same model has been fitted to all data sets being compared. For example, there is no easy correspondence between the rate parameters of the logistic and Gompertz growth curves. In fact, the four parameters of the Richards family of curves should be considered together as a multivariate set in any experimental comparisons being made. Second, where different transformations have been used for the dependent variable, the R^2 value (coeffient of determination) for a fitted linear model cannot be used to determine which model gives the best fit. This limitation applies whether from biological (Morral and Verma, 1981) or statistical considerations (Payandeh, 1981). The safest procedure would seem to be to test the goodness-of-fit of each competing model when back-transformed to the original scale of measurement (Jeger, 1982a), or to use nonlinear modeling. In some cases where statistical goodness-of-fit is the only concern, polynomial models of disease progress have been applied (Griggs et al., 1978). This remains a sound option for those who wish to avoid the modeling preconceptions associated with growth curves.

A model is best used as a means of calculation when it contains some elements of biological realism yet at the same time gives parameters that can be estimated with reasonable ease from experimental data. The linked equations of Walker and Smith (1984) concerning the dynamics of mycorrhizal infection of roots and discussed in a later section in this chapter are a good example. Jeger and Starr (1985) developed a model of the overwintering dynamics of nematode eggs and juveniles based on simple considerations concerning the life cycle of the populations. The equations that resulted were complex but procedures are being developed to enable estimation of the parameters and, moreover, to indicate where greater experimental precision is required.

Use of Models in Disease Management

The last use of models in epidemiology to be considered here is in disease management. Disease management is influenced by more than purely scientific considerations; it is aimed not at understanding but rather manipulating disease in agricultural systems. The extent to which management is possible without understanding will not be discussed here (see Butt and Jeger, 1985).

It should immediately be admitted that theoretical models have played little role in practical disease management, except in the broadest sense of suggesting the outcomes of alternative strategies given certain epidemiological scenarios. In practice most models proposed, if not implemented, have been empirical regression models or threshold criteria that forecast infection or some other component of the disease cycle. It has been claimed that complex models, which attempt to mimic many environmental factors and other influences, are more appropriate models for disease management than theoretical models or simple regression models. What is not usually stated is that these "realistic" models are usually based on many empirical relationships albeit linked together in a more logical structure than in the simpler regression models. In one way this is saying little more than that explanation at one level demands only description at a lower level. Thus it is the systems integration of these lower-level descriptions that is important in complex models. I have reservations, however, concerning the relevance of this approach for the understanding of plant disease epidemics. Furthermore, it has not been demonstrated that better disease management results from using complex models even when they are simplified. Indeed, mathematical techniques, which make few or no biological assumptions, have increasingly been used for optimization and decision analysis in disease management (Betters and Schaeffer, 1981; Shoemaker, 1981) rather than complex models.

COMPARISON OF ANALYTIC AND SIMULATION APPROACHES

Early Epidemic Models

The modeling of plant disease epidemics effectively started with the pioneering syntheses of Van der Plank (1963) and, especially in aerobiology, of Gregory (1961). The treatment of Van der Plank was idiosyncratic in that little attempt was made to relate the dynamics of disease epidemics to the mainstream of population biology. Despite this shortcoming, various equations were proposed that have had a major influence on the development of plant disease epidemiology. The derivations of Van der Plank's and other simple models are discussed by Jeger (1984c); their limitations and subsequent development have been discussed by Rouse (1985) and Jerger (1985).

There is now renewed interest in developing models of plant disease epidemics that reflect the work that has occurred in other areas of population biology. For example, it is salutary to find discussion of a differential-difference equations (Wangersky and Cunningham, 1956) to which the Van der Plank (1963) equation reduces in certain cases. These single-equation models do have a place in population biology and epidemiology but are not the best means for comparing the two main approaches to modeling: analytic and simulation.

The starting point used here will be the case in which linked differential equations are used to describe the dynamics of a disease epidemic. The discussion will be oversimplified necessarily in order to compare the two modeling approaches. In the analytic approach, I will discuss how the qualitative behavior of the system can be investigated to obtain results of general intuitive value. In the simulation approach I will discuss how the equations are used to mimic, quantitatively, the response of the system to external variables such as environment, management practice, and other variables. The main characteristics of the two approaches are summarized in Table 1 and form the basis for the comparisons made and the discussion that follows. The differences can be subtle and are certainly more than whether explicit (closed-form) solutions can be found, or whether numerical solutions must be sought. Pedersen (1984) has contrasted the two approaches in the context of pesticide resistance models.

Table 1   Comparison of analytic and simulation models

Characteristic	Analytic model	Simulation model
Objective	Problem-oriented	Goal-oriented
Logical status	Deductive	Inductive
Empirical status	Data-independent	Data-dependent
Mathematical form	Differential, difference, and algebraic equations	Differential, difference, and "structured" regression equations
Structure	Simple	Complex
Mathematical sophistication	High	Low
Variables	Few	Usually many
Parameters	Few	Many
Inputs	Few	Many
Interpretation of parameters	Indirect	Direct
Solution	Closed-form or numerical methods	Numerical methods (by definition)
Qualitative analysis	Usually possible	By definition not possible
Calibration ("fine tuning")	Not relevant	Necessary
Verification (Does model equal concept?)	Easy	Difficult
Validation (Does model fit data?)	Usually subjective	Objective, but never complete

A Common Starting Point for Simulation and Analytic Approaches

Waggoner (1977, 1978, 1981), in a comprehensive description of the simulation approach, discussed how systems of differential equations may be "composed" by considering the life cycles of fungal plant pathogens, by compartmentalizing the various components of the life cycle (e.g., sporophores, spores, dispersed spores, lesions), and by specifiying the rates of change of each variable. If the variables in the system are denoted by $X_1, X_2, \ldots, X_n$, then the rate of change of the ith variable can be written as

$$\frac{dX_i}{dt} = f_i(X_1, X_2, \ldots, X_n; E_1, E_2, \ldots, E_m) \tag{1}$$

where $E_1, E_2, \ldots, E_m$ are variables reflecting the influence of external variables, such as environment, on the dynamics of the epidemic. Waggoner showed how life cylces can be mimicked by specifying the exact forms of the functions, f_1, $f_2, \ldots, f_n$ and solving the equations numerically with a computer. In this sense the model is serving to synthesize all known or assumed elements of the life cycle, and their relationships with the environment, in order to mimic an epidemic. This is the concept of a simulation model that has grown up in the context of plant disease epidemiology and is perhaps unduly restrictive.

This concept of modeling is also incomplete. In population biology, there are other approaches that can be taken to analyze systems of equations such as Eq. 1. For example, the effects of environment can be either taken as constant or reduced to time-dependent functions. With a constant environment, autonomous differential equations of the following form result:

$$\frac{dX_i}{dt} = f_i(X_1, X_2, \ldots, X_n) \tag{2}$$

With time-dependence, nonautonomous equations of the following form result:

$$\frac{dX_i}{dt} = f_i(X, X_2, \ldots, X_n; t) \tag{3}$$

where time t appears explicitly on the right-hand side of the equations, result. As the functions are all likely to be nonlinear, the question is not whether numerical techniques are required to obtain a particular solution of the equation or not (they almost certainly will), but whether the system is analyzed in terms of its general dynamic properties. Even where the system includes time on the right-hand side, it is desirable to analyze the corresponding autonomous system as a first step. In my view such analysis of dynamic systems specified for plant disease epidemics has been sorely neglected and should always be attempted even when simulation is the ultimate goal. There is a rich body of theory that has been developed in population dynamics as a whole, but this makes very few references to the microbial pathogens of plants.

Differences in Approach

Waggoner (1978) outlines an interesting example of how a dynamic system can be composed for *Erwinia stewartii* epidemics in maize. Linked equations were proposed describing the dynamics of infested vectors, the numbers of diseased plants, increases in leaf area, and the proportions of leaf area that are healthy, infected, or dead. The analytic approach would then assemble these equations into a coherent system, analyze the system for its general dynamic properties. These include the existence of equilibrium values, at which time there is no change in any of the variables, the stability of these equilibria, and the dynamics of the system near equilibrium. For example, in the case of *E. stewartii*, does the epidemic run to completion with all leaf area diseased, or does it level-off ("equilibrate") at some intermediate and stable level of disease?

It may be argued that the concepts of equilibrium and stability are of little value in considering epidemics in agricultural systems. This view, however, reflects a bias toward intensive agriculture and the small number of crops grown with high levels of fungicide input, a view that is hardly representative of economic crops worldwide. Even where crops are subject to repeated inputs affecting pathogens, there is not reason to suppose that these cannot be analyzed in a nonautonomous system (such as given by Eq. 3). Disease incidence on growing vegetative shoots of apple trees treated with weekly sprays of fungicide approached equilibrium during the time of application (Jeger, 1984b); when applications ceased (for the epidemic, a perturbation) there was divergence from the equilibrium. In terms of the analysis to be described later in the chapter, this situation represents an unstable equilibrium. It would be invaluable to characterize this equilibrium in terms of epidemiological parameters.

Other examples of the usefulness of the concepts in agricultural systems can be found are: selection-mutation equilibria in pathogen populations (Leonard and Czochor, 1980); the efficacy of eradication or sanitation practices (for a pragmatic approach, see Allen, 1983). In natural or quasi-natural systems, the usefulness of the concepts has been demonstrated many times in ecosystem analysis.

The simulation approach, by contrast, would exhaustively relate each variable to environmental or other factors and obtain particular solutions according to each possible combination of parameter values. The range of particular solutions is limited by the availability and cost of computer time, and there is no guarantee that the global dynamic behavior of the system will be exposed by the simulations.

In passing it should be noted that there are other approaches to simulation modeling than to use linked differential equations (Bloomberg, 1977a, 1977b; Bruhn and Fry, 1981; Sall, 1980; Zadoks, 1971). Also, the point that systems analysis and simulation modeling are not synonyms should be further stressed

(Kranz and Hau, 1980). Systems analysis is simply one means of conceptualizing multiple cause-and-effect relationships and, at its most useful, need only involve qualitative analysis of the system under consideration. Teng (1985) has reviewed the various interrelationships between systems analysis and simulation and other model types, and notes difficulties in comparing approaches even among modelers.

Thus we can see that essentially the same starting point can lead to two quite divergent approaches to modeling. Of course both have their place and are by no means mutually exclusive. It can be instructive, however, to compare the conclusions obtained from complex simulations with those obtained by analysis (usually neglected) of the basic equations.

USES OF ANALYTICAL MODELS IN PLANT DISEASE EPIDEMIOLOGY

With few exceptions, the analysis of plant disease epidemics has followed the treatment outlined more than 20 years ago by Van der Plank (1963). Very rarely have analogies with similar phenomena in population biology been drawn. Van der Plank (1975), for example, in discussing the derivation of his differential-difference equation, commented in a footnote that rather different approaches had been taken in medical epidemiology. In fact, in medical epidemiology there is a vast literature on the theory and mathematics of epidemics (Bailey, 1975).

Leaving aside important stochastic considerations, the purely deterministic models of medical epidemiology have reached a level of sophistication as yet not sought in the modeling of plant disease epidemics. Similarly, Jowett et al. (1974) consider the use of differential equations in population ecology and yet do little more than indicate two possible applications of this wide body of theory to plant disease epidemics. There is a vast literature on the use of linked differential equations, usually of the Lotka-Volterra type, in population ecology (Pielou, 1977; Wangersky, 1978).

The lack of cross reference between disciplines is curious but not altogether unexpected given the totally different backgrounds and settings of plant pathology, medicine, and ecology. Indeed until recently very few studies attempted to reconcile the different strands in medical epidemiology and population ecology. There is some evidence that the situation is changing with respect to plant disease epidemiology. Jeger (1982b) specified linked differential equations for the different categories of disease (latent, infectious, and postinfectious) commonly present in a plant disease epidemic. In medical epidemiology the so-called SIR epidemic (susceptible, infective, and removed) is the usual categorization.

Despite the obvious differences between plant disease epidemics and epidemics in animal populations (e.g., mobile individuals, recovery and immunity, "superspreaders" of disease), there seems considerable scope for exploring the use of linked differential equations in studies of plant disease. A good example of how a quite general epidemic model can be used to examine a specific animal disease epidemic is given by Hahn and Furniss (1983). In particular the spatial component of animal epidemics is well developed compared to the almost non-existent theory in plant disease epidemiology other than that for the airborne dissemination of fungal spores. A further unexplored area is where a vector is responsible for pathogen dissemination. This is potentially an area in which cross-disciplinary work would be both intellectually stimulating and applicable.

Some of the differential equations used in population ecology have recently been applied in plant disease epidemiology (see also Rouse, 1985). Fleming (1980) developed a general model of biological control of cereal rust by incorporating a term developed in the modeling of predator-prey relationships. A single equation describing the population density of cereal rust pustules subject to biological control resulted. The key feature of the analysis was the identification of critical values of population density necessary to overcome the effect of natural enemy control. This sort of conclusion, it will be seen, is quite common to some analytic approaches. Although the existence of critical population densities may seem self-evident, their definition in terms of model parameters does at least formalize intuition. Also quite typical of some analytic approaches is the lack of biological detail; for example potential natural enemies such as the cereal rust hyperparasite *Eudarluca caricis* (= *Darluca filum*) are only discussed in passing. This is in quite marked contrast to the article by Hau and Kranz (1978) in which the effectiveness of *E. caricis* is modeled by simulation, but in a way that makes it very difficult to obtain any general theoretical insights. Their model however does remain commendably simple. The merits and demerits of each modeling approach obviously have to be decided by the would-be modeler, although there is a considerable area in between these two approaches in which advances could be made.

The Lotka-Volterra competition equations were used explicitly by Skylakakis (1982) with respect to the dynamics of competing pathogen races sensitive and insensitive to a fungicide. The magnitude and direction of the change in the race populations was noted to be dependent on the model parameters but the parameter combinations were not given or interpreted in the context of plant disease epidemics. Skylakakis (1982) then incorporated epidemiological relationships between Van der Plank's so-called apparent and basic infection rates, and thus the latent period, to reach conclusions stated clearly and, it was claimed, supported by the data examined. The explicit time-dependence of the relationship between the rate parameters was shown by Jeger (1984a) and may affect the conclusions reached.

Levy et al. (1983), using essentially the framework of Lotka-Volterra competition equations, have recently proposed the most complex model for the development of fungicide insensitivity. They elaborate on the models developed by Skylakakis (1982) by incorporating linear functions of time representing the effects of fungicide weathering and loss of efficacy. Their model is based on subpopulations insensitive and sensitive to a systemic fungicide. The dynamics of the subpopulations and the overall level of disease severity are examined with respect to various combinations of the systemic and protectant fungicides. They take a simulation approach to solving their equations, but with few details of how repeated spray applications are incorporated into the simulation, and show that the relative outcome depends not only on intrinsic rates of increase and other epidemiological parameters, but also on the rates of weathering and loss of fungicide efficacy. In fact, analysis of the equations without any recourse to a computer shows quite clearly that the linear functions describing weathering and efficacy loss are critical in determining the dynamics of disease given any of the strategies. One must then ask whether complex simulation is justified when based on the improbable assumptions of linearity.

Barrett (1983) posed the general problem of estimating relative fitness in plant pathogens and used the Lotka-Volterra equations as a means of demonstration. He showed succinctly that the subtraction of intrinsic rates of increase of two pathogen races to give a measure of relative fitness is inappropriate when there is intraspecific or interspecific competition or different carrying capacities in the two races. Barrett also showed that for each model given in terms of population numbers, rare in plant pathology, an equivalent model can be formulated in terms of disease severity but with different consequences due to host growth. Putter (1982) used competition equations to examine the outcome of the competition between *Phytophthora infestans* and *Alternaria solani* causing late and early blight, respectively, of potatoes and tomatoes. There was little analysis of the basic equations; they were solved numerically using values considered appropriate for each pathogen.

Predator-prey equations have been less used in epidemiological studies. Brittain (1983) discussed a model system for four species consisting of two competing tree species and two host-specific fungal pathogens. Although not tightly linked (e.g., no interaction between the two pathogens), the models enabled examination of equilibrium values in the long term. All analysis was done by numerically solving their equations. Their procedure of establishing equilibrium values and then perturbing the dynamics near to the equilibrium to examine the subsequent behavior of the system (i.e., whether the populations return to the same or different equilibria) is characteristic of both the analytic and simulation approaches to modeling.

Walker and Smith (1984) formulated equations describing the dynamics of mycorrhizal infection of root systems. The two variables described were the numbers of infection units per plant, and the length of infected root per plant. The parameters were the infection frequency per unit length root per time unit, the rate of spread of mycelium within the root, and an assumed constant length of root. Walker and Smith (1984) formulated these equations in terms of the proportion of root length infected and the frequency of infection units. They then obtained steady-state values for the variables, which they assumed were asymptotically stable (i.e., at arbitrarily large time values), but without following a formal procedure to show, in fact, that this was so. Further, Walker and Smith (1984) show how these models not only offer theoretical insight but can also be used as a means to calculate parameters from experimental data. These and other models of the dynamics of mycorrhizal infection (see references in Walker and Smith, 1984) represent an advance in the epidemiology of soil microbes, an area that has long been bedevilled by perhaps outmoded concepts and controversies and presents an exciting challenge to those working with soilborne pathogens.

The area in which rapprochement of ideas in population biology and plant disease epidemiology is most likely to occur is the area of the dynamics and genetics of plant pathogens and populations (Jeger and Groth, 1985). In a broader context, there have been attempts to reconcile epidemiological and population genetic ideas into models of host-parasite interactions (e.g., May and Anderson, 1983). Leonard and Czochar (1980) review the literature on the genetic inter-actions between plant and pathogen populations. They also use many of the techniques and concepts relating to equilibria and stability introduced in the next section, although the equations used and cited are mainly stated in discrete time. It is usually possible to reformulate discrete-time models as continuous-time models, such as the differential equations used in the next section, but this will not be discussed in this chapter.

EXAMPLE OF AN ANALYTIC APPROACH: HOST GROWTH AND DISEASE DYNAMICS

This chapter is directed to the would-be modeler who has familiarity with calculus and differential equations, and possibly an acquaintance with linear algebra and complex numbers, but who is probably not equipped to deal with mathematical papers in specialist journals. With this individual in mind, this section is devoted to showing how models involving linked differential equations can be formulated, analyzed, and used to extract the essential features of disease dynamics. The best exposure to this sort of modeling that can then be obtained is to formulate a problem of concern and to repeat the procedure for oneself.

Consider a simple logistic model of host growth in which a host population P grows sigmoidly from an initial population size P_0 to approach a maximum size P_{max} for arbitrarily large time values,

$$\frac{dP}{dt} = k_1 P \left(1 - \frac{P}{P_{max}} \right) \tag{4}$$

where k_1 is the intrinsic rate of increase of the host population. The measure of P may be numbers of plants or plant parts (e.g., tillers, shoots, or leaves) or a dimensioned quantity such as leaf area. The term "plant" will be used quite generally to indicate any one of these measures, but is unlikely to be biomass as is the case in most crop growth studies. The effects of disease on host growth can be investigated by making simple suppositions concerning their interaction. For example, suppose that the host population can be partitioned into healthy H and diseased X components; that new host growth is proportional only to the healthy component; and that the healthy component becomes diseased at a rate k_2 proportional to the product HX, the interaction between the healthy and diseased components. Then modifiying Eq. 4 gives

$$\frac{dH}{dt} = k_1 H \left(1 - \frac{H}{P_{max}} \right) - k_2 HX \tag{5}$$

and the equation for the diseased component is

$$\frac{dX}{dt} - k_2 IIX \quad k_3 X \tag{6}$$

where $k_3 X$ indicates the additional mortality or loss of diseased plants. The rate parameter k_1 is the intrinsic rate of host increase, k_2 can be considered a contact or exchange rate between diseased and healthy plants (e.g., as mediated by spores or vectors), and k_3 is the mortality rate of diseased plants. If the measure of the host population is in leaf area, then the measure of the diseased population should be in equivalent units, such as lesion area. Equations 5 and 6 will be reformulated later with a more direct interpretation of the parameters. For simplicity of presentation Eqs. 5 and 6 will first be analyzed.

Equations 5 and 6 are a linked pair of differential equations and, unlike Eq. 4, cannot be solved explicitly. That is, it is not generally possible to find a pair of functions f and g such that $H = f(t)$ and $X = g(t)$. This is not a problem is particular starting values of H_0 and X_0 and the rate parameters k_1 k_2, and k_3 are known, as numerical methods can generally be used to obtain values of H and X for any value of time t. Provided that the relative range of parameter values is known, or there are only certain parameter combinations of interest, then numerical methods are perfectly acceptable as a means of obtaining quantitative results. It is not possible, however, to obtain all qualitative aspects of dynamics of equations such as 5 and 6 by numerical methods alone and this is where further analysis becomes indispensible. Qualitative dynamics are usually determined by the existence of critical values of H and P at which there is neither increase nor decrease in the populations (i.e., at equilibria) and by the behavior of the populations near to these critical values. In the case of Eqs. 5 and 6 we may ask if there are equilibria in the healthy and diseased components of the host population; if so, are those equilibria stable? It should be noted that equilibria may be unstable in that, unless the populations are initially at the equilibrium values, they depart from rather than approach the critical points. A wider question may be to what extent disease can regulate a host population to a certain population size.

There are various analytic and graphic techniques for obtaining these critical values, and for investigating stability near the equilibria (Pielou, 1977; Roughgarden, 1979). In some cases it is possible to examine the global stability of equations such as 5 and 6 by using more advanced mathematical techniques than can be considered here (Leighton, 1966).

The critical values of Eqs. 5 and 6 are obtained by setting both equations to zero and solving for H and X. The critical values of H and X are denoted by H^* and X^*. In Eq. 3 this gives $dx/dt = 0$ when $k_2HX = k_3X$, that is when

$$H^* = \frac{k_3}{k_2} \tag{7}$$

Substituting this value of H^* into Eq. 5 gives $dH/dt = 0$ when $(k_1k_3k_2)$ $(1 - k_3/k_2P_{max}) = k_3X$, that is when

$$X^* = \frac{k_1}{k_2}\left(1 - \frac{k_3}{k_2P_{max}}\right) \tag{8}$$

Note that H^* by definition is less than P_{max}, and X^* is positive; hence we have the constraint $k_3 < k_2P_{max}$. Apparently then, the critical value of the healthy component of the host population depends only on the relative values of the k_2 and k_3 and not on its intrinsic rate of increase k_1 or its maximum size P_{max}. The influence of disease certainly is to regulate the host population. The question can now be asked: At what level of pathogen attack (the value of the rate parameter k_2) is the critical value for X^* maximized? Differentiating Eq. 8 with respect to k_2 gives

$$\frac{\delta X^*}{\delta k_2} = \frac{k_1}{k_2^2}\left(\frac{2k_3}{k_2P_{max}} - 1\right) \tag{9}$$

Setting Eq. 9 to zero gives the conditions $k_2 = 2k_3/P_{max}$ for X^* to be a maximum. Inserting this value of k_2 into Eq. 7 gives a value of H^* of $P_{max}/2$. Effectively this means that pathogen "success," as given by the equilibrium value X^*, is maximized when the healthy host population is regulated to half the maximum size P_{max}. Intriguingly the maximum rate of host growth in the absence of disease (Eq. 4) also occurs at precisely $P_{max}/2$.

The nature of these critical values H^* and X^* now must be examined in more detail. Under what conditions do H and X both approach H^* and X^*? A standard technique that can be used to determine the behavior of H and X in the neighborhood of the critical values is local stability analysis. This technique consists of obtaining a special matrix, the Jacobian matrix (Roughgarden, 1979), in which, for a system of two equations such as 5 and 6, partial derivatives are taken with respect to H and X and evaluated at the critical values. Of course the critical values must always be found first (not always an easy task) for the technique to be of use. In this case note first that the right-hand sides of Eqs. 5 and 6 can be written as the functions

$$f_1 = k_1 H \left(1 - \frac{H}{P_{max}}\right) - k_2 HX$$

$$f_2 = k_2 HX - k_3 X$$

Taking partial derivatives of each function with respect to H and X, and evaluating at the critical values H^* and X^* gives

$$
\left.
\begin{aligned}
\frac{\delta f_1(H^*, X^*)}{\delta H} &= k_1 - \frac{2k_1 H^*}{P_{max}} - k_2 X^* \\
&= -\frac{k_1 k_3}{k_2 P_{max}} \\
\frac{\delta f_1(H^*, X^*)}{\delta X} &= -k_2 H^* \\
&= -k_3 \\
\frac{\delta f_2(H^*, X^*)}{\delta H} &= k_2 X^* \\
&= k_1 \left(1 - \frac{k_3}{k_2 P_{max}}\right) \\
\frac{\delta f_2(H^*, X^*)}{\delta X} &= k_2 H^* - k_3 \\
&= 0
\end{aligned}
\right\} \quad (10)
$$

The jacobian matrix **J** is then formed from these partial derivatives to give the 2×2 square matrix:

$$
\begin{bmatrix}
-\dfrac{k_1 k_3}{k_2 P_{max}} & -k_3 \\[3ex]
k_1 \left(1 - \dfrac{k_3}{k_2 P_{max}}\right) & 0
\end{bmatrix}
$$

Elementary matrix theory (Coulson, 1965) informs that for a square matrix such as **J** there are quantities, eigenvalues, that are of great importance in determining the solution of a system of equations. In essence, local stability analysis consists of finding the eigenvalues of the jacobian matrix and classifying these according to their numerical values. In our case the eigenvalues turn out to be

$$\lambda_1 = \{(-k_1 k_3/k_2 P_{max}) + [(-k_1 k_3/k_2 P_{max})^2 - 4k_1 k_3 (1 - k_3/k_2 P_{max})]^{1/2}\}/2$$

and

$$\lambda_2 = \{(-k_1k_3/k_2P_{max}) - [(-k_1k_3/k_2P_{max})^2 - 4k_1k_3(1 - k_3/k_2P_{max})]^{1/2}\}/2$$

There are three possibilities in terms of the numerical values of these eigenvalues:

1. λ_1 and λ_2 are distinct real numbers.
2. λ_1 and λ_2 are equal with $\lambda = \lambda_1 = \lambda_2$.
3. λ_1 and λ_2 are a conjugate pair of complex numbers.

Assume, for the present, that λ_1 and λ_2 are real numbers or, equivalently, that the term in square brackets is positive. The critical values of H^* and X^* are stable equilibria if λ_1 and λ_2 are distinct and both are negative. Now $-k_1k_3/k_2P_{max}$ is certainly negative and hence the second root λ_2 is negative. Hence only the sign of λ_1 need be considered. We require that

$$\frac{-k_1k_3}{k_2P_{max}} + \left[\left(-\frac{k_1k_3}{k_2P_{max}}\right)^2 - 4k_1k_3\left(\frac{1 - k_3}{k_2P_{max}}\right)\right]^{1/2} < 0$$

After some manipulation of the inequality, this is equivalent to

$$0 < 4k_1k_3\left(\frac{1 - k_3}{k_2P_{max}}\right) \tag{11}$$

This condition is ensured because of the constraint $k_3 < k_2P_{max}$, already noted, giving a positive value for the right-hand side of Eq. 11. Thus we conclude that the critical values H^* and X^* are stable equilibria and that H and X approach equilibrium. The way in which H and X approach equilibrium can also be specified. The deviations from equilibrium, with time, decay exponentially at rates dependent on λ_1 and λ_2.

Let us now consider the case where $\lambda = \lambda_1 = \lambda_2$. This is a rather special case in that the term in square brackets must now exactly equal zero to satisfy this condition. In this case $\lambda = -k_1k_3/k_2P_{max}$ which is certainly negative and H^* and X^* are again stable equilibria. The case where λ_1 and λ_2 are a conjugate pair of complex numbers leads to a special kind of dynamics close to the critical values. In this case the term in square brackets is a negative number and the roots are written in the form $\lambda_1 = \alpha + \beta i$ and $\lambda_2 = \alpha - \beta i$ where α and βi are the real and imaginary parts, respectively, of the complex numbers. For this to occur the condition

$$\left(\frac{-k_1k_3}{k^2 P_{max}}\right)^2 < 4k_1k_3\left(1 - \frac{k_3}{k_2P_{max}}\right) \tag{12}$$

must hold. After some manipulation an equivalent condition is obtained that depends on finding the roots of a quadratic equation P_{max}. If P_{max} is greater than a given value (defined by k_1, k_2, and k_3), then the inequality given by Eq. 12 holds and λ_1 and λ_2 are a conjugate pair of complex numbers. However, the real part of the pair is again negative and the critical values are again stable equilibria. The approach to equilibria, however, is markedly different from the case in which λ_1 and λ_2 are both real numbers. When H is plotted against X a spiral around the equilibrium point (H^*, X^*) is formed. This is because damped oscillations in H and X are occurring as each approach their equilibrium values (Fig. 1) and provides a contrast to smoothly increasing sigmoid curves.

Several conclusions follow from the discussion of the host-disease system described by Eqs. 5 and 6. The existence of critical values of H and X at which there is no change in either population has been demonstrated. Moreover these critical values are stable in the sense that all trajectories of H and X in the neighborhood of the equilibria converge to the critical values. Finally, the approach to equilibria, whether oscillatory or not, can be determined from the relative parameter values. In this sense the potential for regulation of a host population by disease has been quantitatively described.

The basic suppositions made in deriving these conclusions are now reexamined to investigate the robustness of the conclusions. The basic flaw in the equations is that time delays in disease development are not included. This can be accounted for by partitioning the disease component into disease that is latent, infectious, and postinfectious and specifying equations for each category of disease to give a more complex system of Eqs. than 5 and 6. Alternatively, time delays can be introduced into Eq. 6 to indicate the lag due to a latent period. The effects of this are to introduce a further aspect of stability: that is the existence of a trajectory called a stable limit cycle to which, eventually, all trajectories converge (Wangersky, 1978). The mathematical techniques involved in analyzing those cases are complex and are not considered further.

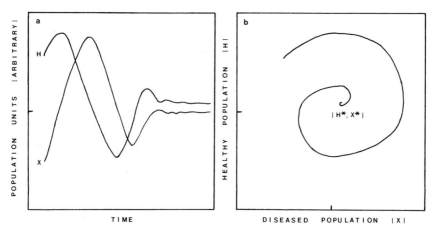

Figure 1 (a) Time trajectories, and (b) phase diagram of healthy (H) and diseased (X) populations in which both populations oscillate and approach equilibrium values (H^*, X^*) in the form of a spiral.

Two improvements to the basic equations can be made without increasing unduly the complexity of the analysis. First, the parameter k_2 at present relates the healthy and diseased component of the host population in a "mass action" term. This is commonly done in medical epidemiology where diseased and healthy individuals freely intermingle in a community; the rate parameter k_2 is then a contact rate. For our purposes it is better to consider k_2 to be an intrinsic rate of disease increase rather than a contact rate between plants and propagules. This can readily be done by incorporating the term $H/(H + X)$, which is the proportion of the plant population healthy, rather than H alone. Thus the term is a correction factor to the intrinsic rate (just as the term $1 - x$ is a correction factor to the rate parameter r in Van der Plank's equation). When H is large relative to X then the correction is minimal, but as X increases the limitation due to lack of healthy plants becomes more important.

The second improvement that can be made is to the term giving the increase in healthy plants. Equation 5, at present, uses the correction factor $1 - H/P_{max}$. This is effectively an underestimate because plants that become diseased are not included. A more appropriate correction factor to host increase would be the term $1 - (H + X)/P_{max}$ which reflects the cumulative production of all host tissue in relation to the theoretical maximum P_{max}. Incorporating both improvements gives

$$\frac{dH}{dt} = k_1 H \left(1 - \frac{H + X}{P_{max}} \right) - \frac{k_2 X H}{H + X} \tag{13}$$

and

$$\frac{dX}{dt} = \frac{k_2 X H}{H + X} - k_3 X \tag{14}$$

We then follow exactly the same procedure as before. Setting Eqs. 13 and 14 simultaneously to zero gives critical values of

$$H^* = \frac{k_3}{k_2} P_{max} \left(1 - \frac{k_2 - k_3}{k_1} \right) \tag{15}$$

and

$$X^* = \left(1 - \frac{k_3}{k_2} \right) P_{max} \left(1 - \frac{k_2 - k_3}{k_1} \right) \tag{16}$$

It is rather more difficult to establish the critical values than it was previously but they can readily be checked by substituting into Eqs. 13 and 14. Note also that

$$H^* + X^* = P_{\max}\left(1 - \frac{k_2 - k_3}{k_1}\right) \tag{17}$$

In other words, the total plant productions, relative to $P_{\max}$, is given by the proportion $1 - (k_2 - k_3)/k_1$, where of course we require $k_1 > k_2 - k_3$. The actual proportions of the total population healthy u and diseased v are then given by $u^* = H^*/(H + X^*) = k_3/k_2$ and $v^* = X^*/H^* + X^* = 1 - k_3/k_2$.

Thus the proportions of the host population that are healthy and diseased are determined entirely by the rate of disease increase k_2 and the mortality rate of diseased plants k_3, and are independent of both the intrinsic rate of host increase k_1 and the maximum population size $P_{\max}$. This certainly confirms the earlier results and the conclusion that disease can regulate the size of host populations. The critical values have to be evaluated for stability. Taking partial derivatives of the new system of equations and evaluating at the critical values gives a new jacobian matrix. The eigenvalues of the matrix can be obtained as before and, although more complex, lead to the same conclusions: that the critical values are stable equilibria, and may be approached with or without oscillations in either component of the population.

In fact the basic conclusions to be drawn from the model remain essentially the same as before. For example, the improved version of the correction factor incorporated into Eq. 14 leads again to the conclusion that X^*, the equilibrium value of disease, is maximized when the healthy component $H^* = P_{\max}/2$. This procedure, of changing the suppositions in simple models and evaluating whether the change fundamentally alters the conclusions drawn, is of great value in showing that simple models, even when limited by an obvious lack of realism, may distill enough of the essense of a dynamic system to offer insight into its solution.

INTEGRATIVE SIMULATION MODELS

Having considered in some detail what analytic approaches can achieve in modeling plant disease epidemics, I will now consider two areas in which I feel valid, if limited, use can be made of simulation models. In both examples the defining feature will be that the models integrate phenomena at different levels of organizations. A more complete evaluation is given by Teng (1985).

Host Physiology, Crop Loss, and Disease

Much interest has been expressed recently in the coupling of plant disease models with crop or plant physiological models that provide estimates of either crop yield or biomass (Boote et al., 1983; Loomis and Adams, 1983; Rouse, 1983). In my view there are unresolved difficulties with this coupling. First, there are the technical problems of coupling. Many plant models are strictly defined at the whole plant rather than the population level and it is not clear how the effects of disease epidemics should be incorporated. The crop model, for example, should refer to yield in populations of plants.

Second, the backgrounds of crop modeling and epidemiological modeling are very different. The former basically are concerned with physiological integrations involving carbon allocation and other processes within single plants, whereas the latter are population dynamic models deriving from quite different traditions. I believe a more subtle view of disease in populations, as a determinant of yield, is needed than simply to consider "disease" as a submodel in an overall single-plant crop model. Only then can meaningful rapprochement occur.

Third, crop models are increasingly complex and cover a wide range of levels of integration. I remain unconvinced of the usefulness of hierarchical models spanning more than two levels of integration, either within crop physiology, or when coupled with disease dynamics in populations.

Passioura (1973) commented on what he considered to be the sense and nonsense of crop simulation. He discussed the reasons often proposed for composing complex crop models: that models are testable hypotheses; that they serve as frameworks within which to direct research, generate hypotheses, and mobilize knowledge; and that they lead to multidisciplinary spinoffs. Passioura, in general, concluded that each of these reasons was more nonsense than sense, and reasoned that simple models—in which the parameters were few enough to be measured directly or indirectly, and in which components of the system and their interaction were modeled two at a time—represented more the "sense."

Loomis et al. (1979) discussed Passioura's article without really considering his criticisms. In their view, the reasons for complex simulation models is that no other means for studying the integrative physiology of hosts exist. Originally, the phytotron was considered the major integrative tool in crop physiology but now its major use seems to be to provide data for complex simulation models. In my view this rather inverts the relationship that is the defining feature of science: experiments are done to test hypotheses and models rather than to provide flesh to a rather bare skeleton. In fact Loomis et al. (1979) estimate that 50 to 80% of their research effort goes into experiments to fill information gaps.

The claim is made that combined models coupling crop growth and disease can be reliable predictors of pest and disease development but the point of reference is always yield rather than some epidemiological consequence and no substantive cases are cited. As far as I am aware, the one successful example of where crop and disease models have been coupled is with verticillium wilt in cotton (Gutierrez et al., 1983). The host model is unusual in that population dynamic aspects of the cotton boll population are more explicit than usual, and are expressed in terms of both time and the age structure of the boll population. The model predicts cotton yields, or biomass, as number of bolls per plant for different inoculum densities and pathotypes of *Verticillium dahliae*. Importantly, the coupling appears successful because it appears that the pathogen has no dynamics within a growing season. There is one generation of infective propagules to infect the cotton population.

The main reservation against too many resources being allocated to coupling host and disease models is that in many cases the models resulting are unlikely to be cost effective. As Loomis and Adams (1983) admit, if the occurrence of disease at some level is known to have serious economic consequences then epidemiological (that is, population dynamic) models are quite sufficient. In my opinion this is more often the case than not, and we should resist the temptation to become so much grist to the crop modeler's mill unless it is unequivocally demonstrated as necessary. In particular, epidemiological models deal increasingly with population genetic and evolutionary considerations with respect to pathogen populations. These effects on epidemics and thus on crop loss cannot currently be incorporated into standard crop models.

To return to the distinction made between description and explanation at different levels of integration: if explanation of an epidemic (its economic consequences being taken as given) is desired, then a description of host dynamics, and possibly phenology, is required to couple with a description of the pathogen population. According to this view it is not necessary to use explanatory hierarchical models of host growth and development to explain the dynamics of epidemics. For example, many epidemics caused by biotrophic pathogens such as powdery mildews depend on the continual production of new susceptible foliage. In many cases a simple growth model describing leaf area increases is sufficient to couple with pathogen dynamics (Rouse, 1983). Similarly the infection of roots by soil-inhabiting fungi largely depends on the probability of contact between spatially distributed inoculum and a developing root system (Huisman, 1982). Accordingly descriptions of the distribution of inoculum and of root dynamics are required to explain, in part, the occurrence of initial infections. And yet, in the writer's experience, to ask a crop modeler what root models are available for this purpose is to become immersed in matters of shoot:root partitioning, photosynthate allocation and other physiological processes that are largely irrelevant to the question at hand. One area where host physiological models may prove important in modeling epidemics is the effect of the nutrient status of host tissues on disease dynamics.

Disease Epidemics and Management Practices

It has been claimed that the most appropriate means of using models in disease management is to construct a complex simulation model, evaluate the model (see Teng, 1981), and then simplify the model to a form that can be implemented (Loomis et al., 1979; Rabbinge and Carter, 1983). This approach has appeal but there are several drawbacks in addition to those already cited (Passioura, 1973). First, there are large discrepancies in what is meant, philosophically and practically, by evaluating a complex hierarchial model (Jeger, 1983; Teng, 1981). Further, there is no universally accepted algorithm for determining the validity of any complex simulation model and it is unlikely that there will be one in the foreseeable future. Second, I know of no examples in which a complex simulation model has been constructed, fully evaluated, simplified by some means, and

implemented in practical disease management. Even if there were such an example, to be convincing it would require demonstration that a simplified model
could not have been obtained by more conventional means.

Despite these qualifications there is some intuitive appeal in the concept
and there are some experimental and modeling programs that have this direction.
Bruhn and Fry (1981) used a simulation model to analyze the epidemiology of
potato late blight (*Phytophthora infestans*). The simulation model was constructed rather differently from those involving differential equations. The model
was explicitly discrete and formulated in terms of transition matrices with lesions
categorized into a large number of age classes. Both the number of lesions and
the mean surface area of lesions in each age class were simulated in response
to environment, host-factors including cultivar resistance, and fungicide application. The model successfully mimicked epidemic development in four cultivars
over three growing seasons and analysis of the simulations indicated several
strategies to be used in optimizing late blight control. In subsequent models
(Bruhn and Fry, 1982a, 1982b), the deposition and weathering of the fungicide
chlorothalonil, interactions with plant growth and architecture, cultivar, application method and dosage, and environment effects such as rainfall, were incorporated. These two fungicide models were based largely on the use of a
statistical distribution, the gamma probability function, to describe the variation
in residues, and a rainfall-dependent decay model to describe the dynamics. The
various components in the model were thus of rather diverse form.

Fry et al. (1983) subsequently evaluated the complex simulation model by
comparison with the late blight forecasting criteria specified in BLITECAST
modified to incorporate host resistance. A forecasting scheme was developed
from analysis of the simulation results and this compared favorably in all respects
with the BLITECAST-based forecasts. In fact BLITECAST tended to cause
more sprays to be applied than was actually necessary, whereas the simulation
model gave better forecasts with resistant cultivars. Neither method of forecasting, however, led to any difference in the yields of potato tubers or the
proportion of tubers blighted when sprays were applied on the basis of the
forecasts. Hence, as neither the benefits associated with the reduced number of
sprays nor the costs associated with either forecasting scheme were presented,
the actual economic gains cannot be commented on. However, this example of
an original simulation model, combined with practical objectives and the eventual
aim of simplified forecasts, does give some confidence in the potential of the
procedure.

CONCLUSIONS

The broad conclusions to be drawn from this chapter can be considered by
returning to the categories of model use introduced earlier: in development of
theory, as an aid to calculation, and in disease management. Analytic approaches

are eminently suitable for the development of theory but there are dangers in these models being considered as explanatory and hence being used for direct quantitative prediction. The main use is in suggesting the possible range of outcomes that may be expected and, in so doing, providing qualitative features of epidemics to be compared. The extended example presented in this chapter is intended to illustrate the analytic approach. By contrast simulation approaches are, by definition, concerned only with particular, or a set of particular, behavior and are not suited for the development of theories concerning the real system.

Simple descriptive models that make no pretension to explanation are often useful as a means for calculation. There are dangers where the parameters of such models take on a mechanistic meaning never originally intended. The very best models to use as calculation tools possess some biological realism and they extend into the range of analytic approaches. Simulation models, however, can rarely be used as a means of calculation. They can only be fitted to experimental data and unknown parameter values can only be estimated if sophisticated optimization packages are available. The most common link of simulation models with experimentation is that experiments are done to provide information for a model rather than to test predictions of a model.

Analytic approaches to modeling are useful in the development of modeling skills and in avoiding an excessive reliance on the digital computer. Undeniably, however, the power and flexibility of the computer lends itself to the development of instructional programs based on simulation models which can be of direct use and heuristic value to students and practitioners of disease management.

Analytic approaches are most useful for analyzing a range of management options, provided the problems can be posed in simple but realistic terms. Thus, typically, threshold criteria are derived in terms of model parameters according to which broad strategies can be determined. Simulation approaches may also be useful for determining strategies as, for example, in the use of Monte Carlo techniques to determine probabilities of outcomes given a range of management options. (This kind of approach has not really been discussed in this chapter.) The main value of the simulation approach lies in integrating across traditional boundaries and this has been discussed in the examples given. The application of such integrations lies very much in the future. For the present it should be recognized by crop modelers that epidemics of plant disease have a direct impact on yield in crop populations through purely population dynamic considerations and that sophisticated coupling at different levels of integration may not be necessary for many diseases.

Neither the analytic nor the simulation approach to modeling will have much impact on practical disease management by comparison with more direct empirical means of forecasting. Undoubtedly, hybrid models with a combination of analytic and simulation approaches will continue to emerge in the future; it is important that the potential of each approach be explored and exploited to the fullest.

REFERENCES

Allen, R. N. 1983. Spread of banana bunchy top and other plant virus diseases in time and space, in: *Plant Virus Epidemiology* (R. T. Plumb and J. M. Thresh, eds.), pp. 51–60. Blackwell Scientific Publications, Oxford.

Bailcy, N. T. J. 1975. *The Mathematical Theory of Infectious Diseases and Its Applications*. Charles Griffin & Company Ltd, London.

Barrett, J. A. 1983. Estimating relative fitness in plant parasites: Some general problems. *Phytopathology* 73:510–512.

Betters, D. R., and Schaefer, J. C. 1981. A generalized Monte Carlo simulation model for decision analysis illustrated with a Dutch elm disease control example. *Can. J. For. Res.* 11:342–350.

Bloomberg, W. J. 1979a. A model of damping-off and root rot of Douglas-fir seedlings caused by *Fusarium oxysporum*. *Phytopathology* 69:74–81.

———. 1979b. Model simulations of infection of Douglas-fir seedlings by *Fusarium oxysporum*. *Phytopathology* 69:1072–1077.

Boote, K. J., Jones, J. W., Mishoe, J. W., and Berger, R. D. 1983. Coupling pests to crop growth simulators to predict yield reductions. *Phytopathology* 73:1581–1587.

Brittain, E. G. 1983. A model system of four species with a unique equilibrium point. *Ecol. Model.* 19:199–211.

Bruhn, J. A. and Fry, W. E. 1981. Analysis of potato late blight epidemiology by simulation modeling. *Phytopathology* 71:612–616.

———. 1982a. A statistical model of fungicide deposition on potato foliage. *Phytopathology* 72:1301–1305.

———. 1982b. A mathematical model of the spatial and temporal dynamics of chlorothalonil residues on potato foliage. *Phytopathology* 72:1306–1312.

Butt, D. J., and Jeger, M. J. 1985. The practical implementaton of models in crop disease management, in: *Advances in Plant Pathology,* vol 3: Mathematical Modelling of Crop Disease (C. A. Gilligan, ed.) pp. 207–230. Academic Press, London.

Coulson, A. E. 1965. *An Introduction to Matrices.* Longmans, Green & Co. Ltd. London, 152 pp.

Fleming, R. A. 1980. The potential for control of cereal rust by natural enemies. *Theor. Pop. Biol.* 18:374–395.

Fleming, R. A., and Bruhn, J. A. 1983. The role of mathematical models in plant health management, in: *Challenging Problems in Plant Health* (T. Kommedahl and P. H. Williams, eds.), pp 368–378. The American Phytopathological Society, St. Paul, Minnesota.

Fry, W. E., Apple, A. E., and Bruhn, J. A. 1983. Evaluation of potato late blight forecasts modified to incorporate host resistance and fungicide weathering. *Phytopathology* 73:1054–1059.

Gregory, P. H. 1961. *The Microbiology of the Atmosphere.* Leonard Hill, London, 251 pp.

Griggs, M. M., Nance, W. L., and Dinus, R. J. 1978. Analysis and comparison of fusiform rust disease progress curves for five slash pine families. *Phytopathology* 68:1631–1636.

Gutierrez, A. P., DeVay, J. E., Pullman, G. S., and Friebertshauser, G. E. 1983. A model of Verticillium wilt in relation to cotton growth and development. *Phytopathology* 73:89–95.

Hahn, B. D., and Furniss, P. R. 1983. A deterministic model of an anthrax epizootic: Threshold results. *Ecol. Model.* 20:233–241.

Hau, B., and Eisensmith, S. P. 1983. The crop ecosystems approach: A new challenge, in: *Abstracts of Papers, Fourth International Congress of Plant Pathology,* Melbourne, Australia, August 17–24, p. 72.

Hau, B., and Kranz, J. 1978. Modellrechnungen zur Wirking des Hyperparasiten Eudarluca caricis auf Rostepidemien. *Z. Pflanzenkrankh. Pflanzenschutz* 85:131–141.

Huisman, O. C., 1982. Interrelations of root growth dynamics to epidemiology of root-invading fungi. *Annu. Rev. Phytophathol.* 20:303–327.

Jeger, M. J. 1982a. Using growth curve relative rates to model disease progress of apple powdery mildew. *Prot. Ecol.* 4:49–58.

———. 1982b. The relation between total, infectious and post-infectious diseased plant tissue. *Phytopathology* 72:1185–1189.

———. 1983. Mathematical models: Their nature and development to practical ends in crop protection. *Bull. IOBC/WPRS NS (New Series)* 1983/VI/2: 3–17.

———. 1984a. The relation between rate parameters and latent and infectious periods during a plant disease epidemic. *Phytopathology* 74:1148–1152.

———. 1984b. Relating disease progress to cumulative numbers of trapped spores: Apple powdery mildew and scab epidemics in sprayed and unsprayed orchard plots. *Plant Pathol.* 33:517–523.

———. 1984c. The derivation and use of epidemic models in plant pathology. *Sci. Horti.* 35:11–27.

———. 1985. Modelling the dynamics of pathogen populations, in: *Populations of Plant Pathogens: Their Dynamics and Genetics* (M. S. Wolfe and C. E. Caten eds.), Blackwell Scientific Publications, Oxford (in press).

Jeger, M. J., and Groth, J. V. 1985. Pathogenicity and resistance: Epidemiological and ecological mechanisms, in: *Mechanisms of Resistance to Plant Diseases.* (R. S. S. Fraser, ed.), pp. 310–372. Martinus Nijhoff/Dr. W. Junk Publishers, The Hague.

Jeger, M. J., and Starr, J. L., 1985. A theoretical model of the winter survival dynamics of *Meloidogyne* spp. eggs and juveniles. *J. Nematol.* 17:257–260.

Jowett, D. A., Browning, J. A., and Haning, B. C. 1974. Non-linear disease progress curves, in: *Epidemics of Plant Diseases: Mathematical Analysis and Modeling* (J. Kranz, ed.), pp. 115–136. Springer-Verlag, Berlin.

Kranz, J., and Hau, B. 1980. Systems analysis in epidemiology. *Annu. Rev. Phytopathol.* 18:67–83.

Leighton, W. 1966. *Ordinary Differential Equations,* 2d ed. Wadsworth Publishing Company, Inc., Belmont, California, 246 pp.

Leonard, K. J., and Czochor, R. J. 1980. Theory of genetic interactions among populations of plants and their pathogens. *Ann. Rev. Phytopathol.* 18:237–258.

Levy, Y., Levi, R., and Cohen, Y. 1983. Build up of a pathogen subpopulation resistant to a systemic fungicide under various control strategies: A flexible simulation model. *Phytopathology* 73:1475–1480.

Loomis, R. S., and Adams, S. S. 1983. Integrative analyses of host-pathogen relations. *Ann. Rev. Phytopathol.* 21:341–362.

Loomis, R. S., Rabbinge, R., and Ng, E. 1979. Explanatory models in crop physiology. *Ann. Rev. Plant Physiol.* 30:339–367.

Madden, L. V. 1980. Quantification of disease progression. *Prot. Ecol.* 2:159–176.

May, R. M., and Anderson, R. M. 1983. Parasite-host coevolution, in: *Coevolution* (D. J. Futuyma and M. Slatkin, eds.), pp. 186–206. Sinauer Associates Inc., Sunderland, Massachusetts.

Morral, R. A. A., and Verma, P. R. 1981. Disease progress curves, linear transformations, and common root rot of cereals. *Can. J. Plant Pathol.* 3:182–185.

Passioura, J. B. 1973. Sense and nonsense in crop simulation. *J. Aust. Inst. Agr. Sci.* 39:181–183.

———. 1979. Accountability, philosophy and plant physiology. *Search* 10:347–350.

Payandeh, B. 1981. Choosing regression models for biomass prediction equations. *For. Chron.* 57:229–232.

Pedersen, O. C. 1984. Models of pesticide resistance dynamics. *Acta Agr. Scand.* 34:145–152.

Pielou, E. C. 1977. *Mathematical Ecology*. John Wiley & Sons, Inc., New York, 385 pp.

Putter, C. A. J. 1982. An epidemiological analysis of the Phytophthora and Alternaria blight pathosystem in the Natal Midlands. Ph.D. thesis, Unviersity of Natal, South Africa, 192 pp.

Rabbinge, R., and Carter, N. 1983. Application of simulation models in the epidemiology of pests and disease: An introductory review. *Bull. IOBC/WPRS NS (New Series)* 1983/VI/2:18–30.

Roughgarden, J. 1979. *Theory of Population Genetics and Evolutionary Ecology: An Introduction*. Macmillan Publishing Co., Inc., New York, 634 pp.

Rouse, D. I. 1983. Plant growth models and plant disease epidemiology, in: *Challenging Problems in Plant Health* (T. Kommedahl and P. H. Williams, eds.), pp. 387–398. The American Phytopathological Society, St. Paul, Minnesota.

———. 1985. Construction of temporal models: I. Disease progress of air-borne pathogens in: *Advances in Plant Pathology,* vol 3, Mathematical Modelling of Crop Disease (C. A. Gilligan, ed.), pp. 11–29. Academic Press, London.

Sall, M. A. 1980. Epidemiology of grape powdery mildew: A model. *Phytopathology* 70:338–342.

Schoemaker, C. A. 1981. Applications of dynamic programming and other optimization methods in pest management. *IEEE Trans. Automat. Control.* 26:1125–1132.

Skylakakis, G. 1982. The development and use of models describing outbreaks of resistance to fungicides. *Crop Protection* 1:249–262.

Teng, P. S. 1981. Validation of computer models of plant disease epidemics: A review of philosophy and methodology. *Z. Pflanzen. Pflanzenschutz* 80:181–187.

———. 1985. A comparison of simulation approaches to epidemic modeling. *Annu. Rev. Phytopathol.* 23:351–379.

Thornley, J. H. M. 1980. Research strategy in the life sciences. *Plant Cell Environ.* 3:233–236.

Van der Plank, J. E. 1963. *Plant Diseases: Epidemics and Control*. Academic Press, London, 349 pp.

———. 1975. *Principles of Plant Infection*. Academic Press, London, 216 pp.

Waggoner, P. E. 1977. Contributions of mathematical models to epidemiology. *Ann. N. Y. Acad. Sci.* 287:191–206.

———. 1978. Computer simulaton of epidemics, in: *Plant Disease: An Advanced Treatise,* vol 2: How Disease Develops in Populations (J. G. Horsfall and E. B. Cowling, eds.), pp. 203–222. Academic Press, Inc., New York.

————. 1981. Models of plant disease. *BioScience* 31:315–319.

————. 1983. Quantifying the effect of the physical environment, in: Challenging Problems in Plant Health (T. Kommedahl and P. H. Williams, eds.) pp. 215–225. The American Phytophathological Society, St. Paul, Minnesota.

Walker, N. A., and Smith, S. E. 1984. The quantitative study of mycorrhizal infection. II. The relation of rate of infection and speed of fungal growth to propagule density, the mean length of the infection unit and the limiting value of the fraction of the root infected. *New Phytol.* 96:55–69.

Wangersky, P. J. 1978. Lotka-Volterra population models. *Annu. Rev. Ecol. System.* 9:189–218.

Wangersky, P. J., and Cunningham, W. J. 1956. On time lags in equations of growth. *Proc. Nat. Acad. Sci. U.S.* 42:699–702.

Zakoks, J. C. 1971. Systems analysis and the dynamics of epidemics. *Phytopathology* 61:600–610.

Part Five

Disease Spread

Chapter 12

Disease Gradients and the Spread of Disease

K. P. Minogue

The development of disease in a population of plants requires not only that a pathogen can infect and colonize individual plants, but also that it can be transmitted from one plant to another. The mechanisms of dispersal, and the means by which it is studied, are the subject of Chap. 13. This chapter discusses how dispersal of the pathogen gives rise to the spread of disease, and how this process can be described, quantified, and modeled.

The contagious nature of biotic plant diseases results in the development of a distinct class of spatial patterns of disease, on scales ranging from leaflets to continents. Various methods for describing and quantifying these patterns have been developed (Strandberg, 1973; Gross et al., 1980; Campbell and Pennypacker, 1980; Gilligan, 1982; Madden et al., 1982; Martin et al., 1983). A pattern of particular importance is the disease gradient, which occurs whenever there is a general increasing or decreasing trend in disease level that is related to distance. Such a trend can develop in response to a gradient in environmental favorability for disease, or to topographical or vegetational features that influence spore deposition patterns, but our concern in this chapter will be almost exclusively with gradients that develop as a result of dispersal around a localized source of inoculum. Such gradients are the result of a complex dynamic process, and their properties are influenced by all of the factors that affect the development and reproduction of the pathogen. The exact nature of this influence is still poorly understood, although some recent advances have been made.

Gradients related to dispersal are of a number of types. The *dispersal gradient,* discussed in detail in Chap. 13, refers to the number of spores or other propagules that are deposited on natural or artificial surfaces as a function of distance from a source. More useful for modeling gradient development is the closely related *dispersal function,* which is the probability that a dispersing propagule will travel a given distance from its point of origin before being deposited. The dispersal function is essentially the same as a dispersal gradient around a point source; the dispersal gradient is a more general concept that also applies to area sources.

After propagules have been deposited, infection occurs, giving rise to an *infection gradient.* If all propagules producing infection originate from outside the plot or field of interest, or from primary inoculum sources within the field, it is referred to as a *primary infection gradient.* In a uniform environment the infection gradient has the same form as the dispersal gradient, although it differs in level because not all propagules are successful in initiating infection.

Infection gradients are never observed directly, but only after the infections have produced visible disease in the host. At this point a *disease gradient* exists, which is called a *primary disease gradient* if it develops from a primary infection gradient. The disease gradient may or may not have the same form as the infection gradient, depending on the level of disease, and on whether disease is assessed in such a way that individual infections are distinguishable. For polycyclic diseases, gradients that develop from secondary infections are called *secondary gradients.*

Apart from its intrinsic interest and its contribution to our understanding of epidemiological dynamics, the study of gradients has a number of important practical applications. Primary disease gradients in a uniform environment are closely related to dispersal gradients, and thus give information about the distance over which a significant amount of dispersal occurs (e.g., Kable et al., 1980). The shape and steepness of a gradient can give clues about the mode of dispersal (Gregory, 1968; Thresh, 1976) and even about the behavior of vectors (Hampton, 1967). Comparison of gradients has been used to make inferences about the role of secondary inoculum in epidemics (Johnson and Powelson, 1983), to determine the spatial effects of host resistance (Mackenzie, 1976) and multilines (Fried et al., 1979), to test models of disease spread (Minogue and Fry, 1983b; Jeger, 1983), to infer etiology (Ries and Royse, 1978), and to locate sources of inoculum (Bonde and Schultz, 1943).

This chapter begins with a discussion of the methods currently available for quantifying and comparing gradients. The following section describes approaches to modeling the spread of monocyclic and polycyclic diseases, with consideration of the development of the gradient and the factors determining the velocity of spread. The next section outlines some problems that are commonly encountered in gradient experiments, and the chapter concludes with a look at some of the outstanding questions in this field. Although many of the concepts discussed are applicable to a wide range of diseases, the reader will notice a definite bias toward those that affect aboveground portions of the host.

DESCRIPTION AND COMPARISON OF GRADIENTS

To compare gradients it is necessary first to quantify them. The simplest and most direct measure of gradient steepness is the ratio of a change (Δy) in the amount of disease (incidence or severity) to the distance (Δs) over which the change occurs. For a sufficiently small distance $\Delta y/\Delta s$ approximates the mathematical definition of the gradient, the derivative dy/ds. This ratio is rarely used, however, because it is usually not constant at all points along the gradient. Primary gradients are typically steep near the source (implying a large magnitude of dy/ds), and become flat farther from it (small dy/ds), while well-developed secondary gradients are flat near the focal center, become steeper with distance, and finally flatten out again. As will be seen later, it is possible to use the derivative to compare gradients in spite of this inconsistency, and in fact this is commonly done in an informal way when gradients are compared visually. The usual approach, however, is to transform disease intensity or distance (or both) in such a way that the transformed gradient approximates a straight line. The slope of this line is then used as the basis for comparing gradients.

The use of transformations is based on the assumption that the gradients under study can be adequately described by a particular equation, one of whose parameters can be interpreted as a reflection of gradient steepness. A variety of equations have been proposed for this purpose (Gregory and Read, 1949), but only two have been widely used:

1. *The inverse power law:* $y = a/s^b$, popularized by Gregory (1968)
2. *The negative exponential law:* $y = ae^{-bs}$

where e is the base of natural logarithms. The parameters a and b are interpreted differently for the two equations (see later). The negative exponential equation is often called Kiyosawa and Shiyomi's (1972) equation, but in fact it had been used by Frampton et al. (1942) 30 years earlier.

The Log-Log Transformation

Gregory's equation is the basis of the log-log transformation,

$$\log y = \log a - b \log s$$

which describes a straight line of slope $-b$, a dimensionless constant. This slope is interpreted as the steepness of the gradient, and the estimate of b (which can be obtained by standard linear regression techniques) is used as the statistic for comparing gradients. Experimentally, b is most often in the range of 0.1 to 5.0; larger values are taken to indicate a steeper gradient.

The application of the log-log transformation has been discussed in detail by Gregory (1968), but some confusion over the interpretation of b persists in the literature (Barrett, 1981). There are three aspects, in particular, that seem to need clarification.

Response to a Uniform Change in Disease Intensity   If the level of disease along two gradients differs by a constant proportion at all distances (i.e., if $y_1(s) = ky_2(s)$), then b is unaffected even though on arithmetic axes one gradient will appear steeper than the other (Fig. 1). Thus, fungicides, host resistance, multilines, and so on, cannot directly affect the value of b (although they may have indirect effects, as will be discussed below). The expectation that multilines can affect primary disease gradients (Fried et al., 1979) is therefore unfounded (Barrett, 1981).

Effects of Measurement Scale   The parameter b is dimensionless; it therefore has the same value whether distance is measured in centimeters or kilometers. This lack of scale information introduces some hazards to the comparison of gradients, which become more severe as the scales of measurement diverge. For example, values of b have been measured at 0.2 to 1.0 for *Cercosporella* foot rot (Rowe and Powelson, 1973) and up to 4.3 for beet mosaic (Shepherd and Hills, 1970). Given only this information, the latter would be classified as steeper. This interpretation changes, however, when the scales of measurement are taken into account: the foot-rot gradients were contained entirely within a few meters of the source, while those of beet mosaic covered many kilometers.

When scales of measurement are closer this kind of interpretation problem is less severe. Users of the log-log transformation should be aware, however,

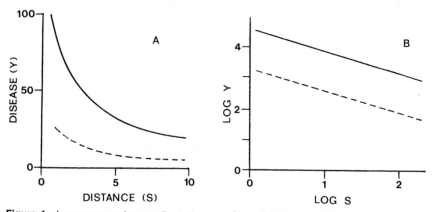

Figure 1   Inverse power law gradients ($y = as^{-b}$) on (A) linear and (B) logarithmic axes. For both curves $b = 0.7$; $a = 100$ (——) or $a = 25$ (----).

that the parameter b contains no distance information. The sense in which it can be considered as a measure of gradient steepness (change in disease per unit distance) is therefore often obscure.

Choice of Base for Logarithms It is nearly universal to use common (base 10) logarithms for the log-log transformation, but this is not a requirement of the method. The estimate of b will be the same whether common, natural (base e), or any other logarithms are chosen. It is preferable that natural logarithms be used, in the interest of consistency with other transformations in common use.

The log-log transformation is often quite successful in linearizing gradients, and perhaps for this reason there have been a few attempts to use the inverse power law as a gradient model, that is, to propose a conceptual framework for the equation within which the parameters can be given a mechanistic interpretation (Dimond and Horsfall, 1960; Lambert et al., 1980). The equation has, in fact, often been referred to as Gregory's model (e.g., see Mackenzie, 1976), even though Gregory himself was emphatic that its use was a matter of convenience, and that it was intended for the description of gradients, not their interpretation (Gregory, 1968). Attempts to give it a mechanistic interpretation, such as the geometric arguments of Dimond and Horsfall (1960), have failed, because they do not take into account a major determinant of gradient steepness, spore deposition. The fact that the equation predicts infinite disease at the origin ($s = 0$) seems sufficient to rule out the possibility that any reasonable conceptual basis for it will be found. In the light of this, it seems best to use the log-log transformation only when the objective is simply to describe and compare gradients; it is unlikely to be useful when gradient differences are to be interpreted in terms of fundamental epidemiological processes.

The Log-Linear Transformation

For many years, the only serious competitor to the log-log transformation has been the log-linear transformation,

$$\ln y = \ln a - bs$$

where ln signifies the natural logarithm. This transformation is based on the negative exponential equation. Again, b is the gradient parameter (it is different from the b in Gregory's equation), and again a larger value of b implies a steeper gradient. The transformation has few of the drawbacks of the log-log transformation: it does not predict infinite disease at the source, and b has explicit units (m^{-1}, km^{-1}, etc.), thus avoiding the scale problems discussed above. Furthermore, the exponential equation has a plausible conceptual basis: it results from the assumption that a spore will deposit in a unit distance with equal probability regardless of distance from the source.

In general, the log-linear transformation is probably as successful as the log-log transformation in straightening gradients, and its superior mathematical and interpretational qualities recommend its use wherever possible. Neither transformation makes any allowance for multiple infection; for this reason the multiple infection transformation (Gregory, 1948) should be applied whenever disease incidence exceeds about 10%. Even with this correction, however, the transformations are usually applicable only to primary gradients, or to secondary gradients at an early stage in the epidemic. They have limited usefulness when the objective is to follow gradient development through an epidemic. A number of transformations have recently been proposed to overcome this limitation. A logit-log transformation was used by Berger and Luke (1979), while Minogue and Fry (1983a) advanced arguments in favor of a logit-linear transformation. These and several others are included in a family of transformations recently proposed by Jeger (1983).

What these transformations have in common is that they apply functions that are normally used to describe temporal disease progress to the spatial pattern of disease. While this practice has been criticized (Imhoff et al., 1982) in the apparent belief that such functions are not appropriate for the description of gradients, it will be seen below that their use has inspired significant new insights into the relationships between space and time in the progress of epidemics. Unfortunately, with many new transformations being proposed, and more likely to come, conclusions about the effects of various treatments on gradient steepness will become increasingly dependent on the chosen transformation. In view of this it seems worthwhile to develop in some detail a definition of gradient steepness that applies directly to arithmetic axes, and does not require recourse to a specific transformation. A proposal for such a definition is presented in the next section.

The Meaning of Gradient Equality

A natural definition for the equality of slopes of two gradients is that the derivative (or some finite approximation $\Delta y/\Delta s$) is the same for both gradients. As noted previously, the derivative changes within a given gradient, so to compare two gradients dy/ds must be measured at equivalent points on both. For this we need a definition of "equivalent points."

Two possibilities suggest themselves, leading to what may be called the "fixed-s" and "fixed-y" definitions of gradient equality (Fig. 2). In the first, measurements of the derivative are taken at equal distances (s) from the focal center. Two gradients are considered equal if dy/ds is the same for both at all distances. Alternatively, measurements may be taken at equal levels of disease (y), and gradients are equal if dy/ds is the same at every value of y.

That these definitions are not equivalent is most easily seen by means of an example. The negative exponential gradient has the derivative

$$\frac{dy}{ds} = -by$$

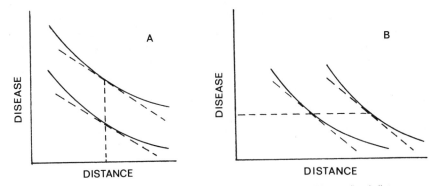

Figure 2 Gradient steepness measured as the derivative (----) (A) at a fixed distance, or (B) at a fixed level of disease.

Two gradients with the same b are equal at equal values of y, and they are therefore equal by the fixed-y definition. At equal distances, however, y will in general be different for the two gradients, so that they will be unequal by the fixed-s definition. The reader may verify that gradients following Gregory's equation and having the same b are not equal by either definition. This is a reflection of the fact that Gregory's b contains no information about the distance scale, and thus does not measure the gradient as the term is being used here.

Since the fixed-s and fixed-y definitions are not equivalent, a choice must be made between them. The first is probably more natural, and is undoubtedly easier to measure in the field; the second, however, has properties that make it much more useful as a tool for understanding the dynamics of gradient development. Examination will show that if two gradients are equal by the fixed-y definition, then one is a horizontal translation of the other (Fig. 2B). This will prove to be a very useful property when we come to consider the stability of gradients during disease spread. By contrast, equality in the fixed-s sense means that one gradient is a vertical translation of the other (Fig. 2A); it is very difficult to imagine a reasonable dynamic process that would cause a gradient to evolve in this way.

What is being proposed, then, is that gradients are to be considered equal if their derivatives are the same when measured at equal levels of disease. Furthermore, a parameter is to be considered a measure of gradient steepness only if equality of the parameter implies equality of the gradient in this sense. By this criterion the slopes of the log-linear and logit-linear transformations are measures of the gradient, but slopes of the log-log and logit-log transformations are not.

This criterion is not advanced as a replacement for the use of transformations, but rather as a means of avoiding the confusion that is bound to result from a proliferation of competing "gradient parameters." Specifically, it provides a way to keep straight what is meant by the steepness of the gradient, and what qualifies a given parameter to be used as a measure of that steepness. For the practical problem of comparing observed gradients, transformations are likely to remain the simplest and most useful method available.

MODELING GRADIENT DYNAMICS

A disease gradient as observed in the field is a dynamic phenomenon, the result of the interplay of a multitude of factors operating in space and time. The spatial scale of the gradient is largely determined by the size and mode of transport of the dispersing propagules. Especially where secondary infection occurs, however, the shape of the gradient, the rate of spread, and other details are strongly influenced by temporal factors as well (for example, the rate of sporulation, and the length of latent and infectious periods). In turn, the steepness of the gradient exerts an influence on the apparent infection rate. To be fully understood, the gradient must be studied in a context that allows spatial and temporal representations of disease progress to be interrelated.

This much has been appreciated for some time, but until recently the necessary tools and concepts have been poorly developed. In a brief but important paper, Berger and Luke (1979) introduced two ideas that have led to a significant improvement in this situation: the use of the logit transformation for disease gradients, and the characterization of disease spread by means of the rate of isopath movement. An isopath is a contour connecting points of equal disease, and as disease increases in a field these contours spread outward from the focal center. In the context of one-dimensional disease spread, an isopath is simply a particular level of disease. The rates at which the various isopaths move, and the way in which those rates depend on disease level, time, and distance, can be used to describe disease spread more thoroughly and precisely than has previously been possible.

In the simplest case, all isopaths move at the same, constant rate. The gradient at any point in time will be a horizontal translation of the gradient at some previous time, and disease will appear to spread as a traveling wave with a constant shape and velocity (Fig. 3). As the wave passes a fixed point, the increase in disease at that point can be viewed either as an increment along the local disease progress curve, or as an increment along a moving gradient. The constant rate of isopath movement makes these points of view equivalent, and as a result the gradient and the disease progress curve can be described by the same mathematical function (Fig. 4). This is the basis for the use of the logit transformation to linearize gradients: if local disease progress is logistic and all isopaths move at the same constant rate, then the gradient must be logistic as well, with a steepness parameter that is related to the apparent infection rate through the velocity of spread. Some examples of measured velocities for a variety of pathosystems are given in Table 1.

There is, of course, no reason to suppose that all isopaths will move at the

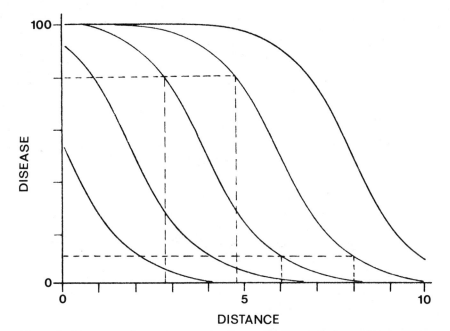

Figure 3 Disease gradients at successive times after the start of an epidemic. When isopaths move an equal distance per time interval regardless of disease level, spread occurs as a traveling wave.

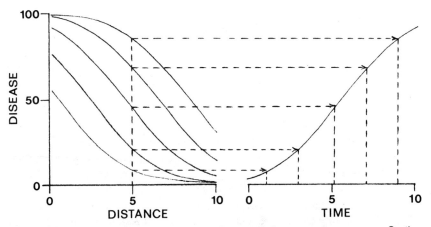

Figure 4 Relationship between disease gradients and disease progress curves. On the left, disease readings are taken at a fixed point as the wave of disease passes at constant velocity. When these readings are plotted over time (right), the disease progress curve has the same functional form as the gradient (in this case, logistic).

Table 1   Velocities of spread for several diseases

Pathogen	Host	Velocity		Source of data
Cercosporella herpotrichoides	Wheat	0.2–0.3	m/day	Rowe and Powelson (1973)
Tomato ringspot virus	Raspberry	2	m/yr	Converse and Stace-Smith (1971)
Ceratocystis wageneri	Pine	0.1–1.8	m/yr	Cobb et al. (1982)
Puccinia coronata	Oats	0.2–1.2	m/day	Berger and Luke (1979)
Phytophthora infestans	Potato	3–4	m/day	Minogue and Fry (1983b)
Septoria nodorum	Wheat	0.3	m/day	Jeger (1983)

same rate, although there is experimental evidence (to be discussed later that this sometimes occurs in the field. More generally, isopathic rates may vary in response to a number of factors, and the way in which they vary determines the way in which the gradient changes over time (Fig. 5). In addition, the rate of movement of a given isopath may change over time, even in a constant environment. To illustrate this, and to show how spatial and temporal factors interact in the development of gradients, a simple model for the spread of a monocyclic disease will be presented in the following section.

Modeling Monocyclic Disease

Monocyclic diseases provide a convenient starting point for understanding gradient development, because in the absence of secondary infection the disease gradient is essentially just a superposition of primary infection gradients. The spatial development of disease is completely determined by the rate of propagule emission from the source, the proportion of "successful" propagules, the geometry of the source, and the geometry of dispersal. Within this sample framework, however, a surprising variety of phenomena can occur.

Consider the monocyclic (*simple interest*) disease model of Van der Plank (1963):

$$\frac{dy(t)}{dt} = QR(1 - y(t)) \tag{1}$$

where Q is the amount of inoculum and R is the infection rate. In the context of disease spread, it is more convenient to replace the product QR by the equivalent product EP, where E is the rate of propagule emission by the source and P is the probability that an emitted propagule will cause an infection (in the absence of density effects). As a temporal model, Eq. 1 ignores spatial effects: QR (or EP) is assumed independent of position. In a spatial representation, E, P, and the carrying capacity K (maximum number of lesions per unit area) may be functions of position. In addition, the probability distribution of the dispersal distance for propagules needs to be considered; this is the dispersal function, denoted $f(s)$.

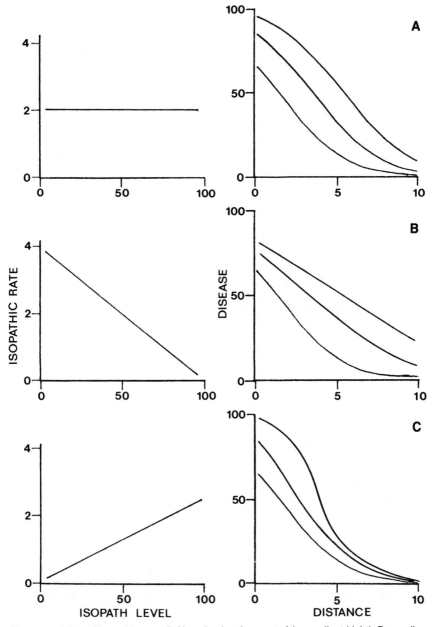

Figure 5 Effect of isopathic rates (left) on the development of the gradient (right). Depending on whether the rate of movement of the isopaths is (A) constant, (B) decreases, or (C) increases with disease level, the gradient respectively remains stable, flattens, or steepens with time.

To simplify the model and focus attention on gradient dynamics, a uniform environment and a point source of infection will be assumed. This allows E, P, and K to be combined in a single constant $C = EP/K$, which is zero everywhere except at $s = 0$. Letting $y(s,t)$ be the number of lesions per unit distance as a proportion of K, Eq. 1 can be extended as a simple spatial model in one dimension,

$$\frac{dy(s,t)}{dt} = Cf(s)(1 - y(s,t)) \tag{2}$$

which has the solution

$$y = 1 - \exp(-Cf(s)t) \tag{3}$$

if $y(s,0) = 0$ for all s (exp signifies the exponential function: $\exp x = e^x$).

Gradients at successive time intervals for an epidemic developing according to Eq. 3 are shown in Fig. 6. A negative exponential dispersal function has been

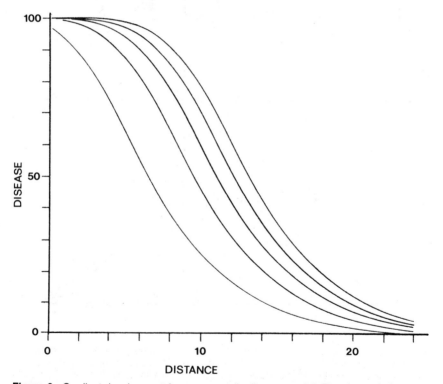

Figure 6 Gradient development for a monocyclic disease model. Time intervals between gradients are equal. Parameters are $C = 3.0$, $k = 0.25$.

assumed ($f(s) = k \exp(-ks)$). To the eye the gradients appear stable, with disease spreading as a wave whose velocity decreases with distance from the source. Notice, however, that the gradient slopes are the same only when compared at the same level of disease, not when compared at a fixed distance from the source.

The concept of isopathic rates can be used to verify gradient stability and to characterize the change in velocity with distance. This is done simply by considering y to be fixed at some value y^* (the level of the isopath), and s to be the position of the isopath at time t. Differentiating Eq. 3 with respect to time and rearranging gives

$$\frac{ds}{dt} = \frac{1}{kt} \tag{4}$$

when $f(s) = k \exp(-ks)$. The derivative ds/dt is the velocity of the isopath, and for this choice of $f(s)$ it is independent of the level of the isopath. This means that the gradient is simply translated horizontally with time, and is therefore (by the fixed-y definition of equality) a stable gradient. The rate of spread is faster as the dispersal gradient becomes flatter (smaller k), but decreases with time: the longer spread has been occurring, the farther the wave of disease will have spread from the source of infection, and the fewer spores will be dispersed far enough to drive the wave onward.

Perhaps surprisingly, Eq. 4 seems to predict that the rate of spread is not directly dependent on C. However, solving for s introduces a constant of integration that includes C, and shows that epidemics with a higher C are in fact farther advanced at any point in time than those with lower C.

At the beginning of this section it was stated that gradients of monocyclic diseases are a superposition of primary infection gradients. This point will perhaps be clearer with a transformation of Eq. 3:

$$-\ln(1 - y) = Ctf(s) \tag{5}$$

The left side of Eq. 5 is the multiple infection transformation, and estimates the number of infections that have occurred. The right side is a multiple of the dispersal function, which increases linearly with t.

By further transformation, in the case of a negative exponential dispersal function,

$$-\ln[-\ln(1 - y)] = ks - \ln Ctk$$

which shows that gradients of monocyclic diseases may under some conditions be linearized by applying the Gompertz transformation to $(1 - y)$. The slope k is a measure of gradient steepness by the fixed-y definition, since

$$\frac{dy}{ds} = k(1 - y)\ln(1 - y)$$

At a fixed level of disease, dy/ds is the same when k is the same.

The primary purpose in presenting this model is to illustrate several points made earlier: the utility of the isopathic rate concept, the evaluation of the velocity of spread, the meaning of gradient equality and the stability of gradients, and the use of the slopes of transformed gradients to summarize gradient steepness. The model could be extended in several ways to make it more realistic: environmental gradients, spatially distributed sources, and time-varying sources are readily accommodated, as are the more complex dispersal functions needed to describe "skip-distance" (Thresh, 1976) and elevated-source (Strand and Roth, 1976) dispersal. It is clear that gradients of monocyclic diseases provide a fertile field for both theoretical and experimental investigation, and the ease with which they can be modeled makes them suitable for testing our understanding of the factors that shape gradients.

One approach to describing the development of gradients of monocyclic diseases was recently proposed by Jeger (1983). He noted first that the rate of isopath movement could be expressed as the ratio of the local rate of disease progress at a particular point to the gradient at that point:

$$\frac{ds}{dt} = \frac{dy/dt}{dy/ds}$$

(The idea expressed in this equation is very close to that shown diagrammatically in Fig. 4.) Minogue and Fry (1983a) have also presented this equation in a slightly different form. Using Van der Plank's (1963) models for dy/dt, gradient equations based on the inverse power law and the negative exponential law for dy/ds, and making certain assumptions about ds/dt, Jeger was able to develop a series of equations to describe disease development in space and time. For a monocyclic disease with a gradient that follows Gregory's equation, Jeger proposed

$$-\ln(1 - y) = a - b\ln(s) + ct$$

where b is the gradient parameter, c is the infection rate, and a is a constant. Applying our earlier methods, the isopathic rate is

$$\frac{ds}{dt} = \frac{cs}{b}$$

and the gradient is

$$\frac{dy}{ds} = \frac{-b(1 - y)}{s}$$

Both of these equations are given by Jeger. The first says that the farther the isopath is from the source of the spores that drive the epidemic, the faster it moves. This rather odd prediction is a consequence of the use of an inverse power law for the gradient, and illustrates the difficulties that arise when such equations are used as gradient models.

Modeling Polycyclic Disease

As shown in the previous section, gradients of a monocyclic disease can be viewed as a superposition of primary infection gradients. The shape and steepness of the gradient are entirely determined by the characteristics of dispersal; temporal factors like the rate of spore production have no influence on the gradient, but affect only the rate of disease spread.

A gradient in a uniform environment can be modified in only two ways, either by changing the pattern of spore dispersal around individual point sources, or by changing the shape of the source. For a monocyclic disease, the shape of the source (that is, the spatial distribution of infectious material) is fixed throughout the epidemic. Ultimately, this is the reason that temporal factors do not affect gradients of monocyclic diseases: they have no effect on spore dispersal patterns, and without secondary infection they cannot modify the shape of the source.

When secondary infections occur, this is no longer true; the source shifts its position constantly in response to the population dynamics of the pathogen. The position and spatial distribution of the source in relation to the advancing wave of disease depend on how fast the wave is moving, the length of latent and infectious periods, and how the reproduction rate changes as a function of the age of the infection (Fig. 7). As a result, the gradient is affected by all of these factors, and may in the end be quite different from the gradient of spore dispersal.

Such phenomena make the analysis of gradients and disease spread much more difficult for polycyclic than for monocyclic diseases. Analytic solutions are usually not possible, so that more reliance must be placed on simulation. The resulting difficulty of drawing general principles from the analysis is to some extent mitigated by the fact that simulation results have so far tended to support more intuitive approaches to modeling, from which principles are more readily derived.

The first attempt to use simulation specifically to investigate disease spread seems to have been the model presented by Zadoks and Kampmeijer (1977), although a number of models incorporating a spatial dimension have been described (Kiyosawa and Shiyomi, 1972; Kiyosawa, 1976; Schrum, 1975; Strand and Roth, 1976; Bloomberg et al., 1980). In Zadoks and Kampmeijer's model

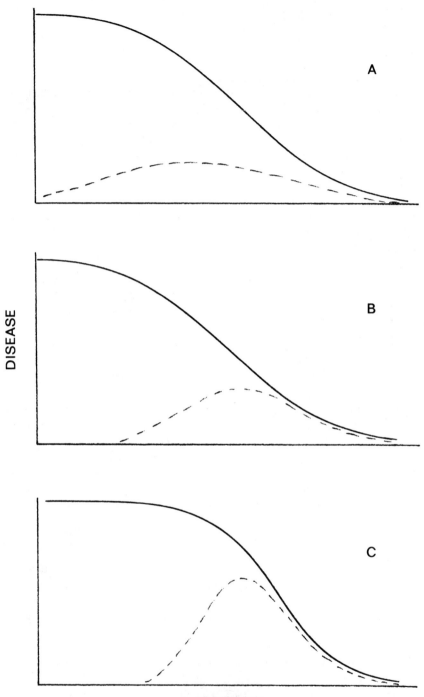

Figure 7 Gradients for polycyclic diseases: (A) with long latent and infectious periods; (B) with short latent period; (C) also with short latent period but with higher sporulation rate or higher infection efficiency. (——), disease gradient; (----), distribution of infectious lesions. (*Based on unpublished simulations using the model described by Minogue and Fry, 1983a.*)

(EPIMUL76), disease progress at any location was assumed to follow Van der Plank's general equation for epidemic development, characterized by a basic infection rate, a latent period, and an infectious period (Van der Plank, 1963, Eq. 8.3). Dispersal between locations occurred in accordance with a two-dimensional normal distribution function. The model was intended to investigate how the rate of disease spread is affected by the components of resistance, the median distance of spore dispersal, field size, and multilines.

The epidemics produced by EPIMUL76 spread outward at a constant rate from the initial focus. Contrary to the widespread belief that secondary infection flattens gradients, the simulated gradients became steeper with time, until a steady state was reached. Thereafter the gradient was stable, and disease spread as a traveling wave. Gradients became flatter as the median distance of spore dispersal increased, but the relationship was not linear: a 10-fold increase in median dispersal distance caused a 100-fold decrease in gradient steepness. The rate of disease spread was approximately proportional to the median dispersal distance, and to the logarithm of the "daily multiplication factor" (the equivalent of Van der Plank's corrected basic infection rate). Zadoks and Kampmeijer did not report on the effects of latent and infectious periods on disease spread, nor did they attempt to characterize the shape of the gradient, or relate it to the rate of spread.

EPIMUL76 presented a significant challenge to several established concepts concerning disease spread, most notably in its predictions that secondary spread does not flatten gradients and that the rate of spread is relatively unaffected by the infection rate. Zadoks and Kampmeijer attributed the opinions of previous authors to misinterpretation of experiments and faulty reasoning, but the flattening of primary gradients by secondary spread seems well established in the literature (Cammack, 1958; Rowe and Powelson, 1973; Johnson and Powelson, 1983; Minogue and Fry, 1983b; Imhoff et al., 1982). The failure of EPIMUL76 to reproduce this effect is worrisome. It is also not clear to what extent the spatial patterns produced by EPIMUL76 were the result of the specific assumptions that went into the model, for example, the assumption of a normal distribution function for spore dispersal. However, the model produced the first explicit recognition of the traveling wave as a mode of disease spread, even if it did not define the conditions under which such a wave may occur, and thus provided an important conceptual foundation for subsequent work.

A very different approach to modeling disease spread was adopted by Fleming et al. (1982). They rejected simulation models as unwieldy and unrealistic, and instead adopted an analytic approach in which pathogen spread was treated as a diffusion process. The model they developed was applied to a number of spatial problems, notably the effects of field size and shape, and spread between fields. They predicted, among other things, that there would be a critical field size below which the pathogen population will die out rather than produce an epidemic, and that less disease will occur in a crop planted in many small fields than in a few large ones.

The model presented by Fleming et al., as a general model of disease spread, could be applied to the study of gradients and the velocity of spread; since the authors did not address these questions their model will not be discussed in detail here. The notion of a critical field size can, as a practical matter, be dismissed out of hand: if it were important, small plot experiments would be impossible in plant pathology. The diffusion model adopted by these authors is an adaptation of models developed for animal populations, whose members retain the ability to move throughout their lives. This is not true for most plant pathogens, and therefore the structure of the model does not seem appropriate as a description of disease spread.

An approach that seems more promising has been developed in great mathematical detail by Mollison (1972, 1977). Conceptually, his models are very simple. Individuals, some of which are infective, are distributed at constant density along a line. Infectious individuals emit "germs," which travel some distance and then come in contact with another individual. The distance traveled by a germ is taken from a probability distribution referred to as the contact distribution. (In our context, a germ is a spore or other disseminule and the contact distribution is the same as the dispersal function.) If the individual contacted is not already infected it becomes so, and itself begins to emit germs. In this way the epidemic spreads down the line.

The case examined in the most detail by Mollison is the simplest, in which individuals become infectious immediately on contact with a disseminule, and remain so for the rest of the epidemic. In this case the epidemic spreads as a wave at constant velocity, with a waveform well described by the logistic function. The velocity of the wave was determined by the rate at which infectious individuals emit germs, and by the dispersal function; the initial distribution of infected individuals had only a transient effect.

Mollison investigated in some detail the effects of the dispersal function on the form of the wave and its velocity of spread (he did not address the question of gradient steepness). His simulation results indicated that the velocity was linearly dependent on the standard deviation of the dispersal function (similar to the behavior of EPIMUL76), but the exact shape of that function had relatively little effect. That is, whether dispersal followed a normal, exponential, or even uniform probability distribution was relatively unimportant as long as the variance was the same in all cases.

Mollison delineated the conditions under which wavelike spread of disease can be expected: in essence, the moments of the dispersal function are required to be finite. In particular, functions with an infinite variance produce a continuously accelerating and flattening wave that never achieves a stable state. Such behavior is sometimes observed in real epidemics when the range over which dispersal takes place is much larger than the scale of observation. He also simulated epidemics that spread as a wave most of the time, but occasionally produced infections far in advance of the front. These "great leaps forward," as he called them, are typical of many plant pathogens with windborne spores, and it may be that dispersal functions of the type described by Mollison will be increasingly useful in the description of disease spread.

These concepts were extended and applied to plant disease epidemics by Minogue and Fry (1983a, 1983b). Our model differed primarily in the inclusion of latent and infectious periods, and in the use of a point source for the initial infection. Mollison was concerned almost entirely with the velocity of spread as influenced by the dispersal function; we considered also the shape and steepness of the gradient, and the effects of the components of resistance.

The inclusion of latent and infectious periods had relatively little effect on the qualitative properties of disease spread. After a period of rapid flattening, a stable wave developed that spread at a constant velocity. Because of the finite infectious period, such a wave must eventually come to a stop (see Kelly in the discussion of Mollison, 1977), but we did not observe this. The shape of the wave was described well by the logistic function, so that the logit transformation linearized the gradients. For this reason we used the absolute value of the slope of this line (denoted g) as a measure of gradient steepness, and the rate at which the wave spread along the distance axis as a measure of velocity of spread v. Using these parameters it was possible to tie together spatial and temporal representations of disease progress, in the form of an equation

$$g = \frac{r}{v}$$

where r is the apparent infection rate at any location in the plot. The same relationship (with different symbols) was arrived at independently in a more intuitive way by Jeger (1983).

The characteristics of the gradient, as expressed in the parameters g and v, could be modified by altering the components of resistance, sometimes with surprising effect. An increase in either the infectious period or the sporulation rate caused the gradient to steepen, while an increase in the latent period had little effect. Since factors that affected g usually also affected r, there was no fixed relationship between gradient steepness and the velocity of spread.

Some of these effects were tested in a small experiment with potato late blight (Minogue and Fry, 1983b). Epidemics started at one end of long, narrow field plots developed steep gradients that rapidly became flatter until a stable

Table 2 Values of the gradient parameter g, apparent infection rate r, and velocity of spread v for Phytophthora infestans on potato

Treatment	g (m^{-1})	r (day^{-1})	v (m · day^{-1})
	Value[a]		
Untreated	0.156 (a)	0.465 (a)	2.98 (a)
Horizontally resistant	0.037 (b)	0.163 (b)	4.44 (a)
Fungicide treated	0.065 (b)	0.235 (b)	3.62 (a)

[a]Values in each column followed by the same letter are not significantly different at $P = 0.05$ by Fisher's LSD test.
Source: K. P. Minogue and W. E. Fry, Phytopathology 73:1173–1176, 1983b.

slope was attained. Thereafter spread was at a more or less constant rate through the plots. Gradients were linearized by a logit transformation of disease severity. Fungicide treatment and horizontal resistance were expected on the basis of the modeling work to reduce r and flatten the gradient (reduce g); these effects were observed (Table 2) but, unexpectedly, neither treatment affected the velocity of spread.

DESIGN OF GRADIENT EXPERIMENTS

Field Plots

The observation and manipulation of gradients in the field present a number of problems to the experimenter, especially if the object is to study the development of the gradient over time. For many pathogens the scale of dispersal dictates that very large plots must be used if a significant portion of the gradient is to be observed. As we have noted before (Minogue and Fry, 1983b), the scale of experiments must match the scale of the questions being asked. Because of the expense of large plots, the number of treatments and the number of replicates must usually be severely restricted. The problem is compounded by the need to isolate the plots, both from each other and from outside sources of infection, in order to prevent contamination of the gradients.

Large plots introduce an additional complication when it comes time to collect disease data. Since the process of dispersal is responsible for the development of gradients, it is clearly vital not to interfere with or modify that process during the course of the experiment. For many pathogens, however, it is difficult to avoid this when the plot is entered to make disease readings. As a result, observations have often been taken only once or twice, making it impossible to study the development of the gradient over time.

A number of solutions to these problems have been attempted. Replication has occasionally been achieved (Mackenzie, 1976; Lim, 1978), not by using separate plots, but by taking several transects through the same plot (Fig. 8). The transects radiate from the point of inoculation (usually either the center of the plot or the upwind corner). These are not, of course, true replicates; because of dispersal the gradients will be more highly correlated with each other than would be the case for gradients in separate, isolated plots. This will affect statistical tests, underestimating the experimental error and making it more likely that chance differences in gradients will appear significant. Whether these effects are important under normal experimental conditions is not known.

Another approach, adopted by Minogue and Fry (1983b), is to use long, narrow plots, inoculated at one end. Such plots provide a relatively inexpensive way to study gradients of moderate scale; they also solve the problem of disturbance during disease assessment, since the plots are narrow enough to be assessed from outside the plot boundaries. The plots are sensitive to the direction of the prevailing wind, and they cannot be used to study spread in two dimensions. However, they seem well suited to studying some of the factors that govern

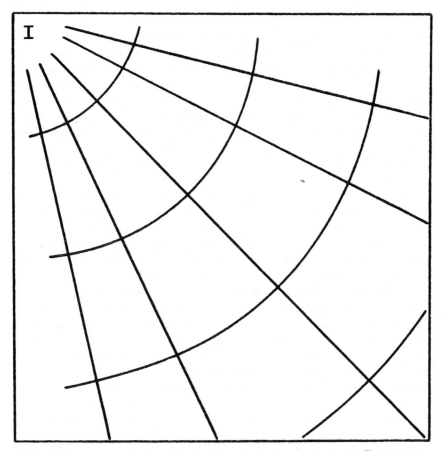

Figure 8 Use of transects to obtain pseudoreplicates from a single plot. Five transects radiate from the point of inoculation I, which is ideally in the windward corner of the plot. Disease readings are taken where transects intersect arcs concentric about I. (*Adapted from Mackenzie, 1976; Lim, 1978.*)

gradient development, and to testing one-dimensional disease spread models. A number of other plot designs are described by Thresh (1976).

Plot isolation is usually achieved either by providing large distances between plots (Imhoff et al., 1982), or by interposing a nonhost between the experimental plots (Mackenzie, 1976; Minogue and Fry, 1983b). Neither approach is satisfactory: the first may be difficult to accomplish without introducing significant environmental variability, and the second is often rather ineffective. Johnson and Powelson (1983) recently used benomyl-tolerant pathogen isolates to distinguish experimentally induced gradients from the natural background infection. This or other kinds of genetic marker (Garber et al., 1983) may be useful in avoiding cross contamination between treatments. Whenever possible, uninoculated checks should be included to ensure that plot isolation is sufficient.

Data Analysis and Interpretation

Analysis of data has usually been carried out in one spatial dimension, either by taking measurements in only one direction or by averaging measurements from several directions. Lambert et al. (1980) recorded the compass direction of their disease observations, and included this as a second spatial dimension in some equations. This provided only a small increase in the coefficient of determination compared to their one-dimensional equations, but nevertheless this approach seems preferable to averaging the gradient in different directions. When spread is anisotropic, averaging will distort the functional form of the gradient. It is probably better either to include a second spatial dimension in the analysis or to analyze gradients in each direction separately.

When the development of the gradient over time is of interest, the usual practice has been to analyze data for each observation date separately (Imhoff et al., 1982; Mackenzie, 1976). Occasionally, data have also been analyzed over time, with a separate disease progress curve for each distance of observation (Berger and Luke, 1979; Minogue and Fry, 1983b). Only rarely have both time and space been included in the same equation (Frampton et al., 1942; Jeger et al., 1983), although this seems to be the ideal way to handle data of this type. Part of the reason such analyses are rare is probably the difficulty of devising spatiotemporal equations that are both simple and biologically reasonable. In addition, observations taken from the same plot are correlated over both space and time. The statistical problems this creates in the time dimension are discussed in Chap. 3 of this volume, and these can be expected to become more severe with the addition of one or more spatial dimensions to the fitted equations. The extent of these problems and their possible solutions remain to be studied.

The process of fitting gradient equations to data is much the same as fitting disease progress curves. The procedures and pitfalls associated with this are discussed in Chaps. 1 and 3, and need not be repeated here. There are, however, three points relevant to gradient analysis that should be mentioned.

Treatment of Zeroes   Transformations that involve taking logarithms do not work when some disease readings are zero. This is sometimes avoided by adding a small quantity to each reading (Gregory, 1968) or by more involved procedures (Lambert et al., 1980), but this inevitably distorts the form of the gradient, sometimes dramatically. Since many transformations have a high variance at low disease levels, it seems much better to truncate the data before any zeros occur, or to use nonlinear estimation methods that do not involve transformations.

Spatial Averaging   Gradient equations are usually continuous in the spatial dimension, and data are fitted as if they represent point values of disease severity. In fact, disease must be measured over a finite area, which introduces a spatial averaging effect. This distorts the gradient to some extent, and causes errors in the estimation of parameters. These effects have not been studied either experimentally or theoretically, but they appear to be small for exponential and logistic gradients. For the inverse power law, errors can be large, especially if b is large and observations near the source are included. This seems to be a question that deserves more attention.

Disease Measurement   So far in this chapter there has been no consideration of whether the level of disease is being measured as incidence or severity. The choice between these will affect the form of the observed gradients. Ries and Royse (1978) found that disease incidence gradients had a smaller value of b (Gregory's equation) than severity gradients measured at the same time. Jeger et al. (1983), in examining the utility of a number of transformations, found that the best transformation depended on whether incidence or severity was being measured. Little is known as yet about the relationships between incidence and severity gradients, but the fact that they are distinct should be kept in mind when interpreting observations.

OUTSTANDING QUESTIONS

It will be clear that our understanding of the development of disease gradients and the factors that shape them is still relatively poor. It seems appropriate to close this chapter with a brief discussion of a few topics in which research seems to be particularly lacking.

Spread in Two Dimensions

Although spread in the field is two-dimensional or three-dimensional, and a few simulation models have included a second dimension (Schrum, 1975; Kiyosawa, 1976; Zadoks and Kampmeijer, 1977), all our analytic techniques for describing and quantifying gradients are one-dimensional. If a second dimension is present in the data, it is removed by directional averaging prior to analysis. Theoretical approaches too have been mostly one-dimensional, developed with the implicit hope that such efforts give some insight into two-dimensional spread. The second dimension presents a number of modeling difficulties, including the greatly increased expense of modeling and the problem of extending the concept of the gradient to two dimensions. However, some preliminary theoretical work, at least, seems necessary.

Nonwavelike Spread

Much of the recent work in disease spread has assumed, directly or indirectly, that the gradient eventually stabilizes and that disease spreads as a wave. In many experimental situations this has not been observed, and as previously discussed there is reason to believe that such behavior will occur only under certain circumstances. Furthermore, even when the epidemic as a whole is wave-like, near the front, spread occurs as a series of discrete jumps, forming foci that spread and gradually merge. This is a region of particular importance, since the epidemic is spreading here for the first time into previously healthy areas, but the analogy of the wave seems inappropriate to describing this kind of spread. New concepts and techniques are needed to analyze spread in this region.

Plant Density

As the number of plants per unit area increases, it becomes easier for the pathogen to spread from plant to plant. On the other hand, the increased plant density will decrease the average distance traveled by a dispersing propagule. Since these are competing effects in terms of gradient steepness and the velocity of spread, the net effect of planting density on the spatial development of disease is difficult to predict. Higher plant density may increase the velocity of spread (Cobb et al., 1982; Scott, 1956) or decrease it (Bloomberg et al., 1980). The increase or decrease apparently depends on whether host colonization or plant-to-plant transmission is the rate-limiting process. Much of the relevant literature on plant density has been summarized in an excellent review by Burdon and Chilvers (1982), who point out that the reduction of disease by multilines is primarily a plant density effect.

Spatiotemporal Relationships

Despite scattered theoretical contributions that go back many years, there has been very little experimental investigation into the relationships between the gradient and the velocity of spread, on the one hand, and the latent and infectious periods, sporulation rate, infection efficiency, and dispersal distance, on the other. One of the primary tasks of quantitative epidemiology is to determine the way in which properties of the pathogen population as a whole are determined by the properties of its individual members. From this viewpoint there are many questions that need to be clarified before the behavior of spreading disease will be fully understood.

REFERENCES

Barrett, J. A. 1981. Disease progress curves and dispersal gradients in multilines. *Phytopathol. Z.* 100:361–365.

Berger, R. D., and Luke, H. H. 1979. Spatial and temporal spread of oat crown rust. *Phytopathology* 69:1199–1201.

Bloomberg, W. J., Smith, R. B., and Van der Wereld, A. 1980. A model of spread and intensification of dwarf mistletoe infection in young western hemlock stands. *Can. J. For. Res.* 10:42–52.

Bonde, R., and Schultz, E. S. 1943. Potato refuse piles as a factor in the dissemination of late blight. *Maine Agr. Exp. Sta. Bull.* 416, pp. 229–246.

Burdon, J. J., and Chilvers, G. A. 1982. Host density as a factor in plant disease ecology. *Annu. Rev. Phytopathol.* 20:143–166.

Cammack, R. H. 1958. Factors affecting infection gradients from a point source of *Puccinia polysora* in a plot of *Zea mays*. *Ann. Appl. Biol.* 46:186–197.

Campbell, C. L., and Pennypacker, S. P. 1980. Distribution of hypocotyl rot caused in snapbean by *Rhizoctonia solani*. *Phytopathology* 70:521–525.

Cobb, F. W., Slaughter, G. W., Rowney, D. L., and DeMars, C. J. 1982. Rate of spread of *Ceratocystis wageneri* in ponderosa pine stands in the central Sierra Nevada. *Phytopathology* 72:1359–1362.

Converse, R. H., and Stace-Smith, R. 1971. Rate of spread and effect of tomato ringspot virus on red raspberry in the field. *Phytopathology* 61:1104–1106.

Dimond, A. E., and Horsfall, J. G. 1960. Inoculum and the disease population, in: *Plant Pathology, An Advanced Treatise,* vol. 3 (J. G. Horsfall and A. E. Dimond, eds.), pp. 1–22. Academic Press, New York.

Fleming, R. A., Marsh, L. M., and Tuckwell, H. C. 1982. Effect of field geometry on the spread of crop disease. *Prot. Ecol.* 4:81–108.

Frampton, V. L., Linn, M. B., and Hansing, E. D. 1942. The spread of virus diseases of the yellow type under field conditions. *Phytopathology* 32:799–808.

Fried, P. M., Mackenzie, D. R., and Nelson, R. R. 1979. Dispersal gradients from a point source of *Erysiphe graminis* f.sp. *tritici,* on Chancellor winter wheat and four multilines. *Phytopathol. Z.* 95:140–150.

Garber, R. C., Fry, W. E., and Yoder, O. C. 1983. Conditional field epidemics on plants: A resource for research in population biology. *Ecology* 64:1653.

Gilligan, C. A. 1982. Statistical analysis of the spatial pattern of *Botrytis fabae* on *Vicia faba:* A methodological study. *Trans. Br. Mycol. Soc.* 79:193–200.

Gregory, P. H. 1948. The multiple-infection transformation. *Ann. Appl. Biol.* 35:412–417.

———. 1968. Interpreting plant disease dispersal gradients. *Annu. Rev. Phytopathol.* 6:189–212.

Gregory, P. H., and Read, D. R. 1949. The spatial distribution of insect-borne plant-virus diseases. *Ann. Appl. Biol.* 36:475–482.

Gross, H. C., Patton, R. F., and Ek, A. R. 1980. Spatial aspects of sweet fern rust disease in northern Ontario jack pine–sweet fern stands. *Can. J. For. Res.* 10:199–208.

Hampton, R. O. 1967. Natural spread of viruses infectious to beans. *Phytopathology* 57:476–481.

Imhoff, M. W., Leonard, K. J., and Main, C. E. 1982. Analysis of disease progress curves, gradients and incidence-severity relationships for field and phytotron bean rust epidemics. *Phytopathology* 72:72–80.

Jeger, M. J. 1983. Analysing epidemics in time and space. *Plant Pathol.* 32:5–11.

Jeger, M. J., Jones, D. G., and Griffiths, E. 1983. Disease spread of non-specialized fungal pathogens from inoculated point sources in intraspecific mixed stands of cereal cultivars. *Ann. Appl. Biol.* 102:237–244.

Johnson, K. B., and Powelson, M. L. 1983. Analysis of spore dispersal gradients of *Botrytis cinerea* and gray mold disease gradients in snap beans. *Phytopathology* 73:741–746.

Kable, P. F., Fried, P. M., and Mackenzie, D. R. 1980. The spread of powdery mildew of peach. *Phytopathology* 70:601–604.

Kiyosawa, S. 1976. A comparison by simulation of disease dispersal in pure and mixed stands of susceptible and resistant plants. *Jap. J. Breed.* 26:137–145.

Kiyosawa, S., and Shiyomi, M. 1972. A theoretical evaluation of the effect of mixing resistant variety with susceptible variety for controlling plant diseases. *Ann. Phytopathol. Soc. Jap.* 38:41–51.

Lambert, D. H., Villareal, R. L., and Mackenzie, D. R. 1980. A general model for gradient analysis. *Phytopathol. Z.* 98:150–154.

Lim, S. M. 1978. Disease severity gradient of soybean downy mildew from a small focus of infection. *Phytopathology* 68:1774–1778.

Mackenzie, D. R. 1976. Application of two epidemiological models for the identification of slow stem rusting in wheat. *Phytopathology* 66:55 59.

Madden, L. V., Louie, R., Abt, J. J., and Knoke, J. K. 1982. Evaluation of tests for randomness of infected plants. *Phytopathology* 72:195–198.

Martin, S. B., Campbell, C. L., and Lucas, L. T. 1983. Horizontal distribution and characterization of *Rhizoctonia* spp. in tall fescue turf. *Phytopathology* 73:1064–1068.

Minogue, K. P., and Fry, W. E. 1983a. Models for the spread of disease: model description. *Phytopathology* 73:1168–1173.

Minogue, K. P., and Fry, W. E. 1983b. Models for the spread of disease: some experimental results. *Phytopathology* 73:1173–1176.

Mollison, D. 1972. The rate of spatial propagation of simple epidemics. *Proc. 6th Berkeley Symp. Math. Stat. Prob.* 3:579–614.

———. 1977. Spatial contact models for ecological and epidemic spread. *J. Roy. Statist. Soc. B* 39:283–326.

Ries, S. M., and Royse, D. J. 1978. Peach rusty spot epidemiology: Incidence as affected by distance from a powdery mildew-infected apple orchard. *Phytopathology* 68:896–899.

Rowe, R. C., and Powelson, R. L. 1973. Epidemiology of *Cercosporella* foot rot of wheat: Disease spread. *Phytopathology* 63:984–988.

Schrum, R. 1975. Simulation of wheat stripe rust (*Puccinia striiformis* West.) using EPIDEMIC, a flexible plant disease simulator. *Penn. State Univ. Agr. Exp. Sta. Prog. Rep.* 347, 68 pp.

Scott, M. R. 1956. Studies of the biology of *Sclerotium cepivorum* Berk. II. The spread of white rot from plant to plant. *Ann. Appl. Biol.* 44:584–589.

Shepherd, R. J., and Hills, F. J. 1970. Dispersal of beet yellows and beet mosaic viruses in the inland valleys of California. *Phytopathology* 60:798–804.

Strand, M. A., and Roth, L. F. 1976. Simulation model for spread and intensification of western dwarf mistletoe in thinned stands of ponderosa pine seedlings. *Phytopathology* 66:888–895.

Standberg, J. 1973. Spatial distribution of cabbage black rot and the estimation of diseased plant populations. *Phytopathology* 63:998–1003.

Thresh, J. M. 1976. Gradients of plant virus diseases. *Ann. Appl. Biol.* 82:381–406.

Van der Plank, J. E. 1963. *Plant Diseases: Epidemics and Control.* Academic Press, New York, 349 pp.

Zadoks, J. C., and Kampmeijer, P. 1977. The role of crop populations and their deployment, illustrated by means of a simulator, EPIMUL76. in: *The Genetic Basis of Epidemics in Agriculture* (P. R. Day, ed.). *Ann. N.Y. Acad. Sci.* 287:164–190.

Chapter 13

Spore Dispersal in Relation to Epidemic Models

B. D. L. Fitt
H. A. McCartney

Spores of plant pathogens are transported from plant to plant by many agents, including wind, rain, soil water, insects, and human beings; we shall only consider dispersal by wind and rain splash. To understand how disease epidemics develop, accurate models of spore dispersal are needed. The first step in producing such models is to measure spore dispersal using appropriate spore sampling devices. Empirical models of spore dispersal can be developed directly from these measurements. Alternatively, theoretical models can be derived from the physical principles of particle dispersal and then tested against experimental measurements. This chapter considers the collection of spore dispersal data and the construction and use of spore dispersal models in plant disease epidemiology.

DISPERSAL: THE NEGLECTED PHASE IN PATHOGEN LIFE CYCLES

The life of a plant pathogen may be divided into sporulation, dispersal, germination, infection, and incubation phases. In attempting to describe a pathogen life cycle or construct an epidemic model, it is often the dispersal phase about which least is known (Waggoner, 1983). Sporulation, germination, infection, and incubation all occur on or in the host. Because these phases are localized, they are often readily studied in controlled environment experiments to obtain the information needed for epidemic models.

Dispersal which, by definition, involves more than one location is more difficult to study experimentally. Most spore dispersal data are obtained either from simple laboratory experiments or from field observations under a range of weather conditions. Quite apart from the difficulties in choosing appropriate sampling devices to obtain this information, particularly for splash-dispersed spores, it is only rarely accompanied by adequate meteorological measurements. The data loggers and sensitive meteorological instruments needed to obtain some of the information important in spore dispersal studies (e.g., the occurrence of short-lived wind gusts and rapid changes in relative humidity) have only recently become available (Sutton et al., 1984). The consequent lack of accurate spore dispersal and related meteorological information has hindered the development of epidemic models.

SPORE DISPERSAL PROCESSES: REMOVAL, TRANSPORT, AND DEPOSITION

The spore dispersal phase of pathogen life cycles has three stages: removal, transport, and deposition. Wind and rain may be involved in all three stages. For example, conidia of *Erysiphe graminis* DC (cereal powdery mildew) may be removed from infected leaves by wind gusts (Aylor et al., 1981) or by acceleration forces as leaves flap in the wind (Bainbridge and Legg, 1976). They may also be removed by raindrops, through the puff or tap mechanisms (Hirst and Stedman, 1963). As a raindrop spreads out on a dry leaf it causes a puff of air which may remove dry conidia. The impact (tap) of the raindrop on the leaf may also release dry spores as kinetic energy is transferred to the leaf. Dry, airborne conidia of *E. graminis* are a common component of the air spora during dry weather in summer in the United Kingdom (Hirst, 1953) and are deposited onto new host plants by wind impaction and sedimentation (Gregory, 1973). Rain removes many dry airborne spores, such as those of *E. graminis,* from the atmosphere (Gregory, 1973) but the significance of this rain scrubbing in spore dispersal is difficult to quantify.

Ascospores of *Venturia inaequalis* (Cooke) Wint (apple scab) are actively discharged from infected leaves during periods of rain in spring in the United Kingdom (Hirst et al., 1955) and these dry airborne spores are a component of the air spora in apple orchards during rain. Transport and deposition of *V. inaequalis* ascospores is similar to that of *E. graminis* conidia, although deposition by rain scrubbing is probably more important for *V. inaequalis* ascospores. Dry, airborne spores may also be picked up by rain-splash droplets if raindrops impact directly onto spore-bearing structures (Ramalingam and Rati, 1979).

Splash-borne spores, such as the pycnidiospores of *Septoria nodorum* Berk.

(wheat glume blotch), are generally produced in mucilage which prevents their removal from infected plants by wind. However, the first raindrops dissolve the mucilage to leave a spore suspension available for splash dispersal by subsequent raindrops. The splash droplets carrying spores of *S. nodorum* differ greatly in size (Brennan et al., 1985a). The smallest droplets ($<$100 μm) may become airborne and evaporate to leave dry airborne spores. The largest ballistic droplets ($>$1000 μm), which carry most of the *S. nodorum* spores, follow a trajectory little affected by the wind. The size of intermediate droplets determines how much their trajectories are affected by wind. These examples illustrate the snags in dividing spores into rigid wind-dispersed and splash-dispersed categories. For epidemic modeling it may be possible to select some principal mode of dispersal for a particular spore, but other modes should always be borne in mind.

MEASUREMENT OF SPORE DISPERSAL

Samplers for Wind-borne or Splash-borne Spores

The accuracy of a spore dispersal model depends, to a large extent, on the quality of the experimental data on which it is based or with which it is tested. Unfortunately, some published spore dispersal information is difficult to interpret because the spore samplers used were inappropriate for the spores under study. Some of the samplers available are listed in Table 1. We consider that a device for collecting samples of dry airborne or splash-borne spores is better described as a sampler than as a trap, which is a term more appropriate for describing a device for collecting an organism capable of independent movement.

Artificial spore samplers may collect spores on dry, sticky surfaces or in liquids; they may be passive, collecting the spores that reach them, or volumetric, sampling a known volume of air. However, even volumetric samplers rarely accurately collect the spores in a given volume of air; they generally systematically underestimate or overestimate the spore concentration. Spores can be impacted onto the sampling surfaces of inertial samplers by wind (vertical cylinders, rotorod) or by air suction into the sampler (Burkard, cascade impactor, cyclone, impinger), but they sediment onto horizontal slides and funnels (Gregory, 1973). "Bait" plants of a susceptible cultivar may also be used as samplers.

Table 1   Spore samplers

	Artificial		Plant
	Passive	**"Volumetric"**	**Plant**
Dry	Horizontal slide Vertical cylinder	Burkard Cascade impactor Rotorod	"Bait" plants Crop
Liquid	Funnel	Cyclone Impinger	

Susceptible crops are more likely to collect pathogen spores than any artificial sampler because they sample such large volumes of air.

In collecting spore dispersal data, the experimenter must decide which sampler to use, where to place it, and when to sample. These decisions are affected by factors beyond his or her control, such as the modes of spore removal, transport and deposition, as well as by the objectives of the experiment. Since one objective may be to discover more about spore dispersal, the information required to make the best decisions may not be available. In practice, development of spore samplers often occurs in parallel with epidemiological research.

Choice of Samplers

The choice of sampler should be influenced by biological factors such as the mode of spore dispersal, spore size, and spore concentration in the air. If the number of spores deposited on a plant surface is to be estimated, the nature of that surface should be considered. Generally the sampler likely to collect most spores should be chosen, since larger samples are more representative (i.e., counting errors are smaller).

When sampling above crops, it is necessary to use an inertial sampler to collect dry airborne spores; vertical cylinders and the Hirst sampler (similar to the Burkard) collect many more spores per square centimeter than horizontal slides (e.g., Table 2). As impaction efficiencies of small spores (e.g., those of *Cladosporium,* Table 2) are small (Gregory, 1973), a volumetric sampler (e.g., Burkard, cascade impactor, rotorod) will sample them better than a passive vertical cylinder, especially at low concentrations and low wind speeds; Sutton and Jones (1976) collected more *Venturia inaequalis* ascospores (14×7 μm) with Burkard and rotorod samplers than on sticky cylinders. However, Jenkyn (1974) showed that numbers of *Erysiphe graminis* spores (30×15 μm) collected on cylinders showed the same seasonal periodicity as numbers collected by a Burkard. This suggests that the simpler cylinders are appropriate samplers for spores like those of *E. graminis,* which are large, present in high concentrations and dispersed at high wind speeds. Results for pollen (20 to 80 μm) (Table 2) suggest that volumetric suction samplers may underestimate concen-

Table 2 Collection of dry airborne spores by samplers at 2-m height
June to September 1951

| Organism | Organism size (μm) | No. spores per cm² | | |
		Horizontal slide	Vertical cylinder	Hirst sampler
Cladosporium	12 × 5	59	376	8930
Erysiphe	30 × 15	2	69	100
Pollen	20–80 (diam.)	13	490	181

Source: Gregory, Hirst and Last, quoted in Hirst (1959).

trations of the largest pathogen spores (e.g., those of *Helminthosporium* spp.) and that cylinders may be better samplers for them.

In theory, suction samplers operated isokinetically (i.e., facing into the wind with the air speed at the sampler orifice the same as the local wind speed) should measure airborne spore concentration accurately; in practice it is seldom possible to operate suction samplers isokinetically in the field, where wind speed and direction fluctuate (Chamberlain, 1975).

In choosing samplers for spores carried in splash droplets, the size of the droplets is more important than the size of the spores themselves. The range of particle sizes to be sampled is greater than for dry airborne spores. Spore-carrying droplets may be 10 to 1000 μm in diameter and different samplers are required for the efficient sampling of different sizes of droplet.

Spores of many splash-borne pathogens (e.g., those of *Pseudocercosporella herpotrichoides* (Fron) Deighton) are mostly carried in the large ballistic droplets (Fitt and Bainbridge, 1983b). Samplers appropriate for dry airborne spores are generally unsuitable for collecting large spore-carrying droplets because spores can be washed off the vertical sampling surfaces. Of the samplers tested by Fitt and Bainbridge (1983b), horizontal slides (under rainshields to prevent washoff), funnels (draining into beakers), and impingers collected most *P. herpotrichoides* spores. Slides probably lost some spores through runoff but sampling procedures are simple and easily replicated (an important consideration when sampling a patchy disease). Funnels collected most spores, but the liquid samples required concentration before spores could be counted and it was sometimes necessary to separate spores from soil particles (Fitt and Bainbridge, 1983a). The impinger collected more spores per square centimeter than the funnel, but required power to operate the suction pump and thus was less readily replicated.

Small airborne spore-carrying droplets, which are generally present in low concentrations, would not be collected efficiently by samplers with horizontal sampling surfaces near ground level. Volumetric suction samplers, such as a preimpinger (Faulkner and Colhoun, 1977; Wale and Colhoun, 1979) or high volume cyclone separator (Brennan et al., 1985b), which have been used for collecting airborne *S. nodorum* pycnidiospores, are more appropriate.

Fitt and Bainbridge (1983b) exposed healthy wheat plants to straw infected by *P. herpotrichoides* for the same periods as samplers and then incubated them in a cool greenhouse. These plants developed eyespot symptoms throughout the year, even during February when samplers collected few spores (Fig. 1). Although such bait plants gave no information about numbers of spores dispersed, they provided an extremely useful complement to the artificial samplers. Erroneous conclusions have sometimes been drawn from results obtained with either bait plants or artificial samplers alone.

Positioning of Samplers

The positioning of samplers depends on the position of the source of spores, the mode of spore dispersal (especially whether spores are airborne or in ballistic

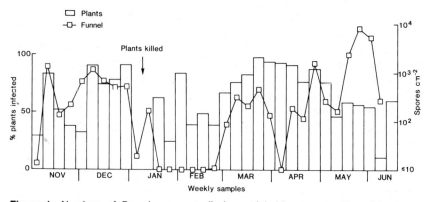

Figure 1 Numbers of *Pseudocercosporella herpotrichoides* spores collected by funnel samplers and the percentages of exposed wheat plants that developed eyespot symptoms after weekly periods in an area of infected wheat straw on the Rothamsted farm.

splash droplets), and the type of sampler used. Bainbridge and Stedman (1979) used different samplers to collect spores of *E. graminis* within and near a strip of barley infected with powdery mildew and demonstrated the change in airborne spore concentration with height above and distance away from a source near ground level. The number of spores collected by suction samplers, a measure of the airborne spore concentration, decreased rapidly with height above the crop (Fig. 2). Similarly the numbers of spores collected on horizontal slides (mostly by sedimentation) decreased with increasing height, whereas the numbers col-

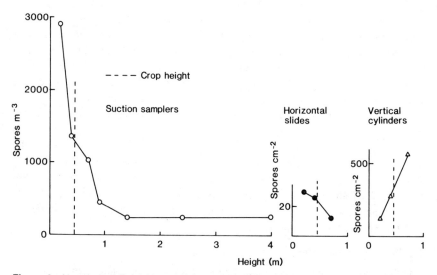

Figure 2 Numbers of *Erysiphe graminis* spores collected by different samplers within and above an infected barley crop. (*Adapted from Bainbridge and Stedman, 1979.*)

lected by vertical cylinders (mostly by impaction) increased. This increase was due to the increase in wind speed with height and these differences between samplers illustrate that it is important to understand the sampling characteristics of a device before interpreting the results it gives.

All three samplers collected fewer spores when positioned farther upwind or downwind of the infected barley (Fig. 3). Such decreases in numbers of spores collected with distance, or spore dispersal gradients, are extremely important in disease spread and hence in epidemiological models. The upwind gradient was steeper than the downwind gradient, which is usually the case for airborne spores (Gregory, 1973). Splash-dispersed spores usually show a steep gradient both with height above (Fitt and Bainbridge, 1983b) and with distance away from a source (Stedman, 1980a).

Therefore, for measurement of background airborne spore concentrations, samplers should be placed above and away from active sources of spores. For measurement of spore dispersal gradients they should be close to the source, particularly if a splash-dispersed spore is being studied. With ground-level sources they must be near the ground. When spore dispersal within crops is of interest they should be positioned so that there is minimum disturbance to the crop, although there must be some gap around the sampler so that infected leaves do not rub against it.

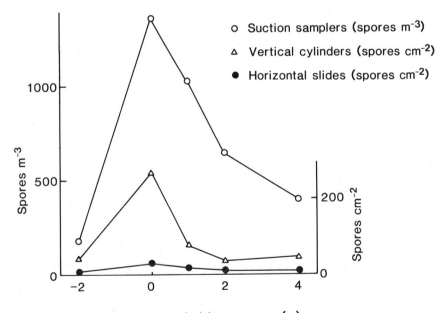

Figure 3 Numbers of *Erysiphe graminis* spores collected by different samplers at 40 cm above ground level upwind and downwind from an infected barley crop. (*Adapted from Bainbridge and Stedman, 1979.*)

Timing of Sampling

If sampling can be restricted to periods when the spores of interest are being released, this may save unnecessary effort and prevent samplers from being overloaded with unwanted spores. Release of many spores shows a diurnal periodicity. On dry summer days the daily average concentration of the dry airborne spores of *E. graminis* is maximal at noon (Sreeramulu, 1964) (Fig. 4), when wind speeds are often great; the daily average concentration of *Botrytis fabae* Sardiña spores is greatest between 7 and 9 a.m., when relative humidity is usually decreasing rapidly (Fitt et al., 1985). *Erysiphe graminis* spore concentrations also show a seasonal periodicity and in the United Kingdom most spores are collected in June (Jenkyn, 1974) (Fig. 4). When such spores are being monitored it may be necessary to sample only for a few weeks of the year and to examine a few hours of the trace of each day if a continuously recording sampler like a Burkard is being used.

Other spores are released in response to specific weather conditions. The splash-borne conidia of *P. herpotrichoides* (Fitt and Bainbridge, 1983b) and dry airborne ascospores of *V. inaequalis* (Hirst et al., 1955) are released during rain. Use of a rain-activated switch (Fitt et al., 1982) will confine sampling to periods of rain and thus reduce the number of unwanted spores collected from the dry air spora.

COLLECTION OF SPORE DISPERSAL DATA UNDER CONTROLLED CONDITIONS

Information required for epidemic models, for example on spore dispersal gradients, is not always easy to obtain in the field, especially if weather conditions favoring dispersal of the spores under study occur only occasionally or if spores of a pathogen are not common in the locality. In these circumstances it may be

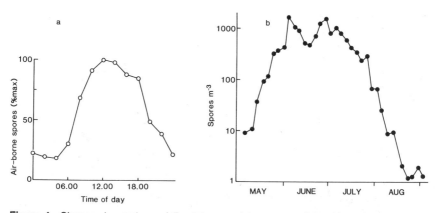

Figure 4 Changes in numbers of *Erysiphe graminis* spores collected by volumetric samplers. (a) Hourly changes; (b) seasonal changes. (*Adapted from (a) Sreeramulu, 1964, and (b) Jenkyn, 1974.*)

useful to study spore dispersal in the laboratory. Such work is often best done in specifically designed systems, such as the Rothamsted raintower and wind tunnel complex (Fig. 5). The equipment can be arranged in two ways: (1) raintower with wind tunnel in "straight-through" position or (2) wind tunnel in recirculating position.

Raintower with Wind Tunnel in "Straight-through" Position

For this system, the height of the raintower (cross section 1 m^2) is 11 m, so that simulated raindrops are falling at approximately terminal velocity when they hit spore-bearing targets. Single drops with diameters of 2 to 5 mm can be produced by a brass drop generator with different nozzles (Stedman, 1979). Simulated rain can be produced by a rain generator (Byass, 1969), which consists of a bank of 150 hypodermic needles, covering an area of 52 × 67 cm, fed by a water reservoir at a rate of about 12 $1 \cdot h^{-1} \cdot m^{-2}$. Air is blown past the needles

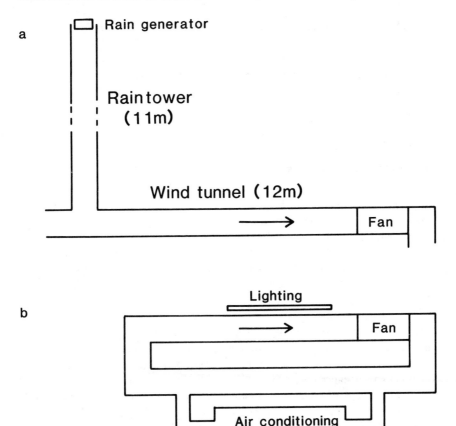

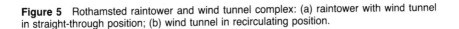
Figure 5 Rothamsted raintower and wind tunnel complex: (a) raintower with wind tunnel in straight-through position; (b) wind tunnel in recirculating position.

to vary the drop size from 1 mm (maximum airspeed) to 3 mm (no air blown past).

This raintower is situated 1 m from the upwind end of a wind tunnel (cross section 1 m^2). Air is drawn through the wind tunnel by a fan 10 m downwind from the raintower; by altering the fan speed, the wind speed can be changed from 10 cm/s to 8 m/s. There are access doors along the sides of the wind tunnel so that plants, samplers, and meteorological equipment can be placed in the air stream.

Wind Tunnel in Recirculating Position

The wind tunnel can also be operated as a closed system with its own internal lighting and environmental control. A proportion of the air circulating in the tunnel passes through an air conditioning unit located in a bypass duct. Temperature and humidity can be controlled over the range 5°C (73 to 85% RH) to 35° C (18 to 45% RH).

Infected spore-bearing plants, sporulating colonies, and spore suspensions may be placed at the base of the raintower (for studying splash dispersal) or at other positions in the upwind section of the wind tunnel. Samplers and meteorological equipment can then be placed at different distances downwind of these sources and the numbers of spores dispersed under different conditions can be measured.

MODELING SPORE DISPERSAL

Spore Dispersal Gradients

The decreases in the concentrations of *E. graminis* spores with distance from (Fig. 3) and height above (Fig. 2) a strip of infected barley are examples of spore dispersal gradients. The decreases in spore concentration with distance away from a source result in spore deposition gradients as portions of the spore clouds are deposited on the ground or crops. These gradients are important in determining the rate of spatial spread of epidemics, and the spore dispersal phase can be incorporated into epidemic models as mathematical descriptions of spore dispersal gradients.

The strip of barley in the experiments of Bainbridge and Stedman (1979) (Figs. 2 and 3) is an example of a line source of spores (Gregory, 1968); an infected plant is a point source, an infected field an area source, and an infected forest a volume source. However the source dimensions must be considered in relation to the distance over which a gradient is measured; an infected field would effectively be a point source if the gradient was measured over several kilometers. Spore dispersal gradients are generally measured in a direction at right angles to a line source or the edge of an area source. Gradients away from a point source may be measured in a particular direction (e.g., downwind) or averaged over all directions.

Empirical Models

Experimental measurements of relationships between numbers of spores deposited y and distance x from a source are usually concave curves, which are difficult to compare between experiments (Gregory, 1968). Empirical models attempt to fit mathematical formulas to experimentally measured spore deposition or concentration gradients. They are descriptive, not interpretative (Gregory, 1968). Since experimental measurements are usually confined to the areas near sources, these models only describe spore dispersal over short distances.

The two most commonly used empirical models are the power law model (Gregory, 1968):

$$y = a\,x^{-b} \tag{1}$$

and the exponential model (Kiyosawa and Shiyomi, 1972):

$$y = c\,\exp(-dx) \tag{2}$$

The constants a and c are related to the source strength and are of interest only if two gradients within one experiment are being compared. The exponents b and d are measures of the dispersal gradient that are independent of source strength, and are more interesting. In the exponential model, the distance in which spore concentration or deposition decreases by half, the half-distance α, is related to d:

$$\alpha = \frac{0.693}{d} \tag{3}$$

Equations 1 and 2 can be put into linear form by taking natural logarithms:

$$\ln y = \ln a - b \ln x \tag{4}$$
$$\ln y = \ln c - d x \tag{5}$$

Linear regressions of $\ln y$ on $\ln x$ and $\ln y$ on x can then be used to estimate values of b and d, respectively.

Comparison of Power Law and Exponential Models

Of 124 dispersal gradients measured for dry airborne spores or pollen and included in the review by Gregory (1968), 59 were fitted better by the power law model and 65 were fitted better by the exponential model (the regression coefficients of determination, r^2, were calculated for each model). This suggests that data on dispersal of dry airborne spores may be fitted equally well by both models, as McCartney and Bainbridge (1984) found in experiments on dispersal

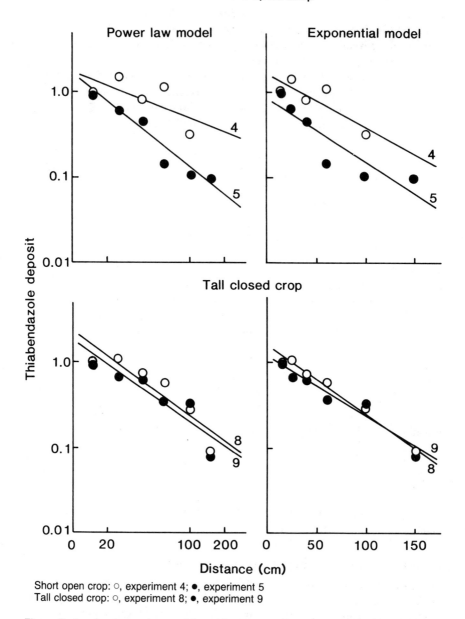

Figure 6 Log-log (power law model) and log-linear (exponential model) plots of thiabendazole deposit (relative to the deposit at 15 cm) against distance from the source in barley crops. Slopes of regression lines are given in Table 3.

of 20 μm droplets (simulating *E. graminis* spores) in barley crops. They added thiabendazole (a chemical tracer) to the droplets and measured deposition at ground level and midcrop height. Deposition of thiabendazole at midcrop height in a short open crop and in a tall closed crop is illustrated in Fig. 6. Of 17 deposition gradients measured (values for eight gradients at midcrop height are given in Table 3), both models gave values of $r^2 > 0.7$ for 15 gradients; the power law model fitted the data better in 7 cases and the exponential model in 10 cases. Gradients of *E. graminis* spore deposition onto vertical cylinders within wheat (Fried et al., 1979) and of *Cronartium quercuum* (Berk.) Miyabe ex Shirai spore concentration over open ground (Schmidt et al., 1982) were both fitted better by the power law than by the exponential model.

However, 15 gradients of splash-dispersed spores, such as those of *P. herpotrichoides* (Fig. 7), measured in still or moving air in the Rothamsted raintower and wind tunnel complex, were all fitted better by exponential than by power law models (Table 4). The gradients were much steeper than those of 20 μm droplets (simulating airborne spores) and the half-distances were smaller.

One advantage of the power law model over the exponential model is that the exponent *b* is independent of the units in which distance is measured (Gregory, 1968), whereas *d* is not. On the other hand the value of the exponential model half-distance is independent of the point from which it is measured, and the half-distance provides an easily visualized measure of the gradient. As *x* tends to 0, the power law equation predicts that *y* tends to infinity (∞); this model generally overestimates deposition near the source (McCartney and Bainbridge, 1984). The exponential model may underestimate deposition near the source (McCartney

Table 3   Deposition gradients at midcrop height

Experiment number	Crop height (cm)	Power law model		Exponential model		
		b	*r²*	*d* (cm⁻¹)	α (cm)	*r²*
Short, open crop:						
3	10	1.11	0.93	0.017	40.8	0.80
4	12–20	0.54	0.47	0.014	49.5	0.68
5	20–30	1.13	0.88	0.018	38.5	0.79
Short, closed crop:						
6	25–30	1.47	0.85	0.025	27.7	0.90
7	30–35	1.27	0.97	0.027	25.7	0.93
Tall, closed crop:						
8	50–58	1.00	0.83	0.018	38.5	0.98
9	85–90	0.95	0.84	0.017	40.8	0.94
10	90	0.91	0.83	0.016	43.3	0.93

Note: Estimated by fitting power law and exponential models to observed thiabendazole deposits. Droplets released at midcrop height in barley. *b* and *d* are gradients estimated by regression analysis; α is the half-distance; r^2 is the coefficient of determination.
Source: H. A. McCartney and A. Bainbridge, *Phytopathol. Z.* 109:219–236, 1984.

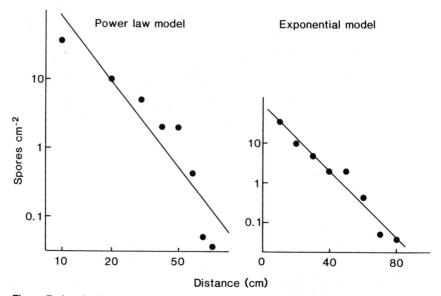

Figure 7 Log-log (power law model) and log-linear (exponential model) plots of number of spores per square centimeter collected, when a 5-mm drop fell onto a suspension of *Pseudocercosporella herpotrichoides* spores, against distance from the target suspension. Slopes of regression lines are given in Table 4.

and Bainbridge, 1984), but predicts finite values of y at all distances and is therefore more easily incorporated into epidemic models. Neither model should be extrapolated to distances beyond the observed gradient (Gregory, 1982).

Interpretation of Spore Dispersal Gradients

Gregory (1968, 1973, 1982) summarized the principles for interpretation of spore dispersal gradients; nevertheless dispersal gradients are frequently misinterpreted. The observation of a spore dispersal gradient implies the existence of a local source of spores. The size, shape and height of the source affect the gradient. Gradients from a point source are generally steeper than gradients from a line, area, or volume source (Gregory, 1968) and ground-level sources often produce steeper gradients than aboveground sources (Stedman, 1980b). Numbers of spores deposited may increase with distance away from an aboveground source when there is a "skip-distance" (a distance before deposition starts) (Gregory, 1968).

Primary gradients result from one source releasing spores for a short time; secondary spore dispersal from new generations of pathogen lesions over a wider area will flatten the primary gradient (Gregory, 1968). Spores from a large number of distant sources (background contamination) will not produce a significant gradient ($b = 0$). Thus, small values of b suggest either proximity to a large area or volume source, spore dispersal from further generations of pathogen lesions, or a large background contamination. A U-shaped gradient suggests

Table 4 Spore dispersal gradients[a]

Experimental details	Power law model		Exponential model			Source
	b	r²	d (cm⁻¹)	α (cm)	r²	
Still air, single drops, spore suspension:						
P. herpotrichoides						
5-mm drop, spores target	3.2	0.82	0.10	7.1	0.95	Fatemi and Fitt (1983)
4-mm drop, spores target	2.8	0.89	0.11	6.1	0.95	Fitt and Lysandrou (1984)
4-mm drop, spores drop	2.0	0.54	0.06	11.0	0.66	
5-mm drop, spores target	3.8	0.78	0.12	5.9	0.93	
5-mm drop, spores drop	3.5	0.87	0.09	7.4	0.94	
P. brassicae						
5-mm drop, spores target	2.6	0.85	0.07	10.0	0.87	Fatemi and Fitt (1983)
S. nodorum						
4-mm drop, spores (dil.) target	2.0	0.89	0.06	10.8	0.95	Brennan et al. (1985a)
3-mm drop, spores (conc.) target	2.1	0.90	0.07	10.2	0.94	
4-mm drop, spores (+ wetter) target	2.6	0.73	0.10	6.8	0.92	
Moving air, single drops, spore suspension:						
S. nodorum						
4-mm drop, downwind	3.2	0.80	0.05	14.8	0.97	Brennan et al. (1985b)
Moving air, simulated rain, straw:						
P. herpotrichoides						
3-mm drops, upwind	7.3	0.99	0.10	6.9	0.99	Fitt and Nijman (1983)
3-mm drops, downwind	4.9	0.77	0.04	16.1	0.89	
3-mm drops, downwind	5.1	0.81	0.05	14.2	0.93	Fitt and Bainbridge (1984)
3-mm drops, downwind (NaOH)	5.2	0.80	0.05	13.1	0.90	
3-mm drops, downwind (cellulose xanthate)	4.1	0.82	0.04	15.4	0.91	

Note: Estimated by fitting power law and exponential models to experimental data for splash-dispersed spores.
[a]Distance in centimeters; number of spores deposited in spores per square centimeter. b and d are gradients estimated by regression analysis; α is the half-distance; r² is the coefficient of determination.

the proximity of a second source. Sometimes a gradient will be steep initially and then flatten out as the spore concentration diminishes to background levels.

Both local weather conditions and crop structure affect spore dispersal gradients. With a strong prevailing wind, gradients downwind of a source are generally shallower than those in other directions (Fig. 3). McCartney and Bainbridge (1984) found that deposition of 20 μm droplets on open ground differed greatly between days with different atmospheric turbulence. Dispersal gradients are generally steeper within crops than over open ground (Bainbridge and Stedman, 1979) and deposition gradients of 20 μm droplets at midcrop height were steeper in short closed crops than in short open crops or in tall closed crops (Table 3).

Since so many factors affect dispersal gradients, it is difficult to interpret small differences in the values of b or d between pathogens or between different experiments with one pathogen. As in wind tunnel experiments (Table 4), field gradients of splash-dispersed spores are often steeper than those of wind-dispersed spores, both over open ground (Stedman, 1979) and within crops (Stedman, 1980a). This may be because most splash-dispersed spores are carried in ballistic splash droplets (Fitt and Bainbridge, 1983b), which are larger than most wind-dispersed spores. However, wind-dispersed spores of different sizes, e.g., spores of *Lycopodium* (32-μm diameter) and *Bovista* (4 μm), may have similar dispersal gradients (Stepanov, 1935).

One common misconception is that a steep spore deposition gradient implies that the source is inactive at long distances (Gregory, 1982). It is a paradox that meteorological factors, such as frictional turbulence and thermal convection, which cause steep gradients, can lead to larger escape fractions (the parts of spore clouds which reach heights where they can disperse over long distances). Stepanov (1935) estimated the source strength and deduced that the fraction of the spore cloud that remained airborne was greater for the smaller *Bovista* spores than for the larger *Lycopodium* spores. Another fallacy is to deduce from a change from a continuous spore dispersal gradient near to a source to a discontinuous gradient further away that there are two different methods of spread; there may or may not be (Gregory, 1982).

Physical Models

Although empirical model formulas may fit observed spore dispersal gradients well, they give little information about the physical processes underlying spore dispersal. Thus, such models cannot be used to predict dispersal under weather conditions different from those of the original experiment. Physical models, however, attempt to describe spore transport and deposition in terms of measurable physical quantities and mathematical relationships based on the laws of physics. Thus they are potentially more versatile, although they are more difficult to construct and use. In principle such models can be applied to the dispersal of any spore in any crop, given appropriate information on spore size, structure, and removal mechanism, crop structure, and wind flow. Furthermore, models based on measurable physical parameters offer a greater insight into the mechanisms of spore dispersal and deposition.

Pollutant gases and small particles carried in the wind are dispersed by turbulent eddies in the air, both above and within crops. Over the last 20 years there has been much work on the construction of mathematical models for application to air pollutant dispersal. Three of these atmospheric dispersion models have also been applied to spore dispersal: the Gaussian plume, gradient transfer theory, and random walk models. They are mainly applied to spore transport but can incorporate spore deposition and spore removal. The construction of physical models describing spore dispersal within and above crops has recently been reviewed by McCartney and Fitt (1985a); thus only a brief outline of the construction and use of each model will be given.

Gaussian Plume Models

The Gaussian plume, or statistical, model is one of the simplest and most widely used atmospheric dispersion models. The model assumes that the time-averaged concentration of spores across the height and width of a plume downwind of a point source (across the height only for a line source) follows a Gaussian curve with standard deviations of σ_z and σ_y, respectively (Fig. 8) (Pasquill, 1962). The mean concentration $C(x,y,z)$ downwind of a point source at height H $(0,0,H)$ is:

$$C(x,y,z) = \frac{Q'(x)}{u} \frac{\exp(-y^2/2\sigma_y^2)}{2\pi\sigma_z\sigma_y)} \left[\exp \frac{-(H-z)^2}{2\sigma_z^2} + \exp \frac{-(H+z)^2}{2\sigma_z^2} \right] \quad (6)$$

and downwind of an infinite line source at right angles to the mean wind direction:

$$C(x,z) = \frac{Q'(x)}{u} \frac{1}{(2\pi)^{0.5} \sigma_z} \left[\exp \frac{-(H-z)^2}{2\sigma_z^2} + \exp \frac{-(H+z)^2}{2\sigma_z^2} \right] \quad (7)$$

where u is the mean wind speed at source height and Q' is the effective source strength allowing for deposition of particles to the ground (Gregory, 1945). Equations 6 and 7 assume that the plume is "reflected" when it reaches the ground and that all deposition processes are accounted for by the Q' term. The model also assumes particles are well mixed within the plume and that deposition does not alter the shape of the vertical concentration profile (it remains Gaussian). Deposition rate to the ground s is often assumed to be proportional to the spore concentration at a reference height z', and the constant of proportionality is called the deposition velocity v_d (Chamberlain, 1975):

$$s = C(z')v_d \quad (8)$$

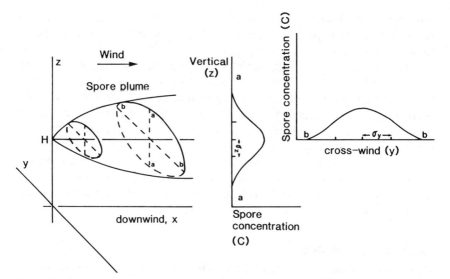

Figure 8 Spore dispersal downwind of a point source at height H. The x axis represents the mean wind direction, the y axis the crosswind direction and the z axis the vertical direction. Gaussian distributions of spore concentration c in the vertical (a–a) and crosswind (b–b) directions, with standard deviations σ_z and σ_y, are illustrated.

If the source is emitting Q spores per unit time (Q spores per unit time per unit length in the case of a line source) then the effective source strength Q' is given by (Horst, 1977):

$$Q'(x) = Q \exp\left[-\int_0^x \frac{v_d}{u} D(x') \, dx' \right] \qquad (9)$$

where

$$D(x) = \frac{1}{(2\pi)^{0.5}\sigma_z} \left[\exp \frac{-(H - z')^2}{2\sigma^2} + \exp \frac{-(H + z')^2}{2\sigma^2} \right] \qquad (10)$$

in which σ_z is a function of x. The value of the integral in Eq. 9 depends on the form of σ_z and may be evaluated numerically. Plant pathogen spores are often released from ground level ($H = 0$) (i.e., crop height for dispersal above crops), and for ground-level concentration (and deposition) Eqs. 6 and 7 simplify to:

$$C(x,y,0) = \frac{Q'}{u} \frac{\exp(-y^2/\sigma_y^2)}{\pi \, \sigma_z\sigma_y} \qquad (11)$$

and

$$C(x,0) = \frac{Q'(2/\pi)^{0.5}}{u \, \sigma_z}$$ (12)

Plume height and width, defined by σ_z and σ_y, vary with distance downwind (x) and with the atmospheric mixing (stability). Generally the less turbulent the atmosphere the more slowly a plume expands with distance downwind. σ_z and σ_y have to be determined experimentally and there have been several recent reviews of their use in atmospheric diffusion modeling (e.g., Mitchell, 1982). However, much use is made of Pasquill atmospheric stability classes, determined from simple meteorological observations, and analytic approximations of σ_z and σ_y for each class are given in Table 5.

Gaussian plume models have been used to examine the dispersal of *Lycopodium* spores released above grass (Gregory et al., 1961; Reddi, 1975). The values of σ_y were fitted to the model of Sutton (1953):

$$\sigma_y = Ax^m$$ (13)

where A and m are constants; m had values between 1.2 and 1.75.

Table 5   Atmospheric stability classes

Stability classes and analytic approximations to Gaussian plume crosswind (σ_y) and vertical (σ_z) standard deviations[a]

Class	Description	σ_y (m)	σ_z (m)
A	Highly unstable	$0.22x \, (1 + 0.0001x)^{-0.5}$	$0.20x$
B	Unstable	$0.16x \, (1 + 0.0001x)^{-0.5}$	$0.12x$
C	Slightly unstable	$0.11x \, (1 + 0.0001x)^{-0.5}$	$0.08x \, (1 + 0.002x)^{-0.5}$
D	Neutral	$0.08x \, (1 + 0.0001x)^{-0.5}$	$0.06x \, (1 + 0.0015x)^{-0.5}$
E	Slightly stable	$0.06x \, (1 + 0.0001x)^{-0.5}$	$0.03x \, (1 + 0.0003x)^{-1}$
F	Stable	$0.04x \, (1 + 0.0001x)^{-0.5}$	$0.016x \, (1 + 0.0003x)^{-1}$

Determination of stability class from meterological observation

Wind speed (m/s)	Daytime solar radiation[b]				Night		
	Strong	Moderate	Slight	Overcast	≤3/8 Cloud	Thinly overcast or ≤4/8 cloud	Overcast
<2	A	A–B	B	D	—	—	D
2–3	A–B	B	C	D	E	F	D
3–5	B	B–C	C	D	D	E	D
5–6	C	C–D	D	D	D	D	D
>6	C	D	D	D	D	D	D

[a]Valid for $100 < x < 1000$ m from source. (Briggs, quoted in Hanna, 1982.)
[b]Strong: sunny midday, midsummer in England (>600 W/m^2). Moderate: average midday, summer in England (300 to 600 W/m^2). Slight: sunny midday, winter in England (<300 W/m^2).
Source: F. Pasquill, *Atmospheric Diffusion*, 1st ed. Van Nostrand, London.

Equations 6 and 7 predict concentrations of spores dispersing above crops and, with slight modification to Q', can be applied to dispersal within crops (McCartney and Fitt, 1985a). However, as there is little information about σ_z and σ_y within crops, their use to predict deposition gradients close to sources is limited. They are most suitable for examination of spore dispersal over distances up to a few kilometers from sources (e.g., dispersal between fields).

The use of a Gaussian model is illustrated in Fig. 9, which shows spore deposition rate plotted against downwind (wind speed 2 m/s) distance from an infinite line source (e.g., a field edge) for different deposition velocities and two stability classes. Stability class A represents a warm sunny day with light winds (Table 5) when there is substantial atmospheric mixing, and class D represents an overcast day when there is less atmospheric mixing. As spore deposition velocity increases, deposition gradients (i.e., slopes of the lines) become steeper and more spores are deposited near the source. The influence of deposition velocity or deposition gradients is less when turbulence is large (class A) as spore concentration is quickly reduced by upward spore transport (large σ_z). In the field the amount of turbulence may also affect the spore deposition velocity (Chamberlain, 1975).

Gaussian plume models cannot accurately predict spore concentrations and deposition rates in complex situations, such as within nonhomogeneous crops where the wind speed changes rapidly with height. However, the models are simple to use because they require minimal meteorological information and are appropriate for situations where ease of use is more important than accuracy.

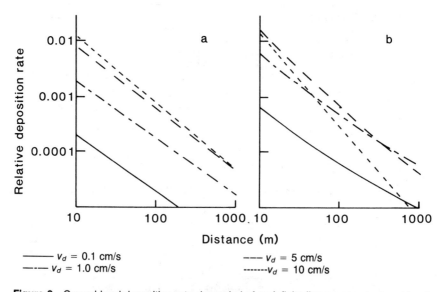

Figure 9 Ground-level deposition rate downwind of an infinite line source at ground level calculated by a Gaussian plume model. σ_z values taken from Table 5. (a) Stability class A; (b) stability class D. Wind speed (2 m/s) and source strength are the same for all cases.

Gradient Transfer Theory Models

Gradient transfer theory is commonly used in mathematical models that describe the dispersal of particles and gases in the atmosphere; these models can also be applied to spore dispersal. The theory assumes that diffusion by eddies in the atmosphere is analogous to molecular diffusion (Taylor, 1915) and obeys Fick's law, i.e., the spores move from higher to lower concentrations at a rate proportional to the concentration gradient. Differential equations can be written to describe dispersal from continuous or instantaneous (puffs) point, line or area sources. If spores are dispersing from a continuous, infinite line source at right angles to the mean wind direction (x direction, Fig. 8), time can be eliminated from the equations and only average concentrations are calculated. Since cross-wind diffusion can be ignored because the source is a line and downwind diffusion is normally much less than downwind advection, only vertical turbulent dispersion need be considered (Pasquill, 1974). The diffusion equation that must be solved to obtain the spore concentration C is then:

$$u \frac{\partial C}{\partial x} + S(x,z) = \frac{\partial}{\partial x} \left(K_z \frac{\partial C}{\partial z} \right) + v_s \frac{\partial C}{\partial z} \tag{14}$$

where u is the mean horizontal wind speed, s is the loss of spores by deposition, K_z is the vertical eddy diffusivity and v_s is the spore fall speed. Equation 14 is a statement of the steady-state conservation of spore numbers (i.e., spore concentration is constant with time). The loss of spores from a volume of air by advection $[u(\partial C/\partial x)]$ and by deposition $[S(x,z)]$ is balanced by the rate of spore diffusion $\partial/\partial x \, (K_z \, \partial C/\partial z)$ and sedimentation $[v_s \, (\partial C/\partial z)]$ into the volume. Equation 14 may be applied to spore dispersal in the atmosphere above crops (where $S = 0$) (Itier and Pauvert, 1979) and within crops (Legg and Powell, 1979; Aylor, 1982). However, before Eq. 14 can be solved the following must be known: (1) wind speed and eddy diffusivity at all values of x and z; (2) the form of the deposition term at all values of x and z; (3) the boundary conditions at the ground and at some upper limit in the atmosphere; and (4) the spore concentration profile at $x = 0$.

Wind speed and eddy diffusivity are generally functions of height and should, in principle, be measured. However, since the variation of u and K_z with height in the atmosphere and within crops has been studied extensively (McCartney and Fitt, 1985a), they may be represented by mathematical functions. Above crops both u and K_z can be represented by power laws (Huang, 1979):

$$u(z) = u(z') \left(\frac{z}{z'} \right)^n \tag{15}$$

$$K(z) = K(z') \left(\frac{z}{z'} \right)^m \tag{16}$$

where z' is some arbitrary height and n and m are constants whose values depend on atmospheric stability. Within many crops the wind speed profile takes the form (Cionco, 1972):

$$u(z) = u(h) \exp\left[\alpha \left(\frac{z}{h} - 1 \right) \right] \tag{17}$$

where h is the crop height and the constant α has a value between 0.3 and 3 depending on crop type and foliage density. K_z, which changes little in the upper two-thirds of the canopy, has been formulated as (Legg and Price, 1980)

$$K_z = K_z(h) \qquad h > z > \frac{h}{3}$$

$$K_z = K_z(h) \frac{3z}{h} \qquad \frac{h}{3} > z \geq 0 \tag{18}$$

Above crops, no spores are lost by deposition ($S = 0$ at all values of x and z) as spores are only deposited in the crop and this is accounted for by the boundary conditions. Within crops, however, spores are removed from the dispersing cloud by impaction and sedimentation onto plant surfaces. Deposition rates (spores per unit volume per unit time) are proportional to spore concentration $C(x,z)$:

$$S(x,z) = r(z) \, C(x,z) \tag{19}$$

where the constant of proportionality r depends on crop structure and density (Legg and Price, 1980), spore size and retention properties (Chamberlain, 1975), and local wind speed $u(z)$.

For dispersal above crops the "ground" ($z = 0$) is taken as the top of the crop, but for dispersal within crops $z = 0$ is taken as the soil surface. If all spores reaching the "ground" were deposited, then the ground would be a "perfect sink" and the boundary condition, $C = 0$ at $z = 0$ could be applied to Eq. 14. This, however, is rarely the case and in practice the flux of spores to the "ground" is usually described by a deposition velocity v_d (Eq. 8) and the boundary condition at z = 0 is written:

$$K_z \frac{\partial C}{\partial z} + v_s C = v_d C \tag{20}$$

The first term on the left represents deposition due to diffusion and the second deposition due to sedimentation. The value of v_d for dispersal above crops depends on spore size, crop structure, and atmospheric turbulence but is generally about 2 to 3 times the spore fall speed (Chamberlain, 1975). Within a few centimeters of ground level under crops, turbulence is usually small and depo-sition can be assumed to be mostly by sedimentation ($v_d = v_s$) (Legg and Powell, 1979; Aylor and Taylor, 1982).

The choice of the upper boundary condition for Eq. 14 generally has little effect on concentrations near the ground (Legg and Powell, 1979; Aylor and Taylor, 1982). For example, upward flux of spores is often assumed to fall to zero at some arbitrary height a (usually the depth of the planetary boundary layer for diffusion above crops), giving the boundary condition at $z = a$:

$$K_z \frac{\partial C}{\partial z} + v_s C = 0 \qquad (21)$$

Equation 14 is difficult to solve analytically for realistic values of u, K_z, and S and solutions, even for simple cases, are complex (Llewelyn, 1983). Numerical methods have been used to solve Eq. 14 for the dispersal of *Erysiphe graminis* spores in barley (Legg and Powell, 1979; Aylor, 1982) and for dispersal of *Peronospora tabacina* spores in tobacco (Aylor and Taylor, 1982). Legg and Powell (1979) found that predicted concentrations of *Lycopodium* spores down-wind of an artificial line source were realistic but that concentrations of naturally released *E. graminis* spores were overestimated. Aylor (1975) has suggested that some spores are only removed from leaf surfaces in gusts of wind above a threshold wind speed (dependent on the strength of attachment of the spores); this will enhance efficiency of impaction close to the source (Aylor et al., 1981). Aylor (1982) therefore changed the deposition term S in the model of Legg and Powell to account for gust removal and consequently improved the model's predictions.

Gradient transfer theory can be used to investigate the relative importance of crop structure, spore size, and turbulence on dispersal. For example deposition gradients predicted by a dispersal model based on Eq. 14 (details in Table 6) are shown in Fig. 10. It was assumed that the leaf area distribution of the crop was independent of height but that the stem/leaf area and leaf angle distribution were those of barley. It was also assumed that spores were continuously released from the lower half of the plants of an infected strip of crop (at $x = 0$) and the model calculated the change in spore deposition onto the whole crop with distance

Table 6   Wind and diffusivity profiles for gradient transfer model used in Fig. 12

	Within crop	Above crop
Wind speed	$u = 0.2 \exp(0.916z)$	$u = 0.375 \ln(10z - 7)$
Diffusivity	$K_z = K_0$	$K_z = K_0 + 0.06 (z - 1)$

Crop height = 1 m.
Units: $z = m$; $u = m/s$; K_z, $K_0 = m^2/s$.
Source: Adapted from a model by Aylor and Taylor (1982).

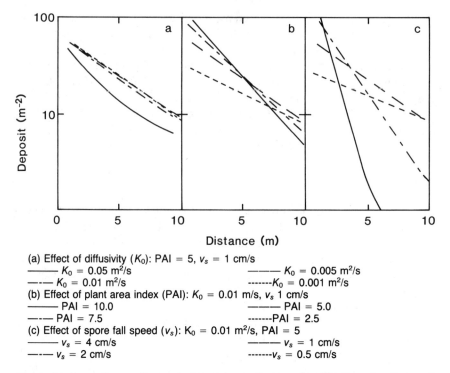

Figure 10 Deposition gradients calculated by a gradient transfer model based on that used by Aylor and Taylor (1982). Details of the model are given in Table 6.

downwind of the source. The figure suggests that the deposition gradients are little affected by turbulence (K_0), except under unstable conditions ($K_0 = 0.05$ m^2/s), but are closely dependent on crop structure (e.g., plant area index, PAI) and spore size (v_s). Both PAI and v_s influence the filtration efficiency of the crop (S term, Eq. 14); thus deposition gradients steepen with increasing crop density and with increasing sedimentation and impaction efficiencies. The wind speed profile was the same for all simulations, a condition that is not strictly realistic as changing crop structure alters the air flow characteristics within crops. However, the results shown in the figure agree qualitatively with those obtained in the droplet dispersal studies of McCartney and Bainbridge (1984).

Although gradient transfer theory has been used extensively to describe dispersal of air pollutants, it has several inherent disadvantages when applied to the dispersal of plant pathogen spores (Legg, 1983). The theory is strictly applicable only when the characteristic size of the wind eddies is smaller than that of the diffusing cloud. This restriction limits the application of the theory close to sources and in regions where the concentration gradient is changing rapidly (e.g., within crops where there are local sources and sinks). Furthermore, the theory has only been tested under conditions of low turbulence where downwind transport by advection is greater than transport by diffusion. This is generally true above crops but may not be true within crop canopies, where air turbulence can be large, despite the low wind speeds. However, gradient transfer theory is a useful tool for investigating spore dispersal in the atmosphere. It can be applied to long distance transport or to dispersal between or within fields and can account for spore deposition and predict realistic spore concentrations.

Random Walk Models

The third and potentially most versatile theory of particle dispersion that has been applied to spore transport is random walk theory, in which individual spore trajectories are simulated (Legg, 1983). This approach offers several advantages over gradient transfer theory of Gaussian models. The paths of individual spores are simulated using realistic mathematical descriptions of turbulent air flow and this means that the theory can be applied close to sources in turbulent flows (e.g., inside crop canopies) and can account for intermittent spore release, such as in gusts. Deposition processes are easily incorporated into such models.

The theory assumes that small particles follow the movements of individual fluid elements in the air flow, except for an additional downward velocity equal to their fall speed. The paths of individual air parcels are simulated as a pseudorandom walk from a knowledge of the turbulence statistics of air flow (Hall, 1975). The trajectories are represented as a series of discrete displacements, determined partly by a correlation between successive velocities and partly by a random component to account for turbulent fluctuations. The nonrandom element in each step is chosen so that the statistical properties of the simulated flow are identical to those expected in the real environment. Such calculations can be treated as a Markov chain process and have been applied to spore dispersal within crops (Legg, 1983) and to the dispersal of spray droplets above crops (Thompson, 1983).

The potential importance of random walk models in spore dispersal studies was demonstrated by Legg (1983), who considered the dispersal of spores from a point source within a barley crop. The model, although restricted to dispersal in a horizontal plane, allowed for continuous spore release or release only when wind speed exceeded a given threshold (1.55 m/s). It assumed that spores were released from the middle of the crop and deposited on the crop by sedimentation or impaction before they could escape from the crop or be deposited on the ground. Characteristics of *Erysiphe graminis* spores ($v_s = 12$ mm/s) and leaf area density and distribution typical of a mature barley crop were used. The model assumed a mean wind speed of 0.5 m/s, with turbulence characteristics typical of those found within a barley canopy.

Figure 11 shows the relative integrated deposition rate (i.e., deposition to an annulus of unit width) by sedimentation and by impaction plotted against distance from the source. Gust release had little effect on deposition by sedimentation, and the gradients were similar in both cases. However, close to the source a smaller proportion of spores were deposited by sedimentation when the spores were released only in gusts than when spores were released continuously.

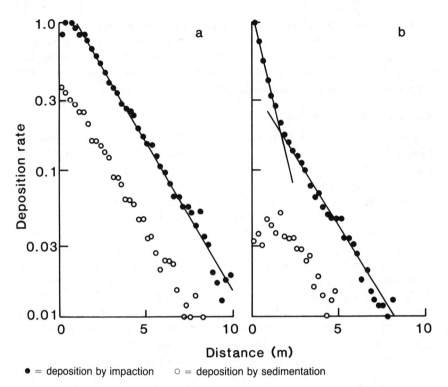

● = deposition by impaction   ○ = deposition by sedimentation

Figure 11 Deposition of *Erysiphe graminis* spores onto a barley crop, integrated around a point source, calculated using a two-dimensional random walk model. (a) Spores released in all wind speeds; (b) spores released only when wind speed exceeded 1.55 m/s, mean wind speed = 0.5 m/s. (*Source: Legg, 1983.*)

The gradient of deposition by impaction was steeper close to the source for spores liberated in gusts. Impaction efficiency, which increases with increasing wind speed, was greater because initial velocities of spores were greater in gusts. As spores are transported away from sources, the influence of their initial velocity is gradually lost and the deposition gradient becomes similar to that of spores released continuously. For a given mean wind speed, horizontal velocity fluctuations could increase deposition gradients close to the source by as much as a factor of four. The gradients predicted by the model were comparable to those of naturally released *E. graminis* spores (Bainbridge and Stedman, 1979) and continuously released 20 μm liquid droplets (McCartney and Bainbridge, 1984).

Before random walk models can be applied fully, further information on the turbulence in the airflow within crops and on the details of spore deposition processes in turbulent airflow is needed. Nevertheless, random walk simulation of spore dispersal is potentially the best model for understanding the dispersal of a wide range of spores in many crops under different meteorological conditions.

USE OF SPORE DISPERSAL MODELS

Incorporation into Epidemic Models

A spore dispersal gradient will result in a disease gradient when dispersal of viable infective spores in a population of susceptible hosts is followed by weather favorable for infection. Thus, for many plant diseases, spore dispersal is an important part of disease spread; it generally needs to be incorporated into disease epidemic models, which attempt to describe the progress of epidemics in both space and time. Epidemic models may be simulation models, which attempt to construct descriptions of epidemics from mathematical descriptions of each of the component phases of pathogen life cycles (Waggoner, 1974) or analytic models, which attempt to describe epidemics as a whole by single mathematical equations (Jeger, 1983).

A simple simulation epidemic model, using an exponential model to describe spore dispersal, has been used to investigate how spore dispersal gradients affect spread of foliar diseases (McCartney and Fitt, 1985b). Values of parameters incorporated into the model were estimated from data for barley powdery mildew *Erysiphe graminis*. It was assumed that a central row of plants in a 20 m wide strip of crop was infected (Fig. 12) and that spores were deposited equally on either side of the central row, with the numbers deposited decreasing exponentially with distance.

The amount of disease on each plant and the total disease in the crop were calculated for two spore deposition gradients, with half-distances $\alpha = 5$ and 50 cm. The disease distribution distance after three, four, five, and six pathogen generations (Fig. 13) suggests that with steep spore deposition gradients disease will intensify rapidly but spread slowly, whereas with shallow gradients disease will intensify more slowly but spread rapidly.

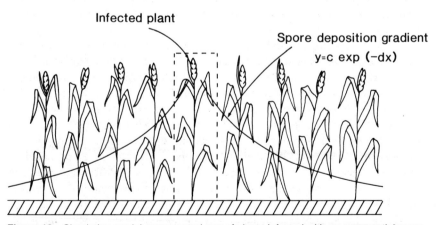

Figure 12 Simulation model crop; central row of plants infected with an exponential spore deposition gradient, $y = C \exp(-dx)$ where y is number of spores deposited and x is distance from the infected row.

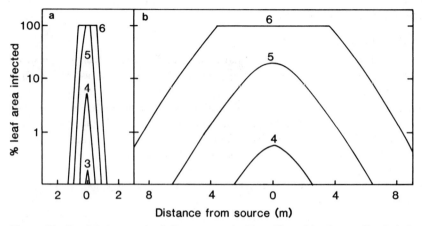

Figure 13 Predicted amounts of disease per plant in a 20-m strip of crop after 3, 4, 5, or 6 pathogen generations with spore deposition gradient half-distances of (a) $\alpha = 5$ cm, (b) $\alpha = 50$ cm. The simulation model assumes values for lesion growth rate, etc., similar to those of barley powdery mildew.

The model also indicated that the steepness of the spore dispersal gradient does not affect the total amount of disease which develops in a crop until about 30% of the leaf area at the focus is infected. Subsequently, multiple infections are more likely; i.e., some spores will land on tissue that is already infected and will not contribute to disease increase. As disease intensifies more quickly when the gradient is steep (α is small), multiple infection effects occur sooner and the total amount of disease in the crop will increase less rapidly than when the gradient is shallow (Fig. 14). When the gradient is very shallow ($\alpha > 100$ cm), less disease develops because spores are lost from the edge of the crop. The model may overestimate disease when gradients are shallow because it does not allow for loss of spores to the atmosphere, which is most likely with shallow gradients.

Analytic epidemic models describing disease development in both time and space have been developed recently (Jeger, 1983). Previously, attempts had been made to describe epidemics by analysis of disease gradients on more than one occasion (Rowe and Powelson, 1973) or by analyzing the same data twice, once for temporal progress and once for spatial spread (Berger and Luke, 1979). Berger and Luke calculated the rate of isopath (i.e., a contour with equal amounts of disease) movement from crown rust foci and Jeger (1983) has now provided the theoretical framework for their calculations. In developing eight models to describe disease increase in space and time from pairs of differential equations describing the rate of isopath movement, he showed how modified versions of either the power law or the exponential dispersal models could be used. Jeger (1983) used his models to evaluate the development of *Septoria nodorum* in pure stands and mixtures of two spring wheat cultivars. These analytic models should provide a powerful tool for the analysis of other epidemic data.

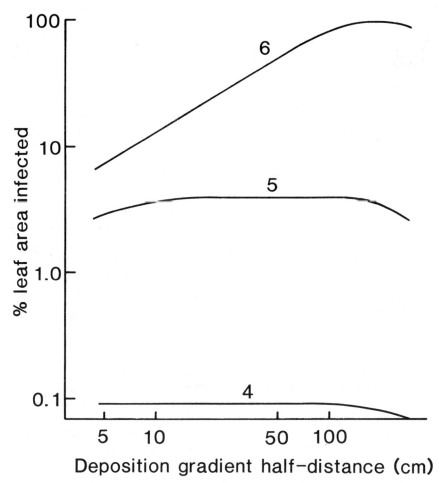

Figure 14   Predicted effect of spore deposition gradient on total amount of disease developing in a 20-m strip of crop after 4, 5, or 6 pathogen generations. The simulation model assumes values for lesion growth rate, etc., similar to those of barley powdery mildew.

Applications

Spore-dispersal-gradient models can be useful in the assessment of the potential of cultivar mixtures and multilines for reducing epidemic development. For example, the simulation model described above was used to assess how the use of cultivar mixtures may modify the effect of spore dispersal gradients on development of a foliar disease spread by airborne spores. Disease development in crops with mixtures of susceptible and resistant cultivars in the ratios 1 : 0, 1 : 1, 1 : 2, or 1 : 3 was calculated. The results suggested that mixtures would be most effective in reducing spread of diseases caused by pathogens with shallow spore dispersal gradients (i.e., large α, Fig. 15), when the rate of disease in-

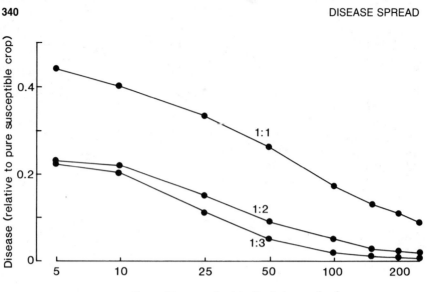

Deposition gradient half-distance (cm)

Figure 15 Predicted effect of spore deposition gradient on amount of disease in mixtures of susceptible and resistant cultivars (1:1, 1:2, 1:3) after six pathogen generations. The simulation model assumes values for lesion growth rate, etc., similar to those of barley powdery mildew.

tensification would be considerably decreased in mixtures through loss of spores by deposition onto resistant plants. When deposition gradients are steep, much of the disease in a focus develops on the initial infected plants (self-infection) and mixtures reduce disease spread less.

The predictions of this model have been compared with the results of a field experiment to investigate the development of powdery mildew (*E. graminis*) foci in mixtures of mildew-susceptible and mildew-resistant barley cultivars in the ratios 1 : 0, 1 : 2, or 1 : 4 susceptible : resistant. Foci were initiated by placing pots of infected plants in the center of plots and amounts of disease were assessed at 2 to 3 weekly intervals. Amounts of disease predicted by the model for spore dispersal gradients with half-distances $\alpha = 25$ and 50 cm (typical values for *E. graminis,* Table 3) after five or six pathogen generations were compared with amounts of disease observed on leaf eight of plants in the two mixtures (Table 7). Although the model does not attempt to simulate the exact conditions of the field experiment, amounts of disease observed, expressed as a fraction of disease in the pure susceptible stand, were similar to those predicted by the model, with $\alpha = 50$ cm.

The spread of *E. graminis* was reduced greatly by multilines in the experiments of Fried et al. (1979) with ratios of 1 : 0, 1 : 1, 1 : 2, 1 : 3 suscepti- ble : resistant winter wheat. Spore deposition gradients were measured with vertical sticky cylinders, and mildew was assessed as numbers of lesions per tiller 15 and 23 days after inoculation. There were no significant differences

Table 7   Amounts of powdery mildew in mixtures of susceptible and resistant barley cultivars (relative to pure susceptible stands), as predicted by a simulation model and observed in infected field crops

Mixture (susceptible: resistant)	Predicted[a]				Observed[b]		
	α = 25 cm		α = 50 cm				
	PG 5	PG 6	PG 5	PG 6	22 June	6 July	18 July
1:2	0.07	0.15	0.06	0.09	0.05	0.08	0.10
1:4	0.04	0.08	0.02	0.03	0.01	0.02	0.03

[a]α is the deposition gradient half-distance; PG = pathogen generation.
[b]Disease assessed on leaf 8, mean of 3 replicates. Infected plants had been placed in the center of each plot on 27 April and on 12 May.

($p > 0.05$) between spore deposition gradients and disease gradients or between disease gradients in multilines and disease gradients in pure susceptible stands, suggesting that the dispersal of spores was similar in all crops and that initially numbers of lesions developing were related to the numbers of spores deposited on suceptible plants. However, much less disease developed in multilines than in the susceptible stands, as predicted by the simulation model when the spore dispersal gradient half-distance was large (Fig. 15).

On the other hand, mixtures of susceptible and resistant cultivars did not greatly reduce the spread of *Septoria nodorum,* a splash-dispersed pathogen with a small spore dispersal gradient half-distance (Table 4) in experiments of Jeger et al. (1983) with spring wheat. They measured disease incidence (proportion of leaves diseased) and disease severity (proportion of leaf area diseased) weekly after inoculation (in the center) of plots with 1 : 0, 3 : 1, 1 : 1, 1 : 3, and 0 : 1 ratios of susceptible : resistant wheat. Although the results suggested that mixtures reduced disease incidence and disease severity slightly, differences from pure susceptible stands were not significant ($p > 0.05$), which agrees, in general terms, with the predictions of the simulation model (Fig. 15).

Spore dispersal gradients may also be useful for tracing sources of spores. The first discovery of *Pithomyces chartarum* (Berk. & Curt.) M. B. Ellis in Britain was made by working up a spore concentration gradient to a site where the fungus was flourishing on grass debris (Gregory and Lacey, 1964). Gradients have also been used to determine the relative importance of primary and secondary inoculum. Rowe and Powelson (1973) inoculated plots of winter wheat with point sources of *Pseudocercosporella herpotrichoides* and measured eyespot disease (as percentage of plants infected) gradients at intervals during the season. Initial disease gradients were steep, as would be expected for a splash-borne pathogen. Gradients subsequently became shallower, possibly because most potential infection sites near the inoculum had been colonized or because secondary rain splash had dispersed spores further from the source. After inoculum was removed from some plots little further disease developed in them, although more disease developed in plots still containing inoculum; these results suggested there was little secondary spread of eyespot.

Johnson and Powelson (1983) measured both spore dispersal gradients and disease gradients in their experiments on gray mold (*Botrytis cinerea* Pers. ex Fr.) of beans (*Phaseolus vulgaris* L.). Plots were inoculated with point sources of *B. cinerea* at blossom initiation. Spore dispersal gradients, measured as numbers of viable spores washed from bean foliage at 5-day to 6-day intervals, became shallower toward harvest as a result of secondary disease spread. However, gradients of pod rot at harvest were similar to those of gray mold incidence on blossoms at full bloom, suggesting that secondary disease spread after flowering did not influence pod rot development.

In small-plot experiments with airborne pathogens interplot interference can be significant and consideration of spore dispersal gradients can help in the interpretation of results. Paysour and Fry (1983) incorporated an exponential spore dispersal gradient model into their model to estimate the amounts of inoculum lost from plots and exchanged between neighboring plots. Their model predicted that the amount of interplot interference was a function of plot size, shape, spacing between plots, and steepness of the dispersal gradient. They used the model to identify plot sizes and spacings that would restrict interplot interference to acceptable levels and tested their predictions in experiments with potato late blight. More generally, the distance from a source in which spore concentration decreases to the background level can be recommended to farmers as the distance away from sources of inoculum (e.g., infected stubble) beyond which new susceptible crops can be planted without the previous cropping influencing disease risk.

Improvements in methods of sampling spores and advances in model construction should result in better spore dispersal models. However, generally applicable spore dispersal models, which can predict spore dispersal gradients from information about pathogen spore characteristics and weather conditions, have yet to be developed. Nevertheless, application of some of the more recent spore dispersal models should help to fill this important gap in our understanding of the life cycles of many pathogens and consequently improve the relevant epidemic models.

REFERENCES

Aylor, D. E. 1975. Force required to detach conidia of *Helminthosporium maydis*. *Plant Physiol*. 66:99–101.

———. 1982. Modeling spore dispersal in a barley crop. *Agr. Meteorol*. 26:215–219.

Aylor, D. E., McCartney, H. A., and Bainbridge, A. 1981. Deposition of particles liberated in gusts of wind. *J. Appl. Meteorol*. 20:1212–1221.

Aylor, D. E., and Taylor, G. S. 1982. Aerial dispersal and drying of *Peronospora tabacina* conidia in tobacco shade tents. *Proc. Nat. Acad. Sci. U.S.* 79:697–700.

Bainbridge, A., and Legg, B. J. 1976. Release of barley-mildew conidia from shaken leaves. *Trans. Br. Mycol. Soc.* 66:495–498.

Bainbridge, A., and Stedman, O. J. 1979. Dispersal of *Erysiphe graminis* and *Lycopodium clavatum* spores near to the source in a barley crop. *Ann. Appl. Biol.* 91:187–198.

Berger, R. D., and Luke, H. H. 1979. Spatial and temporal spread of oat crown rust, *Phytopathology* 69:1199–1201.

Brennan, R. M., Fitt, B. D. L., Taylor, G. S., and Colhoun, J. 1985a. Dispersal of *Septoria nodorum* pycnidiospores by simulated raindrops in still air. *Phytopathol. Z.* 112:281–290.

———. 1985b. Dispersal of *Septoria nodorum* pycnidiospores by simulated rain and wind, *Phytopathol. Z.* 112:291–297.

Byass, J. B. 1969. Laboratory techniques for simulating rainfall. *Chem. Ind.* 1969:1502–1504.

Chamberlain, A. C. 1975. The movement of particles in plant communities, in: *Vegetation and the Atmosphere* (J. L. Monteith, ed.), pp. 155–203. Academic Press, New York.

Cionco, R. M. 1972. A wind profile index for canopy flow. *Bound. Lay. Meteorol.* 3:255–263.

Faulkner, M. J., and Colhoun, J. 1977. An automatic spore trap for collecting pycnidiospores of *Leptosphaeria nodorum* and other fungi from air during rain and maintaining them in a viable condition. *Phytopathol. Z.* 89:50–59.

Fatemi, F., and Fitt, B. D. L. 1983. Dispersal of *Pseudocercosporella herpotrichoides* and *Pyrenopeziza brassicae* spores in splash droplets. *Plant Pathol.* 32:401–404.

Fitt, B. D. L., and Bainbridge, A. 1983a. Recovery of *Pseudocercosporella herpotrichoides* spores from rain-splash samples. *Phytopathol. Z.* 106:177–182.

———. 1983b. Dispersal of *Pseudocercosporella herpotrichoides* spores from infected wheat straw. *Phytopathol Z.* 106:214–225.

———. 1984. Effect of cellulose xanthate on splash dispersal of *Pseudocercosporella herpotrichoides* spores. *Trans. Brit. Mycol. Soc.* 82:570–571.

Fitt, B. D. L., Creighton, N. F., and Bainbridge, A. 1985. The role of wind and rain in the dispersal of *Botrytis fabae* conidia. *Trans. Brit. Mycol. Soc.* 85:307–312.

Fitt, B. D. L., and Lysandrou, M. 1984. Studies on mechanisms of splash dispersal of spores, using *Pseudocercosporella herpotrichoides* spores. *Phytopathol. Z.* 111:323–331.

Fitt, B. D. L., and Nijman, D. J. 1983. Quantitative studies on dispersal of *Pseudocercosporella herpotrichoides* spores from infected wheat straw by simulated rain. *Neth. J. Plant Pathol.* 89:198–202.

Fitt, B. D. L., Rawlinson, C. J., and Smith, C. B. 1982. A comparison of two rain-activated switches used with samplers for spores dispersed by rain. *Phytopathol. Z.* 105:39–44.

Fried, P. M., MacKenzie, D. R., and Nelson, R. R. 1979. Dispersal gradients from a point source of *Erysiphe graminis* f.sp. *tritici*, on Chancellor winter wheat and four multilines. *Phytopathol. Z.* 95:140–150.

Gregory, P. H. 1945. The dispersion of air-borne spores. *Trans. Brit. Mycol. Soc.* 28:26–72.

———. 1968. Interpreting plant disease dispersal gradients. *Annu. Rev. Phytopathol.* 6:189–212.

———. 1973. *The Microbiology of the Atmosphere*, 2d ed. Leonard Hill, London.

———. 1982. Disease gradients of wind-borne plant pathogens: Interpretation and misinterpretation, in: *Advancing Frontiers of Mycology and Plant Pathology* (K. S.

Bilgrami, R. S. Misra, and P. C. Misra, eds.), pp. 107–117. Today and Tomorrow's Publishers, New Delhi.

Gregory, P. H., and Lacey, M. E. 1964. The discovery of *Pithomyces chartarum* in Britain. *Trans. Brit. Mycol. Soc.* 47:25–30.

Gregory, P. H., Longhurst, T. J., and Sreeramulu, T. 1961. Dispersion and deposition of air borne *Lycopodium* and *Ganoderma* spores. *Ann. Appl. Biol.* 49:645–648.

Hall, C. D. 1975. The simulation of particle motion in the atmosphere by a numerical random-walk model. *Quart. J. R. Meteorol. Soc.* 101:235–244.

Hanna, S. R. 1982. Applications in air pollution modelling, in: *Atmospheric Turbulence and Air Pollution Modelling* (F. T. M. Nieuwstadt and H. van Dop, eds.), pp. 275–310. D. Reidel Publ. Co., Dordrecht, The Netherlands.

Hirst, J. M. 1953. Changes in atmospheric spore content: Diurnal periodicity and the effects of weather. *Trans. Brit. Mycol. Soc.* 36:375–393.

———. 1959. Spore liberation and dispersal, in: *Plant Pathology: Problems and Progress 1908–1958* (C. S. Holton, G. W. Fischer, R. W. Fulton, H. Hart, and S. E. A. McCallan, eds.), pp. 529–538. University of Wisconsin Press, Madison.

Hirst, J. M., and Stedman, O. J. 1963. Dry liberation of fungus spores by raindrops. *J. Gen. Microbiol.* 33:335–344.

Hirst, J. M., Storey, I. F., Ward, W. C., and Wilcox, H. J. 1955. The origin of apple scab epidemics in the Wisbech area in 1953 and 1954. *Plant Pathol.* 4:91–96.

Horst, T. W. 1977. A surface depletion model for deposition from a Gaussian plume. *Atmos. Environ.* 11:41–46.

Huang, C. H. 1979. A theory of dispersion in turbulent shear flow. *Atmos. Envir.* 13:453–463.

Itier, B., and Pauvert, P. 1979. Modélisation de transports horizontaux (spores, pollen). *EPPO Bull.* 9:251–264.

Jeger, M. J. 1983. Analysing epidemics in time and space. *Plant Pathol.* 32:5–11.

Jeger, M. J., Jones, D. G., and Griffiths, E. 1983. Disease spread of non-specialised fungal pathogens from inoculated point sources in intraspecific mixed stands of cereal cultivars. *Ann. Appl. Biol.* 102:237–244.

Jenkyn, J. F. 1974. A comparison of seasonal changes in deposition of spores of *Erysiphe graminis* on different trapping surfaces. *Ann. Appl. Biol.* 76:257–267.

Johnson, K. B., and Powelson, M. L. 1983. Analysis of spore dispersal gradients of *Botrytis cinerea* and gray mold disease gradients in snap beans. *Phytopathology* 73:741–746.

Kiyosawa, S., and Shiyomi, M. 1972. A theoretical evaluation of the effect of mixing resistant variety with susceptible variety for controlling plant diseases. *Ann. Phytopathol. Soc. Jap.* 38:41–51.

Legg, B. J. 1983. Movement of plant pathogens in the crop canopy. *Phil. Trans. R. Soc. Lond. B* 302:559–574.

Legg, B. J., and Powell, F. A. 1979. Spore dispersal in a barley crop: A mathematical model, *Agr. Meteorol.* 20:47–67.

Legg, B. J., and Price, R. I. 1980. The contribution of sedimentation to aerosol deposition to vegetation with a large leaf area index. *Atmos. Envir.* 14:305–309.

Llewelyn, R. P. 1983. An analytical model for the transport, dispersion and elimination of air pollutants emitted from a point source. *Atmos. Envir.* 17:249–256.

McCartney, H. A., and Bainbridge, A. 1984. Deposition gradients near to a point source in a barley crop, *Phytopathol. Z.* 109:219–236.

McCartney, H. A., and Fitt, B. D. L. 1985a. Construction of dispersal models, in: *Mathematical Modelling of Crop Disease* (C. A. Gilligan, ed.), *Adv. Plant Pathol.* 3:107–143.

———. 1985b. Spore dispersal gradients and disease development, in: *Populations of Plant Pathogens: Their Dynamics and Genetics* (M. S. Wolfe and C. E. Caten, eds.). Blackwell Scientific Publications, Oxford.

Mitchell, A. E. 1982. A comparison of short-term dispersion estimates resulting from various atmospheric stability classification methods. *Atmos. Envir.* 16:765–773.

Pasquill, F. 1962. *Atmospheric Diffusion*, 1st ed. Van Nostrand, London.

———. 1974. *Atmospheric Diffusion*, 2d ed. Ellis Harwood, Chichester.

Paysour, R. E., and Fry, W. E. 1983. Interplot interference: a model for planning field experiments with aerially disseminated pathogens. *Phytopathology* 73:1014–1020.

Ramalingam, A., and Rati, E. 1979. Role of water in the dispersal of nonwettable spores. *Ind. J. Bot.* 2:8–11.

Reddi, S. C. 1975. Lateral and vertical dispersion of air-borne *Lycopodium* and *Podaxis* spores released from an artificial point source. *J. Palyneol.* 11:111–119.

Rowe, R. C., and Powelson, R. L. 1973. Epidemiology of Cercosporella foot rot of wheat: Disease spread. *Phytopathology* 63:984–988.

Schmidt, R. A., Carey, W. A., and Hollis, C. A. 1982. Disease gradients of fusiform rust on oak seedlings exposed to a natural source of aeciospore inoculum. *Phytopathology* 72:1485–1489.

Sreeramulu, T. 1964. Incidence of conidia of *Erysiphe graminis* in the air over a mildew-infected barley field. *Trans. Brit. Mycol. Soc.* 47:31–38.

Stedman, O. J. 1979. Patterns of unobstructed splash dispersal. *Ann. Appl. Biol.* 91:271–285.

———. 1980a. Splash droplet and spore dispersal studies in field beans. *Agr. Meteorol.* 21:111–127.

———. 1980b. Splash dispersal studies in wheat using a fluorescent tracer. *Agr. Meteorol.* 21:195–203.

Stepanov, K. M. 1935. Dissemination of infective diseases of plants by air currents (in Russian). *Bull. Plant. Prot. Leningr. Ser.2, Phytopathol.* 8:1–68.

Sutton, J. C., Gillespie, T. J., and Hildebrand, P. D. 1984. Monitoring weather factors in relation to plant disease. *Plant Dis.* 68:78–84.

Sutton, O. G. 1953. *Micrometerology*. McGraw-Hill, New York.

Sutton, T. B., and Jones, A. L. 1976. Evaluation of four spore traps for monitoring discharge of ascospores of *Venturia inaequalis*. *Phytopathology* 66:453–456.

Taylor, G. I. 1915. Eddy motion in the atmosphere. *Phil. Trans. R. Soc. London A* 215:1–26.

Thompson, N. 1982. Meteorology and crop-spraying. *Meteorol. Mag.* 112:249–260.

Waggoner, P. E. 1974. Simulation of epidemics, in: *Epidemics of Plant Diseases: Mathematical Analysis and Modelling* (J. Kranz, ed.), pp. 137–160. Chapman and Hall, London.

———. 1983. The aerial dispersal of the pathogens of plant disease. *Phil. Trans. R. Soc. London B* 302:451–462.

Wale, S. J., and Colhoun, J. 1979. Further studies on aerial dispersal of *Leptosphaeria nodorum*. *Phytopathol. Z.* 94:185–189.

Chapter 14

Long Distance Transport of Spores

David E. Pedgley

Many pathogenic fungi release enormous numbers of spores into the air. They are minute, so their free-fall speeds (about 100 m/day) are small compared with updrafts often present in the air. Hence the persistence of these spores in the air is determined more by the updrafts than by fallout. The absence of updrafts on a calm night will allow most of the spores in a layer about 100 m deep to fall to the ground or onto vegetation. In the air just above the ground, the most frequent cause of updrafts is turbulence—irregular fluctuations of wind speed and direction caused by roughness and warmth of the ground. Roughness turbulence can be likened to a jumble of eddies in the form of whirls and waves, similar to those more easily seen in a river flowing over its rough bed, but revealed on a windy day when fallen leaves or snow are lifted into the air. Turbulence due to ground warmth is a form of convective overturning by somewhat more organized eddies in the form of buoyant, rising (and sinking) masses, columns and whirls, similar to those seen in a pan of water heated from below, but revealed in the air in an exaggerated form by the smoke from fires and explosions. Buoyancy develops when the vertical gradient of air temperature (lapse rate) exceeds 10° C/km (in cloud-free air), and the atmosphere is then said to be hydrostatically unstable because initially small upward or downward displacements tend to grow. (By contrast, such displacements tend to be suppressed into vertical oscillations when the atmosphere is stable, i.e., when the lapse rate is less than 10° C/km.) Both kinds of eddies, as they pass by, are expressed as the well-known gusts and lulls superimposed on the mean wind.

Turbulence causes mixing and tends to produce a uniform distribution of particles in the air. The turbulent layer of the atmosphere next to the earth's surface, into which spores are released and in which they spend much of their airborne life, is known as the mixing layer, or the planetary boundary layer. Its depth varies greatly, not only because of spatial variations in ground roughness but also because of temporal variations of ground temperature which lead to large variations of stability from day to night, from day to day, and from hot months to cold. A commonly measured depth in the afternoon is 1 km, and sometimes as much as 5 km, but at night in light winds the depth may be less than 0.1 km.

While airborne in the mixing layer, spores are carried by the wind, some of them for days. In a wind blowing at 20 km/h, such as that often present in the mixing layer, daily displacement is in the order of 500 km. Hence wind-borne spores can be expected to spread across a continent in a few days. The reality of such wind-borne spread is indicated by the capture of spores far from their sources: in the arctic, in deserts, and over the oceans. Of course, comparable spread takes place by other means, e.g., accidentally in planting material brought from one continent to another, or in aircraft; but here we are concerned with the wind as the carrying agent.

A CONCEPTUAL MODEL

It is useful to start with a conceptual model of long distance transport. An instantaneous source of spores can be thought of as producing a puff that drifts downwind and becomes progressively diluted as eddies distort it due to the entrainment of ever greater volumes of clean air. Most sources, however, are continuous, not instantaneous, although they vary in intensity. The result is a plume that may be thought of as a sequence of overlapping puffs. Successive portions of the plume trace out different paths, or trajectories, across country because some passing eddies are larger than plume width and they determine the plume orientation at the source. It is these eddies that cause meandering of the plume downwind, whereas the smaller eddies alter the internal structure of the plume. At any given point downwind of a source there is an intermittency of spore concentration reflecting not only plume meandering but also plume patchiness.

Spores in both puffs and plumes are progressively deposited as they drift downwind, with some of them able to reach susceptible hosts and infect them. Those wind-borne spores able to survive for hours or days in warm, dry air are often released actively when the relative humidity of the air is decreasing, such as happens on sunny mornings, or in sunny weather after rain. Wind gusts also release spores passively, either by temporarily removing the protective film of

viscously slowed air normally present on the substrate and enveloping the spor-
angia, or by inducing leaf flutter or rubbing. Within a crop, sufficiently strong
gusts are rare and come down from above, but they lead to some spore release
in winds whose average speed would otherwise be much too weak. Spores of
some species are knocked into the air by splashing raindrops or by water dripping
from leaves, the smaller spore-laden splash droplets then evaporating before
falling out.

Downwind spread of a plume can be likened to that of smoke, from which
we can get an idea of possible behavior of a spore cloud. The average wind will
stretch the plume while turbulence will broaden and deepen it; both will cause
dilution. At the same time, progressive deposition also causes dilution. Changes
in the average wind direction over time will lead to corresponding changes of
the stretching and meandering of the plume on scales of hours, and therefore of
hundreds of kilometers. These changes contrast with those due to turbulent
eddies, which typically have scales of less than a few kilometers, and whose
nature is poorly described on most occasions. Turbulent eddies are treated statistically, whereas the structure and evolution of larger eddies are revealed by
meterologists' maps of the wind field. These maps show the presence of large,
essentially horizontal circulations, up to several thousand kilometers across and
lasting typically from a few days to a week or two, and moving at speeds of 20
to 40 km/h. Because of their size and persistence they dominate the long distance
transport of spores. They are known as cyclones if winds circulate in the same
sense as that of the earth (counterclockwise in the northern hemisphere), and
anticyclones if in the opposite sense. Between adjacent wind circulations there
can be windshift lines, across which there is an abrupt change of wind direction,
and often also speed. Such windshift lines can also be accompanied by changes
in air temperature, when they are known as fronts. They are of considerable
significance in the long distance transport of spores, not only because there may
be accompanying rains and gusty winds that affect both take-off and landing but
also because the usually sudden change of wind direction leads to a corresponding
change in direction of spore sources. Fronts can therefore define the limit to
which a given airstream may transport its spore load.

Intermediate scale eddies, with dimensions of tens of kilometers, are often
poorly represented on weather maps; hence their effects on plume meandering
are not well known. These latter eddies can be particularly important near mountains, coasts, and convective rain storms. They are most dominant in tropical
latitudes, where the larger-scale eddies are more likely to be weaker or less well-
defined than in higher latitudes.

Downwind spread can be estimated by the use of trajectories. A trajectory
is the calculated track of a windborne particle, and is comprised of a sequence
of joined sections each representing the movement over a fixed time step. Each
section of the trajectory is based on estimates of wind speeds and directions
along the trajectory during that time step that is derived from a map representing
the wind field at the time of midstep. Because the wind field is continually

changing, trajectory accuracy increases with decrease in time step. Wind field maps may be based on wind measurements either measured at fixed times (usually hourly near the ground and smoothed to remove the effects of gusts, but only two or four times daily aloft) or estimated from pressure fields using models based on the theory of fluid dynamics. Time steps shorter than the map interval imply the use of a time-interpolation technique. Also, wind and pressure fields are derived from measurements at a finite number of points spaced tens and sometimes hundreds of kilometers apart, in an irregular network, and they imply the use of a space-interpolation technique. Both techniques introduce errors because they can take little account of eddies smaller than the spacing in the observing network.

Trajectories can be hand-drawn subjectively, or machine-drawn objectively; the former can be tedious but not necessarily if only rough estimates are needed. Trajectory accuracy increases with decreasing separation of observing points or of the grid points in the dynamic model. Trajectory accuracy has been tested by comparison of calculation with observations from the tracking of tracers such as zero-buoyancy balloons. As might be expected, errors increase with duration, and with poorly defined wind fields (Clarke et al., 1983). Because of variations of wind speed and direction with height, further errors can arise from the wrong choice of height. Comparison of trajectories for several heights with observed movements of tracers, however, can suggest heights and times at which movement might have taken place. Pedgley (1982) discusses the various scales of eddies in the atmosphere and their influence on the spread of wind-borne organisms.

Horizontal dilution can be expected to be continuous and lead to plume broadening downwind; typical values are 5 km width at 10, 20 km at 50 km, 30 km at 100 km, and 200 km at 1000 km (Smith, 1979; Smith, 1983). In contrast, vertical dilution is often limited to the depth of the mixing layer, with the result that a spore plume can be expected to have a more-or-less clearly defined top, like the top of a haze layer seen from an aircraft on climb and descent. With such a restricted top, dilution becomes two-dimensional. If the source is very large, as it may be from a crop grown over an area with dimensions hundreds of kilometers, horizontal spread by small turbulent eddies is so slow that dilution effectively ceases. Near a source, where few spores have yet been carried to the top of the mixing layer, a steep vertical concentration gradient will be present. In contrast, far from a source, mixing will have produced a nearly uniform vertical concentration profile, although deposition then leads to a reversed gradient close to the ground, where there is an increase in concentration with height. A variation of wind direction with height in the mixing layer causes a variation of downwind stretching with height. Deposition will dilute the plume: it is unimportant near sources, but its effect increases downwind and after a time becomes comparable to the effect of diffusion by turbulent eddies.

Plume structure is made more complex by intermittency of sources (e.g., spores may be released only when the relative humidity of the air is decreasing), or by a multiplicity of small sources whose individual plumes merge downwind. As with a forest fire, the plume from a spatially and temporally variable source can be expected to have a patchy structure. The plume from a single small source will become insignificantly dilute at a much shorter range than the plume from a large area comprising many individual sources of various strengths. A source persisting for days or weeks gives a meandering and perhaps intermittent plume that streams away in a continuously varying direction due to the passage of a sequence of large eddies (the weather systems shown on meteorologists' maps) across the source.

Wind-borne spores are deposited in two main ways: by impaction on vegetation and by washout in rain. Impaction occurs when the wind approaches an obstacle and passes on either side, but the spores deviate less because of their inertia. It is of greatest importance for large spores and strong winds (both leading to greater inertia), and for small obstacles (leading to small deviation of the wind). For a given spore size and wind speed, impaction efficiency increases from leaf to stem to petiole, but the total number of spores impacting increases in the opposite sense due to the greater area of the leaves. Within a crop, impaction leads to very steep deposition gradients; hence for a plume starting in a crop only the part that is carried above the crop top within about 10 m downwind of a source has a chance of becoming wind-borne over long distances. The ratio of deposition rate (spores per square meter per second) to plume concentration (spores per cubic meter) has the dimensions of a velocity (m/s), and is called the deposition velocity, v_d. For wind-borne spores it is of the order of 1 cm/s, i.e., much faster than typical free-fall speeds. Washout by falling raindrops often causes more rapid deposition than does impaction, especially from dilute plumes, as is demonstrated by spores being found in rainwater before they are caught in traps and before the first infections appear (Rowell and Romig, 1966). Such an occurrence, however, may be due to the presence of a denser plume aloft, as may happen when there is a variation of wind direction with height.

The accumulated deposition from a complex plume, by fallout, impaction, and washout, leads to a complex pattern of inoculum distribution, which may or may not be reflected in disease distribution according to the horizontal patterns of germination and incubation. A disease gradient develops outward from the source, reflecting both dilution with the plume and meandering of the plume as it intermittently crosses any given point. Moreover, the plumes from many different sources will cross a given point from time to time (Fig. 1). Both aerial concentration and density of spore deposition on vegetation decrease approximately exponentially away from a source. Hence, the greater the range at which the deposit will be dense enough to initiate disease with a given probability in a given area, the larger the source must be (either a very intense point source or a large but weaker area source). Fungal species that produce only scattered small sources are therefore unlikely to depend on wind-borne spores for long distance spread because they cannot produce a combined source that is strong

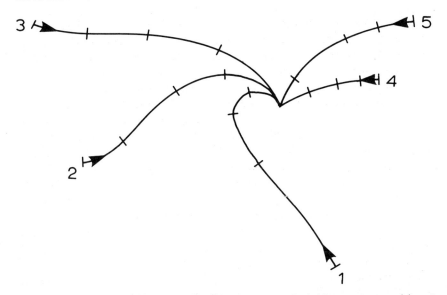

Figure 1 Variation of sources and pattern of movement of wind-borne spores arriving at a given point. The five lines are backtracks of air arriving on five successive occasions. Ticks show positions of the air at equal time intervals. Variations of track shape and tick spacing are due to time changes of the wind field (both speed and direction). Time variations of spore type and aerial concentration at the point will vary in part according to the nature of spore sources over which the backtracks passed.

enough to cause infection at a great distance. A steep spore gradient does not necessarily imply rapid deposition and therefore little long distance spread—it can also imply rapid dilution and hence slower deposition (i.e., a large proportion of the spore cloud is available for long distance spread) (Gregory, 1981).

For problems of controlling spread of disease, or of designing experiments to measure spread, we need to be able to estimate the number of spores likely to reach a given area in a given time. So far, there has been little quantitative work to make such estimates, but we can draw on experience with the numerous studies that have been made on the spread of atmospheric particulates, both natural and man-made. Before coming to that, however, it will be worthwhile to consider some examples of inferred long distance spread of fungal spores and other microorganisms.

EXAMPLES OF LONG DISTANCE TRANSPORT BY SPORES AND OTHER MICROORGANISMS

Several species of rust provide four clear examples. The evidence for long distance spread is both direct and circumstantial. Spores of recognizable races appearing far from known sources provide direct evidence, but circumstantial evidence is more varied. For example, the timing and extent of spread may be consistent with the occurrence of weather systems giving spells of wind blowing from known or likely sources. Such spread may be recognized by inference from subsequent appearance of disease, or by direct trapping of spores. Trapping may be difficult because plume dilution requires large volumes to be sampled, or because the few spores trapped may be masked by numerous other microorganisms, notably pollen grains. Even so, it is sometimes possible to track plumes foward from a source, or backward from an arrival point, using maps of the wind field based either on observations or on fluid dynamic calculations from observed pressure fields (Sykes and Hatton, 1976), but such calculations are still subject to large errors (Clarke et al., 1983). Above the earth's surface, wind observations are often available only two or four times daily, whereas dynamically derived winds can be calculated much more frequently. Plume tracks are in reality three-dimensional because the air meanders in both horizontal and vertical planes, but in practice the vertical component can often be ignored.

These methods were first convincingly applied to spread in North America of *Puccinia graminis* f. sp. *tritici,* the cause of stem rust of wheat, in the 1920s and 1930s (Stakman and Christensen, 1946). The uredial stage of this rust does not survive the winter north of Texas, nor does it survive the summer in Texas. Spread of disease northward in spring and early summer, and then southward in late summer and fall allows continuity. By use of spore trapping, field records of disease, and geographic distribution of races, as well as weather maps, it was possible to provide evidence of wind-borne spread for the first time in 1923, when many barberries (on which the aecial stage develops and provides an alternative source of disease) had been removed. Five successive waves could be recognized from early May to early June, taking spores from Texas to the Canadian border. Similar spread was mapped in later years, although it was not always as clear cut.

In a biogeographic analysis of the spread in Africa of *Puccinia polysora,* the cause of maize rust, in the 1950s (Rainey, 1973), the role of individual weather systems was not examined because of the lack of sufficiently detailed field records of the disease, but it was clear that spread was related to seasonal winds. The fungus seems to have been introduced accidentally into Sierra Leone, where it appeared suddenly in 1949. It then spread eastward, crossing the whole of West Africa by 1951, reaching East Africa by 1952 and Madagascar by 1953 (Fig. 2). The evidence is circumstantial, but spread seems to have been on the wind during the maize growing season, when monsoon southwest winds were blowing over West Africa, and when northeast winds were blowing over East Africa.

The dramatic spread of coffee leaf rust, caused by *Hemileia vastatrix,* in South America during the 1970s provides another example. Disease was discovered in Bahia state of Brazil in January 1970, and it later extended southwestward, reaching São Paulo state by January 1971, and Paraguay and northern Argentina by 1974 (Schieber, 1972; Waller, 1979). This spread was in the direction of the dominant winds, and wind-borne spores were trapped by aircraft up to 1000 m above the state of Paraná before disease was reported there (Schieber, 1975). Such wind-borne spread had already been suggested by Rayner

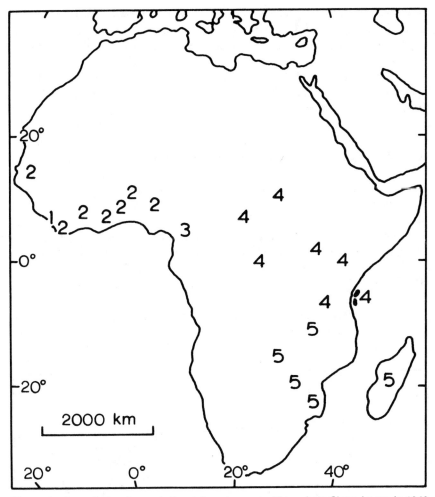

Figure 2 Spread of maize rust, *Puccinia polysora,* in Africa: from Sierra Leone in 1949 (year 1) to Mozambique in 1953 (year 5). (*Reproduced by permission of R. C. Rainey from Weather 28:224–239.*)

(1960), who predicted that uredospores might even be carried from Africa across the Atlantic Ocean to the western hemisphere. Whether the disease reached Brazil that way is still unknown, but Bowden et al. (1971) have provided circumstantial evidence to support the suggestion.

A spread of two species of poplar rust, *Melampsora,* from Australia to New Zealand can be associated with known individual weather systems (Close et al., 1978). Disease was first seen in March 1973 at two places 450 km apart, suggesting an outside source. Field evidence indicated that uredospores were introduced in late February or early March, following widespread disease in New South Wales by February. Winds then were mostly easterly over the Tasman

Sea, but there were west winds from 1 to 3 March. Forward tracks from New South Wales starting on 1 March reached New Zealand in 2 to 3 days, providing strong circumstantial evidence that spores crossed more than 3000 km of open sea (Fig. 3).

Turning now to other species of fungal plant pathogens, the northward spread in North America of southern corn leaf blight, caused by *Cochliobolus heterostrophus*, first became clear in 1970 following the widespread planting of corn varieties susceptible to race T of the fungus, which had been present for years but had not been of major economic importance. Starting in Florida, disease reached the Canadian border by mid-August (Moore, 1970; Wallin, 1970; Fig. 4). A decade later, the northward spread, across the United States, of tobacco blue mold, caused by *Peronospora tabacina*, was studied quantitatively using tracks based on weather maps and a mixing layer of variable depth (Aylor et al., 1982). It was found that spread was not continuous but took place mostly on a few days, and even on a few hours in those days, under the influence of recognizable individual weather systems. The vertical structure of similar clouds of spores was sampled to a height of 1.8 km by aircraft in 1962 over the North Sea, downwind of sources in England (Hirst et al., 1967). Variation in composition could be attributed to changes in sources from day to night.

Among other wind-borne microorganisms, bacteria are smaller (with diameters of a few micrometers) than spores and more difficult to identify but

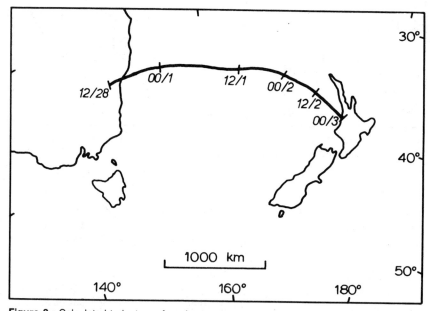

Figure 3 Calculated trajectory of poplar rust spores, *Melampsora*, from Australia to New Zealand, from 12 noon, 28 February to 3 March 1973, assuming drift on winds at a height of 3000 m. (*Adapted from Close et al., 1978.*)

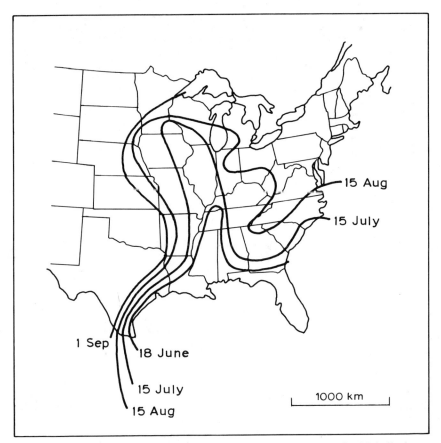

Figure 4 Northward spread of southern corn leaf blight, caused by *Cochliobolus heterostrophus*, across the United States in 1970, represented by approximate isochrones. (*Adapted from Moore, 1970.*)

terrestrial forms have been found in marine air hundreds of kilometers from land (Rittenberg, 1939). A well-documented apperance in Sweden on 18 February 1969 of bacteria with windblown dust deposited with snow was associated with a weather system giving southeast winds blowing from southern Russia, where there had been recent duststorms (Bovallius et al., 1978) (Fig. 5). Accompanying pollen grains were mostly those characteristic of the deduced source area (Lundqvist and Bengtsson, 1970). Blown dust has been shown to carry soil algae into the air (Brown et al., 1964), and spores of some fungi are also carried on windborne soil particles. For example, *Fusarium moniliforme*, a cause of maize rot, was found in blown soil lying on snow in Minnesota 300 to 400 km downwind of the likely source, where little snow was lying (Ooka and Kommedahl, 1977). Arizona desert dust contains a variety of plant components including fungal spores, among which are those of the mold *Coccidioides immitis*, the cause of

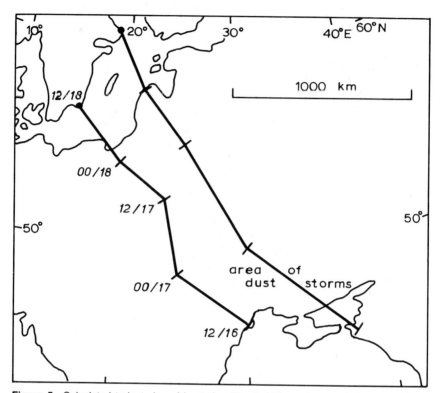

Figure 5 Calculated trajectories of bacteria with wind-blown dust carried from southern Russia to two points in Sweden, 16 to 18 February 1969, assuming drift on winds at a height of 1500 m. (*Adapted from Bovallius et al., 1978.*)

valley fever in human beings and derived from disturbed topsoil (Leathers, 1981). It will therefore be useful to consider the spread of wind-borne dust (and haze, more generally) not only in its role as a potential carrier of spores but also for the light it throws on the long distance spread and dilution of plumes of particles from large sources, because such plumes are visible and therefore more easily tracked than plumes of spores.

WIND-BORNE SPREAD OF HAZE

Meteorologists define haze as visibility less than 10 km due to dry particles in the air. Most of the particles are produced either mechanically (as mineral dust, mainly from soil but some from volcanoes, and with sizes comparable to fungal spores and pollen grains) or chemically (as smoke and fume, mainly from industrial activity but some from natural fires, and with sizes comparable to bacteria or smaller). Long-distance spread of desert dust has been well studied, mainly with satellites and with weather maps showing fields of visibility, and supported by identification of constituent particles (e.g., color, size, or mineral content).

McCauley et al. (1981) provide a case study from February 1977, when dust from the southwestern United States reached well beyond the Atlantic coast in three days. An example of red dust washed out in a trace of rain at Cincinnati, Ohio, in 1965 could be tracked back to a duststorm in western Texas and southeastern New Mexico on the previous day (Cohen and Pinkerton, 1966). (Incidentally, this example provided proof of comparably long distance spread of pesticides.) A similar fall, widespread over England and Wales in 1968, was backtracked to the southern side of the Sahara in West Africa, using winds at heights of 3 to 5 km, where aircraft had reported a dust layer (Stevenson, 1969). In such cases, deposition is made obvious by its spottiness due to individual raindrops; more dilute falls are probably common in heavier rain. The spread of Sahara dust westward across the North Atlantic was extensively studied in the 1970s (Carlson and Prospero, 1972; Schütz, 1980). It was found that transport was continuous, in a plume of up to 5 km deep, taking about a week to reach the West Indies (compare the possible spread of coffee leaf rust from southern Africa to Brazil), but with varying plume concentration reflecting the influence of a sequence of weather systems over the source. The plume was layered due to variation of wind direction with height; in particular, cool, clean, moist northeast winds near the ocean surface undercut the dusty east winds aloft, although fallout (and mixing) took place between the layers and on to the ocean surface (Fig. 6). It is very likely that similar effects occur in spore clouds, and not only on such large, intercontinental scales.

Comparable eastward spread of dust haze from the deserts of China has been tracked to North America and the Arctic. In April 1969, satellites and weather maps clearly showed that dust reached Japan (Ing, 1972), and similarly in April 1979, when layering was demonstrated by use of lidar (laser radar) (Iwasaka et al., 1983). The timing of backtracks was consistent with observed spread, thereby providing a demonstration of the value of backtracking to known sources for invisible plumes of particles such as spores. Layers of this Chinese dust were seen to reach Hawaii in about 9 days (Shaw, 1980; Darzi and Winchester, 1982), and Alaska in about a week (Rahn et al., 1977), both arrivals being confirmed by sampling and by backtracking using wind fields at appropriate heights. The occurrence of such layers above the local mixing layer is due to large-scale weak but persistent updrafts associated with particular kinds of weather systems, or to sources on high mountains, or to local deep convection that takes air from the mixing layer and spreads it aloft. Such effects acting on spore clouds

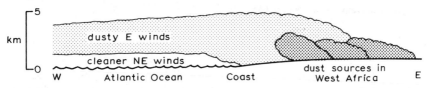

Figure 6 Effect of change of wind direction with height on the plume of Saharan dust crossing the Atlantic.

very likely account for the spores that have been caught by aircraft at heights of several kilometers.

Chemical haze consists mostly of minute sulfate particles (diameters less than about 2 μm) derived by oxidation of sulfur dioxide from industrial and domestic sources. There may also be some soot from partial combustion of hydrocarbon fuels. As with desert dust, the continental spread of chemical haze has been studied from the 1970s with satellites (Mohr, 1971; Lyons, 1980) and visibility maps, with allowance for other causes of poor visibility (e.g., rain and fog) (Hall et al., 1973; Samson and Ragland, 1977; Husar and Patterson, 1980), and on a somewhat smaller scale by aircraft (White et al., 1976; Komp and Auer, 1978). Arctic haze layers have been backtracked to Europe and North America (Rahn and McCaffrey, 1980), and such origins were confirmed by their rich content of vanadium (Barrie et al., 1981). Deposition of chemical haze in rain was first demonstrated in 1965 to be highly episodic rather than more-or-less evenly distributed in time (Brosset, 1976). It is associated with rain in individual weather systems washing out particles simultaneously over areas hundreds of kilometers across. This deposition follows the feeding of hazy air into and beneath the rain clouds, from sources up to 1000 km or more away. Two main types of weather system can be recognized: slow-moving rain areas into which even only moderately polluted air may be drawn for several days; or more mobile rain areas into which is drawn air that is highly polluted because it had been stagnating in a slow-moving anticyclone over persistent sources. Since fungal spores are known to be deposited in rain, it is likely that their deposition from plumes is also highly episodic (Fig. 7).

Invisible gases in the atmosphere, both natural (e.g., radon, from the ground) and generated by human activity (e.g., ozone, sulfur dioxide, and nitrogen oxides in chemical hazes; various organic vapors such as pesticides and aerosol propellants; and motion tracers such as SF_6 and CD_4, introduced experimentally), provide additional supporting evidence for windborne displacements over thousands of kilometers. They are detected by networks of sampling stations on the ground and by aircraft. As with haze, the observations reveal not only the contrast between the polluted mixing layer below and the cleaner air above, but also the presence of layering. Studies of the night-to-day variation in concentration of ozone show the importance of the increasing depth of the mixing layer by day in bringing down night-time layers from aloft, following depletion near the ground (a process known as fumigation). Sudden arrivals of increased concentrations of gaseous pollutants have been backtracked to known sources, both in North American (Lyons and Cole, 1976; Chung, 1977; Samson and Ragland, 1977; Reisinger and Crawford, 1982) and in Europe (Cox et al., 1975; Cox, 1977), and even over periods of one or two weeks into the winter Arctic, where transformation of sulfur dioxide to sulfate is slow (Rahn et al., 1980).

In all these studies, whether of chemical haze or of trace gases, concentrations tended to be greatest on occasions when the mixing layer was shallow. This was usually in association with slow-moving anticyclones persisting for a

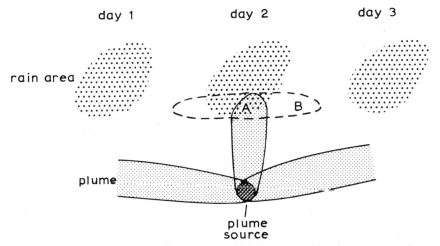

Figure 7 Diagram of development of an area of washout as the direction of a plume from a persistent source changes from day 1 to day 3, resulting in a temporary flow of the plume on day 2 into a rain area moving eastward. Positions of the plume and rain area are shown at the same time on each day. At that time on day 2 the area of washout is shown as A; B is the total area affected by washout due to movement of both plume and rain area.

few days. Such systems are associated with warm, dry weather, and are therefore likely to trigger the release of the kinds of fungal spores we have been considering. Northward spread (such as has been mentioned for stem rust of wheat, southern corn leaf blight, and tobacco blue mold, as well as the bacteria-carrying dust falling with snow in Sweden), would be in the southerly winds on the western sides of anticyclones (in the northern hemisphere), and may be particularly important on cloudy days due to the likely increased survival rate. As these southerly winds spread to higher latitudes they are often lifted, so that any ensuing rain would deposit the wind-borne spores. Whatever the mechanisms of spread and deposition may be, it is clear that there is a dominant role of weather systems; the same is true in relation to the readiness of spores to be released. Such a role must be recognized in modeling the long distance displacement of spores.

MODELING THE WIND-BORNE TRANSPORT OF SPORES

The previous sections provide us with a detailed but descriptive, conceptual model of what very likely happens to a plume of spores that is carried downwind. It would be of great practical value if we could quantify the model in order to be able to calculate the instantaneous plume concentration, and hence also the accumulated deposition, or dose, at any point within the whole area where disease is likely to reach. In principle this can be done if we know not only the geographic distribution and strengths of spore sources and their variations with time, but also the theory of dilution of plumes as they are carried downwind.

Let the accumulated deposition at a given place be S spores per unit area. This can be expressed as the sum of dry deposition by turbulence and of wet deposition by rain:

$$S = \int_0^{T_1 - T_2} C(t) v_d(t) dt + \int_0^{T_2} C(t) v_w(t) dt \qquad (1)$$

where $C(t)$ is the aerial concentration of viable spores (which varies with the previous history of deposition by, addition to, dilution of, and death in that part of the plume passing over at any given time); $v_d(t)$ is the dry deposition velocity (which varies with turbulence structure, and therefore particularly with atmospheric stability); and $v_w(t)$ is the wet deposition velocity (which varies with the raindrop size spectrum). All three vary with time. T_1 is the period when the plume is overhead; and T_2 is that part of T_1 when it is also raining, assuming that dry deposition is negligible during T_2. Deposition by settling (sedimentation) is ignored here; it is frequently small compared with dry deposition due to turbulence.

Of the factors determining S, concentration changes due to turbulence dilution can be estimated from the properties of the turbulence by using the diffusion equation for a Gaussian puff:

$$C = \frac{Q}{(2\pi)^{3/2} \sigma_x \sigma_y \sigma_z} \cdot \exp \frac{-x^2}{2\sigma_x^2} \cdot \exp \frac{-y^2}{2\sigma_y^2} \cdot \exp \frac{-z^2}{2\sigma_z^2} \qquad (2)$$

where Q is the source strength (spores per unit time); x, y, and z are distances measured in the forward, leftward, and upward directions from the puff center (i.e., following the puff); and σ_x, σ_y, and σ_z are standard deviations of the puff concentrations in those directions, and whose relationship to turbulence (expressed as the standard deviations of wind direction changes in the horizontal, σ_θ, and the vertical, σ_ϕ) is still not well understood but can be measured experimentally. Equation 2 applies only to an instantaneous, point source on the ground and in a steady-state, homogeneous atmosphere (such as might be found over open, uniform country) with neutral stability. Moreover, it gives a time-meaned concentration that can bear little resemblance to observed concentration, and its applicability at downwind distances, large compared with the depth of the mixing layer, is still unclear.

In practice, we are more often concerned with a continuous, if varying, source leading to a plume; with a stable or unstable atmosphere; and with distance downwind so large that mixing in the vertical may be considered complete. Only lateral diffusion is then important, and the first and last terms on the right of Eq. 2 can be ignored. For an area source, with dimensions much larger than the diffusing eddies, the plume is so large at source that even lateral growth can be ignored and concentration becomes simply:

$$C = \frac{Q}{lua}$$

where l and a are the width and depth of the plume, and u is the wind speed.

The diffusion equation describes the effects of dilution due to the turbulence in a plume from a single source. To take account also of dilution due to deposition, of meandering, and of a multiplicity of sources, it is clear that analytic solutions of Eq. 1 become impossible unless considerable simplifications are made. For example, Aylor et al. (1982) have used a simple model of this kind to calculate the dose (spores per square meter) on tobacco leaves by blue mold spores windborne over several hundred kilometers. Furthermore, the method used by Gloster (1982) to calculate the fraction R of a plume of foot-and-mouth virus particles remaining airborne at distance x from a source may be applied to a plume of spores. He assumed a simple exponential expression

$$R = \exp\left(-\frac{v_d}{h}\frac{x}{u}\right)$$

where v_d = deposition velocity; h = depth of plume (assumed uniformly mixed in the vertical); and u = wind speed. With v_d = 1 cm/sec, h = 500 m, u = 10 m/sec, an e-fold decrease in R will take place in 500 km.

A more general approach is to consider the time and space variations of all sources and sinks, as well as of the wind field. For this, we must turn to regional numerical models, in which sources and sinks are represented more or less precisely on a grid: sources from field observations of disease or of aerial spore concentration; and sinks from estimates of dry and wet deposition rates based on measured ground roughness, atmospheric stability, and rainfall intensity. So far, such models have not been used, no doubt largely because the necessary field records are not gathered routinely. However, regional numerical models describing the spread of atmospheric pollutants are potentially applicable. Most of these models are kinematic; they are based on observed sources and sinks, as well as measured wind fields, and they assume a uniform concentration profile in the mixing layer. The models are used to calculate aerial concentration and dose at regular intervals, daily or even hourly, either at a grid of fixed points or following the air along a selection of tracks, where pollutants are continuously being taken up from the various sources over which the air passes, and also continuously being deposited.

Models can be used to calculate deposition both during particular episodic events and accumulated over long periods (say weeks or months). Such models were first used in the early 1960s, but they were confined then to short ranges, where deposition and meandering could be ignored, and where the passage of weather systems could be allowed for by the continuous variation in plume orientation at source. Long distance, continental scale models were introduced in the 1970s. Pollutant displacement is dominated in these models by weather systems, as is the spread of heat and moisture in dynamic models of the atmosphere itself. These models need to take account of deposition wherever the average residence time of pollutants in the atmosphere is comparable with transport time. The average residence time is less in middle latitudes (especially in winter, when it may be only a few days), and greatest in the subtropics (where rain is infrequent). The models also need to take account of plume meandering as well as day-to-night variations in source strength, mixing depth, and wind field. Despite advances in the complexity of models, it is still not possible to estimate dilution, vertical or horizontal, with any precision, both in dry weather (where dilution varies with flux rate to the ground and the efficiency of capture there) and in rain (where dilution varies with the distribution and intensity of rain). Pollutant sources are well mapped in North American and Europe, and seasonal variations in the proportions from domestic and industrial sources can be taken into account. Among improvements still needed in these models are: mapping of the wind field on scales of tens and hundreds of kilometers, especially in hilly regions; estimating vertical diffusion, and weighting the observed winds to produce an average over the depth of the mixing layer; developing a theory of concentration fluctuations at any given point, from both point and area sources. More extensive introductions to these models and their operational application are given, for example, by Fisher (1983) and by Pasquill and Smith (1983).

CONCLUSIONS

There is clearly great potential for adapting and applying these pollution models to the spread of spore plumes, but various parts of the sequence—take-off, displacement, dispersion, survival, and deposition—are poorly quantified. Moreover, each species will have its own peculiarities. Maps of spore sources need to be updated sufficiently frequently that they reveal significant time variations. This implies a system of gathering, reporting and collating field records of spore concentration and dose, and of disease development. Such records should also be used to test the results of model computations and to suggest improvements to the model. Effects of gustiness on take-off from various crops needs to be quantified. Estimates are also needed of survival rate and washout efficiency. Case studies of deposition on the scale of individual weather systems would enable a climatology of such systems to be built up that could be of value in forecasting deposition episodes. Effects on dilution due to eddy diffusion are likely to be simulated best by models that assume that an air parcel, with its load of spores, moves along its track in steps, each step being dependent partly on the previous step and partly on a random deviation (Gifford, 1982). More emphasis is needed on the effects on diffusion by medium-scale weather systems (dimensions of the order of 10s of kilometers).

Apart from the long-term aim of modeling quantitatively the spread of spores, with implications for monitoring, forecasting, and controling the spread of disease, there is also a need for qualitative studies to determine to what extent long-distance wind-borne spread in fact takes place with a given species. Case studies of particular events, such as a sudden arrival of spores (recorded in traps or rain water, or inferred from subsequent disease appearance and known development rate) provide circumstantial evidence of wind-borne spread, particularly if winds at the time were blowing from known sources.

Only when these and many more aspects of long-distance transport of spores are better understood it is likely that quantitative mathematical models will lead to the improved understanding of epidemiology that is needed to control the spread of wind-borne pathogenic fungi.

REFERENCES

Aylor, D. E., Taylor, G. S., and Raynor, G. S. 1982. Long-range transport of tobacco blue mold spores. *Agr. Meteorol.* 27:217–232.

Barrie, L. A., Hoff, R. M., AND Daggupaty, S. M. 1981. The influence of mid-latitudinal pollution sources on haze in the Canadian arctic. *Atmos. Environ.* 15:1407–1419.

Bovallius, A., Bucht, B., Roffey, R., and Anas, P. 1978. Long-range air transmission of bacteria. *Appl. Environ. Microbiol.* 35:1231–1232.

Bowden, J., Gregory, P. H., and Johnson, C. G. 1971. Possible wind transport of coffee-leaf rust across the Atlantic Ocean. *Nature London* 229:500–501.

Brosset, C. 1976. Air-borne particles: Black and white episodes. *Ambio* 5:157–163.

Brown, R. M., Larson, D. A., and Bold, H. C. 1964. Airborne algae: Their abundance and heterogeneity. *Science* 143:583–585.

Carlson, T. N., and Prospero, J. M. 1972. The large-scale movement of Saharan air outbreaks over the northern equatorial Atlantic. *J. Appl. Meteorol.* 11:283–297.

Chung, Y-S. 1977. Ground-level ozone and regional transport of air pollutants. *J. Appl. Meteorol.* 16:1127–1136.

Clarke, J. F., Clark, T. L., Ching, J. K. S., Haagenson, P. L., Husar, R. B., and Patterson, D. E. 1983. Assessment of model simulation of long-distance transport. *Atmos. Environ.* 17:2449–2462.

Close, R. C., Moar, N. T., Tomlinson, A. I., and Lowe, A. D. 1978. Aerial dispersal of biological material from Australia to New Zealand. *Int. J. Biomet.* 22:1–19.

Cohen, J. M., and Pinkerton, C. 1966. Windspread translocation of pesticides by air transport and rain-out. *Adv. Chem. Ser.* 60:163–176.

Cox, R. A. 1977. Some measurements of ground level NO, NO_2 and O_3 concentrations at an unpolluted maritime site. *Tellus* 29:356–362.

Cox, R. A., Eggleton, A. E. J., Derwent, R. G., Lovelock, J. E., and Pack, D. H. 1975. Long-range transport of photochemical ozone in north-western Europe. *Nature London* 255:118–121.

Darzi, M., and Winchester, J. W. 1982. Aerosol characteristics at Mauna Loa Observatory, Hawaii, after East Asian dust storm episodes. *J. Geophys. Res.* 87:1251–1258.

Fisher, B. E. A. 1983. A review of the processes and models of long-range transport of air pollutants. *Atmos. Environ.* 17:1865–1880.

Gifford, F. A. 1982. Horizontal diffusion in the atmosphere: a Lagrangian dynamical theory. *Atmos. Environ.* 16:505–512.

Gloster, J. 1982. Risk of airborne spread of foot-and-mouth disease from the Continent to England. *Vet. Record* 111:290–295.

Gregory, P. H. 1981. Disease gradients of windborne plant pathogens: Interpretation and misinterpretation, in: *Advancing Frontier of Mycology and Plant Pathology* (K. S. Bilgrami, R. S. Misra, and P. C. Misra, eds.). pp. 107–117, Today and Tomorrow Printers and Publishers, New Delhi.

Hall, F. P., Duchon, C. E., Lee, L. G., and Hagan, R. R. 1973. Long-range transport of air pollution: A case study, August 1970. *Mon. Weather Rev.* 101:404–411.

Hirst, J. M., Stedman, O. E., and Hurst, G. W. 1967. Long-distance spore transport: vertical sections of spore clouds over the sea. *J. Gen. Microbiol.* 48:357–377.

Husar, R. B., and Patterson, D. E. 1980. Regional scale air pollution: sources and effects. *Ann. N. Y. Acad. Sci.* 338:339–417.

Ing, G. K. T. 1972. A duststorm over central China, April 1969. *Weather* 27:136–145.

Iwasaka, Y., Minoura, H., and Nagaya, K. 1983. The transport and spatial scale of Asian dust-storm clouds: A case study of the dust-storm event of April 1979. *Tellus* 358:189–196.

Komp, M. J., and Auer, A. H. 1978. Visibility reduction and accompanying aerosol evolution downwind of St. Louis. *J. Appl. Meteorol.* 17:1357–1367.

Leathers, C. R. 1981. Plant components of desert dust in Arizona and their significance for man. *Geol. Soc. Am. Special Paper* 186, pp. 191–206.

Lundqvist, J., and Bengtsson, K. 1970. The red snow: A meteorological and pollen-analytical study of longtransported material from snowfalls in Sweden, *Geol. Foeren. Stockholm Foerh.* 92:288–301.

Lyons, W. A. 1980. Evidence of transport of hazy air masses from satellite imagery. *Ann. N. Y. Acad. Sci.* 338:418–433.

Lyons, W. A., and Cole, H. S. 1976. Photochemical oxidant transport: Mesoscale lake breeze and synoptic-scale aspects. *J. Appl. Meteorol.* 15:733–743.

McCauley, J. F., Breed, C. S., Grolier, M. J., and MacKinnon, D. J. 1981. The U. S. dust storm of February 1977. *Geol. Soc. Am. Special Paper* 186, pp. 37–70.

Mohr, T. 1971. Air pollution photographed by satellite. *Mon. Weather Rev.* 99:653.

Moore, W. F. 1970. Origin and spread of southern corn leaf blight in 1970. *Plant Dis. Rep.* 54:1104–1108.

Ooka, J. J., and Kommendahl, T. 1977. Wind and rain dispersal of *Fusarium moniliforme* in corn fields. *Phytopathology* 67:1023–1026.

Pasquill, F., and Smith, F. B. 1983. *Atmospheric Diffusion*, 3d ed. Ellis Horwood, Chichester, England.

Pedgley, D. E. 1982. *Windborne Pests and Diseases: Meteorology of Airborne Organisms*. Ellis Horwood, Chicester, England.

Rahn, K. A., Borys, R. D., and Shaw, G. E. 1977. The Asian source of Arctic haze bands. *Nature London* 268:713–715.

Rahn, K. A., Joranger, E., Semb, A., and Conway, T. J. 1980. High winter concentrations of SO_2 in the Norwegian Arctic and transport from Eurasia. *Nature London* 287:824–826.

Rahn, K. A., and McCaffey, R. J. 1980. On the origin and transport of the winter Arctic aerosol. *Ann. N. Y. Acad. Sci.* 338:486–503.

Rainey, R. C. 1973. Airborne pests and the atmospheric environment. *Weather* 28:224–239.

Rayner, R. W. 1960. Rust disease of coffee. II. Spread of the disease. *World Crops* 12:222–224.

Reisinger, L. M., and Crawford, T. L. 1982. Interregional transport: case studies of measurements versus model predictions. *J. Air. Pollut. Contr. Assoc.* 32:629–633.

Rittenberg, S. C. 1939. Investigations on the microbiology of marine air. *J. Marine Res.* 2:208–217.

Rowell, J. B., and Romig, R. W. 1966. Detection of urediospores of wheat rusts in spring rains. *Phytopathology* 56:807–811.

Samson, P. J., and Ragland, K. W. 1977. Ozone and visibility reduction in the Midwest: Evidence for large-scale transport. *J. Appl. Meteorol.* 16:1101–1106.

Schieber, E. 1972. Economic impact of coffee rust in Latin America. *Annu. Rev. Phytopathol.* 10:491–510.

———. 1975. Present status of coffee rust in South America. *Annu. Rev. Phytopathol.* 13:375–382.

Schütz, L. 1980. Long range transport of desert dust with special emphasis on the Sahara. *Ann. N. Y. Acad. Sci.* 338:515–532.

Shaw, G. E. 1980. Transport of Asian desert aerosol to the Hawaiian Islands. *J. Appl. Meteorol.* 19:1254–1259.

Smith, F. B. 1979. The character and importance of plume lateral spread affecting the concentration downwind of isolated sources of hazardous airborne material, in: *Symposium on Long-Range Transport of Pollutants, Sofia 1979*, pp. 241–251. World Meteorological Organization, Geneva.

———. 1983. Meterological factors influencing the dispersion of airborne diseases. *Phil. Trans. R. Soc. London B* 302:439–450.

Stakman, E. C., and Christensen, C. M. 1946. Aerobiology in relation to plant disease. *Bot. Rev.* 12:205–253.

Stevenson, C. M. 1969. The dust fall and severe storms of 1 July 1968. *Weather* 24:126–132.

Sykes, R. I., and Hatton, L. 1976. Computation of horizontal trajectories based on surface geostrophic wind. *Atmos. Environ.* 10:925–934.

Waller, J. M. 1979. The recent spread of coffee rust (*Hemileia vastatrix*) and attempts to control it, in: *Plant Health* (Ebbels, D. L. and King, J. E., eds.), pp. 275–283. Blackwell Scientific Publications, Oxford.

Wallin, J. R., 1970. Preliminary investigation of the southern corn leaf blight epiphytotic of 1970. *Plant Dis. Rep.* 54:1129–1130.

White, W. H., Anderson, J. A., Blumenthal, D. L., Husar, R. B., Gillani, N. V., Husar, J. D., and Wilson, W. E., 1976. Formation and transport of secondary air pollutants. Ozone and aerosols in the St. Louis urban plume. *Science* 194:187–189.

Index